Advances in Material Research and Technology

Series Editor

Shadia Jamil Ikhmayies, Physics Department, Isra University, Amman, Jordan

This Series covers the advances and developments in a wide range of materials such as energy materials, optoelectronic materials, minerals, composites, alloys and compounds, polymers, green materials, semiconductors, polymers, glasses, nanomaterials, magnetic materials, superconducting materials, high temperature materials, environmental materials, Piezoelectric Materials, ceramics, and fibers.

More information about this series at https://link.springer.com/bookseries/16426

Sumanta Sahoo · Santosh Kumar Tiwari ·
Ashok Kumar Das
Editors

Defect Engineering of Carbon Nanostructures

Springer

Editors
Sumanta Sahoo
Department of Chemistry
Madanapalle Institute of Technology and Science
Madanapalle, Andhra Pradesh, India

Santosh Kumar Tiwari
Guangxi University
Nanning, China

University of Warsaw
Warsaw, Poland

Ashok Kumar Das
Department of Chemistry
Madanapalle Institute of Technology and Science
Madanapalle, Andhra Pradesh, India

ISSN 2662-4761 ISSN 2662-477X (electronic)
Advances in Material Research and Technology
ISBN 978-3-030-94374-5 ISBN 978-3-030-94375-2 (eBook)
https://doi.org/10.1007/978-3-030-94375-2

This Springer imprint is published by the registered company Springer Nature Switzerland AG
The registered company address is: Gewerbestrasse 11, 6330 Cham, Switzerland

Dedicated to

Prof. Andrzej Huczko (Late)

Department of Chemistry, University of Warsaw, Poland

Preface

Carbon materials are unique and surprise from time-to-time scientific fraternity is upgrading its uniqueness, whether it is benzene structure, complex polymer behavior, or the hot materials like graphene and their derivatives. The exceptional properties of carbon exist in every phase of matter, and this is the foremost reason that carbon materials are dominating our daily lives. Over the past 3–4 decades, carbon nanomaterials are thriving and almost 60% of research outputs in chemistry, physics, materials science engineering, and chemical engineering are directly or indirectly linked with the carbon nanomaterials. In this aspect, layered nanostructures and materials consisting of heteroatoms (nitrogen, sulfur, phosphorus, selenium, etc.) catch a great attention, especially for the various engineering applications. Several products, devices, and innovative materials based on the same are also available in the market. Among the various class of materials, 2D and 3D carbon nanomaterials have been studied rigorously during the past 10 years. As a whole, these materials have contributed greatly to the modern science and technologies. Nanomaterials consisting of heteroatoms have been explored a lot worldwide, especially for the electrochemical sensing, solar cells, batteries, and catalysis. For energy applications and catalysis, heteroatom-doped carbon nanostructures are considered as potential candidates due to high surface area, low density, good electrical conductivity, thermos-mechanical stability, and densely dispersed tuneable active sites. Several excellent reviews and few books on the key individual carbon materials like graphite, carbon black, graphene oxide, graphene, porous carbon, and carbon nanotubes consisting of nitrogen, sulfur, etc. have been published from the renowned research groups all over the world. Moreover, recent developments on topological defects owing to heteroatoms doping and their effects on the intrinsic activity of carbon nanostructures have been extensively investigated. Various kinds of active sites have been created in nanocarbon catalysts to enhance their activity.

The main objective of the present book is to offer significant insights and up-to-date analysis of the most current advances, the directions of future research, and the techniques used for the synthesis of these innovative functional carbon nanostructures. Besides, specific attention been made on single-doped or co-doped active materials beyond the electrochemical applications. This book not only discusses the impact of defects on carbon nanostructures, but also gives an idea about the future research directions beyond such carbon nanostructures. The book starts with the biomedical applications of defective carbon materials. The next book chapter focuses on the biological applications of doped carbon nanomaterials. Next two chapters have been dedicated to the electrochemical applications like ORR, fuel cell, batteries, and supercapacitor applications. Two separate book chapters have been devoted to forecast the application perspectives of two prime carbon allotropies like carbon nanotube and graphene. The next chapter is based on the structure and properties of graphene-based composites. A separate book chapter demonstrates the creation of defects on carbon nanostructures through electrospinning technique. Finally, the last chapter demonstrates the impact of metal-oxide semiconduction in DSSC, which can be served as the alternative to the carbon allotropes. This book aims to provide in-depth knowledge of the research and development of various layered carbon nanostructures, which will be beneficial for the readers of any disciplines. We hope this book will be extremely valuable for the researchers working in this field. Moreover, each chapter of this book has plentiful references associated with the defect in carbon nanomaterials, which enable further interpretation and exploration.

Madanapalle, India — Sumanta Sahoo, Ph.D.
Warsaw, Poland — Santosh Kumar Tiwari, Ph.D.
Madanapalle, India — Ashok Kumar Das, Ph.D.

Contents

Defective Carbon Nanostructures for Biomedical Application

Arpita Roy and Chandan Kumar Maity

Abstract In the modern era, biomedical research based on carbon materials has drawn significant focus towards the different biomedical applications because of their incomparable physicochemical properties and large specific surface area. Beyond the pristine carbon materials, modification in carbon nanostructures (CNSs), called defective CNSs, to improve their properties as an efficient candidate for biomedical science has become more significant. Therefore, recent research trends are focusing on the synthesis and application of different defective CNSs for several biomedical applications. Herein, a brief overview of different defective CNSs and their potentiality towards biomedical applications have been vividly discussed with the hope of raising readers' awareness of the great opportunities related to these materials. The heteroatom doping and surface functionalization on CNSs are eminent strategies to generate defects within the carbon lattice. Naturally, the defects in carbonaceous materials having radical impurities, donor–acceptor pairs, impurities and vacancies support to enlarge the application potentiality of these materials (CNSs) in the biomedical field. Therefore, this chapter emphasizes different defective CNSs and existing studies of these materials for biomedical applications such as biosensing, biomedicine, bioimaging, antimicrobial activity and other multimode therapeutic utilizations. The different dimensionalities, the synthetic approaches of different CNSs and the generation of defects in different CNNs have been vividly demonstrated in this chapter. Finally, with the reasonable difference between the great possibilities of defective CNSs and the incomplete studies and applications, the future prospects and outlooks of these materials are very promising for next-generation biomedical advancement.

Keywords Carbon nanostructures · Biomedicine · Carbon nanotube · Biocompatibility · Dimensionality

---

A. Roy (✉) · C. K. Maity
Department of Chemistry, IIT(ISM) Dhanbad, Dhanbad, India

S. Sahoo et al. (eds.), *Defect Engineering of Carbon Nanostructures*, Advances in Material Research and Technology, https://doi.org/10.1007/978-3-030-94375-2_1

1 Introduction

In the modern era, biomedical materials research demands some effective, safe and distinct ways of treatment, so this research field is accelerating its horizons in the direction of more advanced therapies. In this regard, commercially available and used materials have some drawbacks like many times they lack some fundamental properties. For implantation purposes, generally metals and silicon materials are used but these materials have some serious concerns like insufficient long-lasting stability, higher inflammation causing tendency and rigid mechanical properties in original physiological conditions. So, these types of problems impose restrictions on using these types of materials for biomedical applications. Hence, it's necessary to perform all the experiments in the biological conditions to find out the actual toxicity and effectiveness of the therapeutics [1]. Therefore, as an emerging field nowadays nanotechnology is contributing much to the advancement of effective and safe materials for biomedical purposes. However, it is still necessary to introduce some advanced novel nanomaterials with desirable properties for the improvement and the growth of the present pace. This search of unique properties and efficiencies of nanomaterials drew the attention of researchers towards carbon nanostructures (CNSs) and their derivatives for a large number of biomedical applications [2–5]. Due to its catenation property, carbon can form various architectures in many dimensions like fullerenes, carbon quantum dots (0D), carbon nanotubes (1D), graphene sheets (2D) and diamond-like carbon (3D). CNSs are used in different applications owing to their admirable thermal and electrical conductivity, extremely high surface area and mechanical strength, extraordinarily good photo luminescent properties [6], transparency and high structural integrity [7]. Therefore, nowadays CNSs are used as materials in biomedical technologies like health care and artificial implants/organs or other prosthetic devices.

Carbon nanotechnology has become promising towards a new discipline relating to the extraordinary properties of nanocarbon. Recently, several CNSs have drawn significant consideration in the fields of medicine and biology. CNSs with various topologies such as ellipsoidal, cages, wire and tube have been effectively used for biomedical purposes, like drug delivery, bioimaging, theranostics agent for cancer treatment, tissue engineering and so on. In the case of biomedical technology biocompatibility and bioactivity are two major factors associated with CNSs for biomedical technologies. Owing to the outstanding belongings of fullerene and carbon nanotube (CNT), since the mid-1990s, they have been employed for several pharmaceutical and therapeutic purposes [8]. Nanocrystalline diamond (NCD), another carbon nanotopography, has become very widespread in the former decade owing to the advancement of various modification and production techniques. In the class of CNSs, graphene and its derivatives have already fascinated the consideration of the researchers for the development of biotechnology in the last few decades. However, the present research trends mainly focus on the modification of carbon nanostructures (CNSs) to improve their properties and enhance their application potential in different areas of nanoscience and nanotechnology towards biomedical science.

Modification can be adopted in various ways; one of them is the creation of defects in these nanostructures. The surface functionalization as well as heteroatom doping in CNSs are also well-known ways to create the defect within the carbon lattice. Naturally, the defects in carbonaceous materials contain radical impurities, donor–acceptor pairs, impurities and vacancies. These types of defects help to expand the application potential of these materials (CNSs) in the biomedical field. Incorporation of such defects may also tailor their biocompatibility by reducing their toxic effects. Surface functionalization and doping can enhance the specific superficial area of these carbon nanostructures, which in turn offer higher interaction sites to the cells. Therefore, the recent research trend is working to explore the potentiality of these defect-induced CNSs, which primarily emphasizes the anticipated criteria for biomedical application. The principal objectives of these studies are to improve the properties of CNSs and to make them effective for biomedical application purpose. This book chapter mainly features the recent trends and developments of defect-induced carbon nanostructure for different biomedical applications. The prime interest of this chapter is to deliver a brief impression on the current trends in the creation of defects in CNSs, properties of defect-induced CNSs and their application in biomedical science. Moreover, we hope that we will be able to inspire the readers about the roots of the exclusive belongings of these growing materials and to inspire their investigation in an assembly of moving areas of the biomedical field.

2 Carbon Nanostructures

2.1 Zero-Dimensional (0D) Carbon Nanostructures

Fullerenes were the first successfully synthesized zero-dimensional (0D) CNS. These can be synthesized via evaporation and recondensation of graphite in macroscopic amounts [9, 10]. These 0D fullerenes due to some of their extraordinary properties can be used in various fields of application. The delocalization of charges inside the sphere-shaped carbon skeleton and the firm, confined assembly of the aromatic p-sphere offer some exclusive openings to stabilize the charged moieties [9]. Moreover, very less amount of reorganization energies of fullerenes in the case of charge transfer reactions has generated a remarkable innovation in the case of synthetic electron donor–acceptor systems. This generally occurs through a faster charge separation as well as via slower charge recombination. Figure 1 shows the schematic of C_{60} and C_{70} fullerene molecules [11].

With the progress of carbon nanostructures science, it has been known about the carbon dots (CQDs) or graphene quantum dots (GQDs). These CQDs or GQDs are brand new members of the zero-dimensional (0D) carbon nanostructures family. Initially, it was reported that CQDs were synthesized from candle soot and by using the laser ablation of a carbon target [12, 13], but as the research advances, various

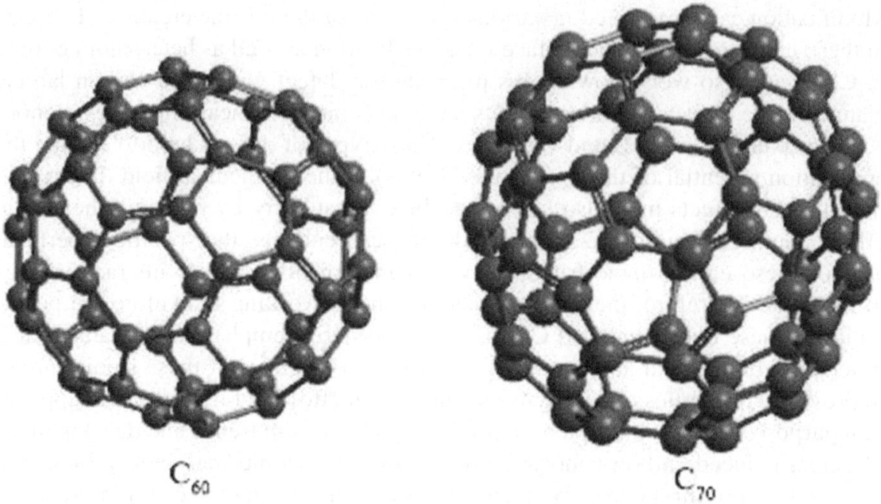

Fig. 1 Fullerene molecules (C_{60} and C_{70}) (Reproduced with permission from Ref. [11])

types of new approaches, as well as techniques, have been introduced for the preparation of CQDs. So, CQDs may be synthesized via various techniques like pyrolysis, electrochemical as well as microwave-assisted approaches and hydrothermal treatments of various carbon sources like household waste materials, biomass or unused foodstuffs [14–17]. There are some advantages associated with these methods like these methods are quite simple, no hazardous chemicals are used and, moreover, they can be used waste materials. Thus, synthesized CQDs give a luminescence spectrum in the visible region. For these luminescence properties of CQDs and their biocompatibility, they are widely used in in vivo biosensing. The CQDs are tiny nano-sized particles (~10 nm) of carbon with different surface functional groups. These various functional groups might be further altered to tune the properties of CQDs to make them more appropriate for application purposes. CQDs can be both crystalline as well as amorphous in nature. The hybridization in CQDs is generally reported to be sp^2 while in some places sp^3 hybridization has also been found. The lattice parameter of CQDs is generally found to be 0.34 nm which resembles (002) graphite interplanar spacing [18].

2.2 *One-Dimensional (1D) Carbon Nanostructures*

Carbon nanotube (CNT), a 1D carbon allotrope, is reported as a ribbon of graphene layers rolled on a cylindrical tube [9]. CNTs demonstrate metallic as well as semi-metallic nature. CNTs also behave as a semiconductor which may also be used for sensing purpose. The various kinds of such properties of CNT are primarily reliant on the inherent belongings of the rolled graphene sheet, with which the CNTs are made up of. The chirality of these materials is also determined by the nature of the rolled graphene sheets present. CNTs may be synthesized using chemical vapour deposition (CVD) of carbon using some transition metal catalyst [9]. CNT can be classified into two parts: one is a single-walled carbon nanotube (SWCNT) and the other one is a multi-wall carbon nanotube (MWCNT). Generally, 1D SWCNTs are said to be like very tiny ribbons of graphene-walled nanocylinders. Synthesis of various kinds of CNT with different geometry, shape and structures as well as different grades of purity may be achieved with various kinds of synthetic techniques like arc-discharge technique or laser ablation [19]. Some extraordinary features of the CNT make them useful for various applications. They have a good capacity of ensuring direct electrical signal, ultra-high sensitivity and superior response. Indeed, the change in nanotubes resistance as a result of chemical interactions between surface atoms and absorbed molecules can be detected in a few seconds. These characteristics urge the use of nanotubes for the next generation of biosensors which require the fabrication of nanotubes with well-controlled morphology [20].

CNT exists in the form of graphene sheet in rolled-up condition; moreover, the CNT has a bonding like that of graphite and has sp^2 hybridization. Among four valance electrons of carbon atom three forms sigma bonds (sp^2 hybridization) with the neighbouring atoms which provide graphene a highly planner rigidity. The left one electron in the valance cell is a delocalized (π-orbital) one and shared with all the atoms. So, this delocalized fourth electron of each carbon atom helps in electronic current transport as these generate conduction bands. SWCNT has a structure with a very close resemblance to the ideal fullerene fibres. These are made up of a single graphene sheet rolled in the form of a tube. These single-layered tubes have a diameter between 0.4 and 2 nm and are joined from one end to another end and forms bundles [19]. They are organized into a larger rope-like structure in a one-dimensional lattice and have a lattice parameter of 1.7 nm and a tube distance of 0.315 nm [10]. Again, MWCNT is somehow different from SWCNT and is made up of concentric graphitic tubes placed on every side of a common central hollow space. MWCNT can be distinguished from SWCNT because they consist of a variety of configurations and shapes. They also have various diameters ranging from 1.5 to 100 nm [19]. MWCNTs have a close resemblance with hollow graphite fibres, but they have a better structural perfection.

2.3 Two-Dimensional (2D) Carbon Nanostructures

2D Graphene is known to be the youngest among all synthetic carbon allotropes. Single-layer graphene was synthesized first from exfoliation of graphite utilizing Scotch Tape in 2004 [20]. However, new modern synthetic strategies are nowadays very important for systematic experiments and technological applications, especially epitaxial growth, solubilization from bulk graphite and so on. Again, high conductance is due to the high mobility for holes and electrons present in the symmetrical structure of graphene. There is one more advantage associated with the monolayer of graphene, i.e., it is completely transparent and has an optical transmittance of 97.7%. Additionally, it is also a low-cost abundant source for multipurpose applications. Graphene is made up of a single-layer sheet of sp^2 carbon atoms [20]. Graphene has an excellent capability to captivate all wavelengths containing light, and along with this property, it has an outstanding electron transportation property which has created massive attention on graphene chemistry. Generally, graphene is synthesized via mechanical as well as liquid-phase exfoliation of graphite [21]. Newly, for the synthesis of graphene on an industrial scale, the usage of several surfactants has also been reported [22]. Additionally, along with pristine graphene, various surface-modified forms, like photo luminescent graphene oxide (GO) and reduced graphene oxide (RGO), have also been synthesized and used for various applications [9, 23]. Graphite upon strong oxidation produces GO. GO is a derivative of graphite with high oxygen concentration and generally contains epoxy, hydroxyl, carboxyl groups and so on. These groups randomly exist on the edges and the basal planes of the GO. Negative surface charge generated by oxy-functional groups on the materials owing to which, they participate in hydrogen bonding like interactions. Thus, during the modification process, some areas remain unmodified and that unmodified portions of the surface keep their π-electrons free, which helps to generate π–π interactions [9]. This unique surface chemistry and moderate hydrophilicity of GO impact the catalytic action and conformational state of a biomolecule.

Another type of chemical derivative of graphene is RGO having a similar basal plane-like graphene and it is different from graphene by the presence of defects on it, created by oxy-functional groups. That is why, a similar kind of π-interaction will happen like graphene; however, there are accompaniments of both hydrogen-bond acceptor and donor moieties from the carboxylate oxygen, carboxylic, epoxides, alcohols and ethers moieties that can induce a supplementary path for interaction [9]. However, graphene is very stable and it exhibits excellent electronic conductivity and a large surface area of 2630 $m^{2}\,g^{-1}$ (theoretically) [24].

In comparison, the 2D shape of graphene is not harmful in small concentration. Hence, graphene the wonder material may be used in various biomedical applications. Mainly, GO is a promising compound for biological requirements due to its cost-effectiveness, amphiphilicity, surface modification ability, compact surface-enhanced Raman scattering (SERS) property and excellent ability for fluorescence quenching [25]. Figure 2 depicts the different types of CNSs according to their dimension.

3 Defects in Carbon Nanostructures

There are various ways to create defects in carbon nanostructures like impurity, vacancy or change in topography. So, in many cases, C atoms are substituted by other different atoms from different elements. The defects thus created are known to be impurity defects. On the other hand, when C atoms (one or more) are eliminated from

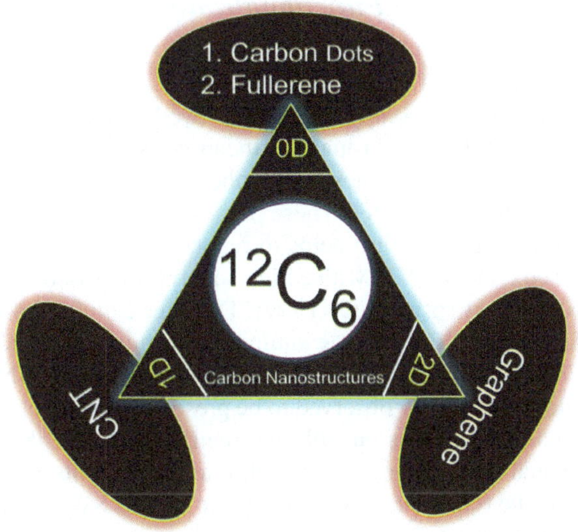

Fig. 2 CNSs corresponding to their dimension

the lattice structure the vacancy defects are generated. However, the topographical defects are mainly created when there is a change in bond angles due to the rotation of C atoms. In case of such defects, the total number of C atoms remains constant in the lattice. These types of defects in lattice sometimes efficiently help in serving chemical or biological applications, as they offer some special active sites for binding or absorption of concerned atoms or molecules. In contrast, sometimes defects also impose many problems for applications of electronic devices like in field-effect transistors or in electrical interconnects as they increase the resistance of carbon. So, defects in CNS are very important for various applications and they can be generated by functionalization, doping or by many other methods.

3.1 Biocompatible Defects in CNTs Through Functionalization

CNTs are normally insoluble materials. So, it is vital to improve their dispersion and solubility in various media like water, serum etc., which in turn improves its dispersion and biocompatibility. Their surface area also must have to be enhanced. These properties make them superior for biomedical applications [26]. Bioactivity and toxic properties for CNTs generally depend on the synthetic route of their modification followed by the purification steps. These also depend on the surface properties of CNTs. Several synthetic processes have been reported for surface functionalization of CNT and respective applications are discussed in the succeeding sections.

3.2 Covalent Functionalization

To use CNTs for biomedical purposes, there are two main key factors: their solubility and cytocompatibility. Improved solubility and lower toxicity in CNT may be accomplished through purification by generating covalent functionalization via multistep acid treatment. A number of purification methods are available to increase the solubility of CNT. These processes sometimes may involve exposure of certain groups which in turn helps to reduce the toxic effects; for example, acid-oxidized, carboxylated CNTs with a high aqueous dispersion ability are safer towards mice as compared to pristine CNT [27]. The inter-cellular uptake study using the acid-treated dispersed SWCNT was examined using human monocyte-derived macrophage cells. From the results it was evident that functionalized CNTs can penetrate into the cells and are generally observed to be present inside lysosomes and cytoplasm. It also has no adverse effects on cells and their structures. Likewise, in the case of pharmacological applications, the potential toxic elements present in carboxylated SWCNTs must be removed, hence the functionalized SWCNTs may be used significantly for mammalian cell cultures also. Sometimes the introduction of some functional groups like –CO, –OH and –COOH enhances the O_2 amount in CNT, which may significantly diminish the toxic effects of CNT [26]. Ultrashort and oxidized SWCNT is nowadays used as nontoxic nonviral vector and as a vehicle for the intracellular delivery of oligonucleotides in human macrophage cells [28]. Currently, it was observed that soluble or dispersed SWCNT are benign towards toxicity of cells. Acute cell death is not responsible due to the functionalization and isolation of SWCNT. Covalent functionalization in CNT was considered to be a good technique influenced greatly by the amount of functionalization. This method improves the biocompatible nature of CNT and reduces cytotoxicity. Additionally, the incapability of functionalization and isolation of CNT can result in severe cell death.

3.3 Noncovalent Functionalization

Surface alteration of CNT can be carried out with useful noncovalent functionalization. It is earlier well known that CNT can interact with different molecules noncovalently using electrostatic interaction, hydrogen bonding, π-electrons, van der Waals force and surface adsorption on its side wall. The dispersibility of CNT can be increased through this noncovalent functionalization, which reduces toxic nature as well [26]. This process has introduced the usage of some surfactants, biomolecules and polymers to gain biocompatibility. For instance, CNT surface is coated with porphyrin derivatives and fluorescein isothiocyanate-conjugated (FITC)-terminated PEG through the π–π interaction between CNT and pyrene, and consequently improves the biocompatible nature and reduces toxicity [26].

3.4 Functionalization Using Protein

Interaction of protein and CNT can influence the biological properties. The π–π interactions of aromatic parts (Tyr, Phe, Trp) of proteins and CNT develop the biocompatible nature and absorptivity. It also reduces the toxicity of bare CNT. CNT–protein nanoconjugates make CNT beneficial for drug delivery, cancer therapy and biosensor fabrication. For instance, when streptavidin was adsorbed on the wall of CNT–protein conjugates, then it was helpful for cancer therapy, and no cytotoxic belongings for proximal cells were introduced. Correspondingly, enhanced cellular actions were tempted by CNT–polycarbonate urethane adsorbed with protein fibronectin. By discriminating physical nano roughness, tissue development also improved. Therefore, a well reorganization of the association of the CNT-based nanomaterials with the serum proteins aided in relatively safer CNT/nanomaterial hybrids fabrication. Bovine serum albumin (BSA), a water-soluble globular protein, consequences the capability of high dispersion towards in vitro by adsorbing onto the CNT surface. BSA-dispersed SWCNT do not affect any important harmful cellular properties when taken up by HeLa cell and mesenchymal stem cell [26]. Correspondingly, albumin-adsorbed SWCNT-induced cyclooxygenase-2 in the RAW 264.7 macrophage cell lines regulate the cytotoxicity and uptake of SWCNT [29]. These reports meaningfully impact the field of biological properties of CNT. These proteins reduce the contacts, modify the cellular mechanism and also help the nanoparticles to improve their biological activity. The interaction of CNT with proteins is testified to mark their uptake, distribution, clearance and delivery to the proposed target site, consequently diminishing toxicity possibly.

3.5 Functionalization Using DNA

Functionalization of DNA with CNT can enhance its stability. The constructions of DNA with CNT make fitted helices around CNT or form noncovalent couples with CNT [26]. The dispersion of nanotubes is enhanced due to the wrapping of CNT with DNA and Flavin mononucleotide. Functionalization of CNT by DNA also have application as biological transporter and biosensor. DNA-functionalized MWCNT shows better thermal removal capability against malignant tissues than normal MWCNT when tested in vivo. Moreover, it was testified that the DNA-CNT reduced the cytotoxic belongings through the penetration of the lymphocytes rapidly by a needle-like mechanism [30]. Similar behaviour of cell-penetrating proteins and functionalized CNT was also studied, and both enter into the cell without endocytosis; however, the incorporation of nanomaterial is influenced by the process of functionalization.

3.6 Functionalization Using Poly (Ethylene Glycol)

Some polymers like PEGylated phospholipids and PEG can form noncovalent bonds with CNT, thus resulting in the enhancement of surface property, biocompatibility and dispersibility of CNT. Recently, it is testified that adsorbing phospholipid (PL)-PEG-functionalized CNT is noncytotoxic [31, 32]. SWCNT-PEG reveals lesser cytotoxicity towards neuronal PC12 cells than bare SWCNT. They possess diminished sensitive oxygen species-facilitated toxicological nature in vitro. Additionally, they also show lesser interaction with the cell membrane than normal SWCNT. Therefore, the usage of SWCNT-PEG in medicine is moderately approaching. SWCNT-PEG helps in multifunctional drugs production and have applications in imaging tools too. Doxorubicin (DOX), an anticancer drug, loaded PEG-functionalized SWCNT, displayed enhanced therapeutic capability and reduced cytotoxicity properties than free drug [33]. When SWCNT-PEG is inoculated intravenously in mice, no evidence of cytotoxicity has been observed [26]. PL-PEG produces a steady dispersed aqueous solution due to the existence of methyl or an amine group. They also excite leading macrophage immune cell and proinflammatory cytokine in culture.

3.7 Functionalization Using Chitosan

Using surface adsorption, CNTs were functionalized by chitosan (CS). CS possesses some unique materialistic properties like better complex formation ability, biocompatibility, outstanding water solubility, nontoxicity and biodegradability, and it is a good choice for the functionalization of CNT. Hence, CS has been extensively considered for pharmaceutical and biomedical applications such as cancer therapy, biosensors and drug delivery [34–36]. For the subtraction of heavyweight metal from an aqueous solution, CS-modified CNT is used as biomaterials [26]. Functionalization of MWCNT with CS coupled with PC (MWCNT–CS–phycocyanin (PC)), a biomaterial for photothermal and photodynamic therapy agent, was experienced on a normal liver cell line (L-O2) and liver and breast cancer cell lines (HepG2 and MCF-7) [26]. The consequences exposed that MWCNT–CS–PC exhibited precise photo-induced toxicity to HepG2 and MCF-7, and the use of CS improved solubility. Cytotoxicity of the CNT was reduced by the use of PC. For targeting tumour cells, modification of SWCNT with CS and coupled with folic acid (FA) (CS–SWCNT–FA) exposed that CS provides an appropriate biological surface for biomolecules immobilization and is successfully able to disperse SWCNT.

3.8 Functionalization Using Other Polymers

For noncovalent functionalization of nanomaterial, the improved dispersible nature of block copolymers like poly(ester), pluronics and poly(l-amino acid) at the time of drug delivery is significantly beneficial [26]. Isopropylene glycol repeating units named pluronic F68 is a linear biocompatible copolymer that stabilized the aqueous dispersion of SWCNT. Additionally, no occurrences of accumulation were revealed by CNT suspensions in two biocompatible dispersants (pluronic F108 and hydroxypropyl cellulose), which continued in a dispersed state in vitro [37]. Correspondingly, the functionalization of SWCNT with polymers like poly(diallyldimethylammonium chloride) and hexamethylene diamine reported the noncovalent conjugation and it was beneficial for the intracellular delivery of negatively charged biomolecule with less cytotoxic properties. Figure 3 schematically depicts some covalent and noncovalent functionalization of CNT [38].

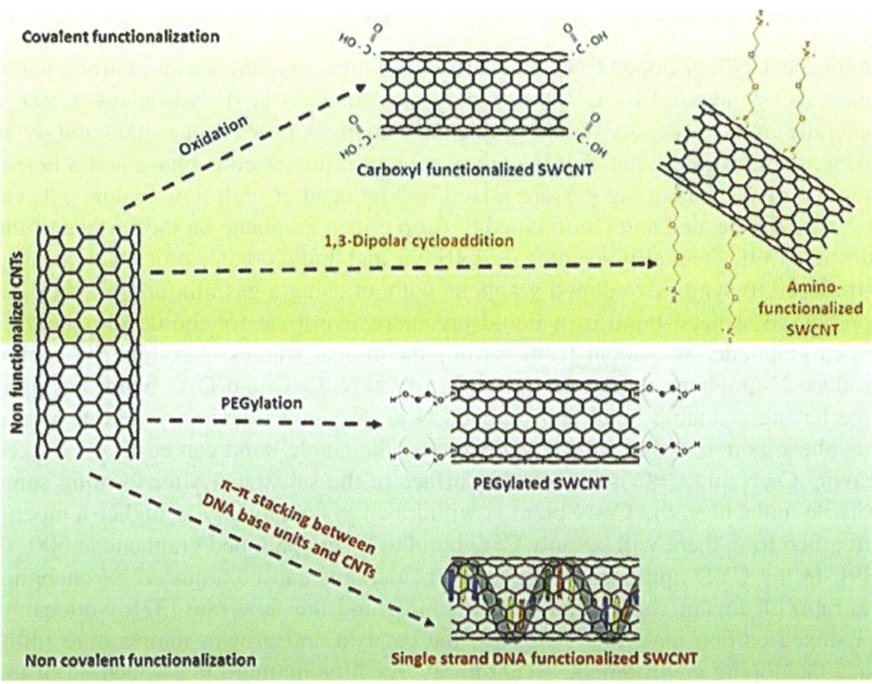

Fig. 3 Schematic representation of covalent and non-covalent functionalization of CNT (Reproduced with permission from Ref. [38])

4 Defects of Graphene

The property of graphene can be modulated using other methods such as chemical doping and functionalization. In different applications, it has been proved effective. Chemically doped graphene is mainly produced in two ways, in general: (1) organic molecules, gas or metal are absorbed on the graphene surface and (2) nitrogen and boron atoms are the two heteroatoms, which are introduced by the substitutional doping within the graphene carbon lattice [39]. The doped graphene is different with respect to various properties from the normal graphene. For instance, the charge distribution, as well as the spin density of carbon atom, might be altered by the neighbouring dopants, which initiate the 'activation area' on the surface of graphene. Graphene doping extends its applications greatly.

4.1 Direct Synthesis

Graphene, CNT, N-doped CNT and carbon nanofibers are the various carbon nanomaterials synthesized by CVD, a widely used method [39]. Nowadays, CVD is generally used to prepare doped graphene. Usually, Cu or Ni, a metal catalyst, is taken as a substrate; after that, a carbon source in the gaseous phase and a hetero atom (B or N) containing gas are mixed and induced at high temperatures. Then, these are dissociated and recombined to form doped graphene on the surface of the substrate [39]. Generally, acetonitrile, pyridine and liquid organic precursors are also introduced to synthesize doped graphene without using a gas mixture [40, 41]. An appropriate skeletal bond of a liquid precursor is critical for the development of doped graphene, as evident from various theoretical studies. Acrylonitrile cannot produce N-graphene due to the presence of C≡N, C=C and C–C bond, but pyridine having a double bond can produce N-graphene. The primary reason behind this phenomenon is that, at low temperature, the single bond can easily be broken leaving C≡N and C=C bonds on the surface of the substrate. After forming some volatile molecules, the C≡N bond is withdrawn systematically at higher temperature; therefore, there will be only C=C bond to form nondoped graphene at 500 °C [39]. In the CVD approach, the number of dopants can be adjusted by changing the ratio of dopant source and carbon source and the flow rate [37]. Noticeably, for doped carbon materials synthesis, the catalyst and growth temperature influence the doping environment. Accordingly, for illuminating the association of the bonding configuration of dopant atoms and the conditions of CVD, further research is essential. The solvothermal method was first successfully applied for the large-scale fabrication of graphene, and recently put on this route at ~300 °C temperature, the large-scale manufacturing of doped graphene has been carried out. Using cyanuric chloride ($N_3C_3Cl_3$) with Li_3N and CCl_4 or mixing tetra chloromethane (CCl_4) with lithium nitride (Li_3N), N-doped graphene having different nitrogen percentages was attained using the hydrothermal method [42]. N-doped graphene contains 1–6 layers of graphene, as evident from HRTEM images. To get CNTs and doped CNTs, an arc-discharge method has been introduced evaporating the source of carbon, usually graphite, at high temperature [38]. Rao et al. effectively synthesized doped graphene using this method [43, 44]. N-graphene, produced from this method, usually contains 0.5–1.5% nitrogen [43, 44].

4.2 Postsynthesis Treatment

The thermal process is generally introduced to produce doped graphene at high temperature. Actually, at tempertaure ≥800 °C, graphene can yield N-graphene in the presence of NH_3 [39]. High temperature is obtained in electrical annealing, which is also useful to synthesize doped graphene [39]. In this method, the content of dopant is low in graphene. Guo et al. attained N-doped graphene at 1100 °C with a doping level of 1.1% and the highest nitrogen level was 2.8% at 900 and 800 °C, according to Geng et al. [45]. The low content of dopants might be accredited to two reasons: (1) the number of defects in high-quality graphene is very low and (2) in doped graphene, high temperature breaks C–dopants bonds [39]. However, in the thermal treatment method, doping can easily happen at the edge of graphene and defects. Graphene oxide (GO) is usually used to produce doped graphene by thermal treatment using several nitrogen precursors. According to Sheng et al., annealing could produce N-doped graphene with the existence of melamine and GO at 700–1000 °C [46]. The mass ratio of melamine and GO and temperature can affect the nitrogen content. Li et al. testified that thermal annealing of GO and NH_3 (gas phase) at 500 °C reduced GO and produced N-doped graphene, the content of N was 5% [47]. Particularly, these two works show that temperature is a vital condition to control the dopant amount. Li et al. attributed the cause to the decreased amount of oxygen functionalities at high temperature. For the construction of the C–dopant bond, these oxygen functionalities are accountable. The reactivity of dopant atoms and GO will diminish after decomposition of the oxygen functionalities at high temperature, which results in lower dopant content. To synthesize N-doped CNT, nitrogen plasma atmosphere is used in the presence of CNT and nitrogen atoms partially replace the carbon atom [39]. Newly, the method has been introduced to prepare N-graphene by revealing graphene or GO in the nitrogen plasma atmosphere [39]. The number of dopants can be changed by varying the exposure time and plasma strength. At the time of plasma treatment, oxy-functional groups and defects are formed. Shao et al. conveyed that N-graphene contains 8.6% oxygen species, while graphene contains 3.5% oxygen species. Wang et al. showed that after doping, the amount of oxygen enhanced from 15 to 26–28% [48]. These consequences designate that the plasma treatment incorporates a substantial quantity of oxygen into graphene.

4.3 Defects in Graphene Through Oxidation and Reduction

Among others, GO has exposed several exceptional physicochemical characteristics like exciting electronic and optical properties, high surface area, small size and extraordinary strength in 2D structure [49]. Due to the hydrophobic nature of exfoliated graphene (from graphite), it is very difficult to disperse it in water; therefore, functionalization is a little bit difficult. However, GO develops hydrophilic nature and becomes soluble in water. Hence, GO achieves outstanding capability for surface functionalization, the ability for fluorescence quenching, amphiphilicity, excellent aqueous dispersibility etc. for hopeful biotechnological requirements [49]. Moreover, nanocomposites of several magnetic nanoparticles, gold and polymers with GO have been directed to several biotechnological applications in antibacterial action, bioimaging, phototherapy, gene and drug delivery, biosensing etc. Mainly, the π–π^* transition of GO reveals the low-energetic electron movement, which helps GO for bioimaging and biosensing. For better biomedical requirements, the preparation and functionalization of GO must be appropriately carried out. Generally, GO is synthesized using Hummer's methods (and its different variations) [49], and it is mostly a chemical oxidation process of graphite, with successive dispersion and exfoliation in different solvents. Manufacture of graphene from GO is now very imperative, as it is produced from low-cost graphite as raw material on a large scale, as well as using a cost-effective chemical process. In addition, due to its hydrophilic nature, it forms a steady aqueous dispersed solution to enable the gathering of macroscopic assemblies through low cost and simple solution methods. For biomedical applications, these two qualities are very significant. To change GO to graphene, reduction of GO is carried out using electrochemical, chemical or thermal processes, which synthesizes reduced graphene oxide (RGO) with a lesser number of oxy-functional groups. Various reducing agents lead to different chemical compositions and carbon-to-oxygen ratios for RGO. However, it is very difficult to completely reduce the GO to obtain pure graphene. The controllability of the oxy-functional group and the cost-effectiveness make RGO promising for biological requirements [49].

4.4 Defects in Fullerene

On the basis of numerous features of fullerene, several theoretical and experimental researches have been executed. Several modern procedures have been advanced to develop the range of nanotechnology research. Through these technologies, it has become easier to investigate the fullerene cluster and its defect. Laser treatment could split C_{60} into C_{58}, C_{56}, and additionally reduced clusters through the loss of C_2 fragment [50]. Through the reactions of C_{60} with O_3, the odd-numbered clusters of C_{59}, C_{57}, C_{55} and C_{53} are generated in laser desorption ionization [50]. The study of defect fullerene clusters has been vastly considered, though several theoretical studies about different fullerene clusters have been carried out. Recently, Ruckenstein and Hu considered the stability and structure of some defect fullerene clusters C_{57}, C_{58} and C_{59} produced by eliminating three, two and one carbon atoms from C_{60} cluster [51]. The exceptional nature of this new form of carbon leads scientists to forecast numerous technical usages. Still, the hard process involved for fullerenes has revealed the main issue for medicinal usage. C_{60} is not soluble in water and could be easily combined. Many attempts have been adopted to reduce the hydrophobic nature of fullerenes. The utmost extensively used methods are: (a) encapsulation or microencapsulation in special carriers like calixarenes, polyvinylpyrrolidone, cyclodextrins, micelles and liposomes [52]. Moreover, the grouping of lipid membranes and fullerenes has been directed to impressive results. Lipid bilayer has vigorous mobile structure, partly ordered and of biopharmaceutical attention for covering biocompatible surfaces or for the sustained release of the drug; (b) suspension using co-solvents by saturating fullerenes in benzene solutions decanted into THF and the subsequent mixture is added dropwise to acetone, and then water is gradually added. A suspension having yellow colour is obtained and solvents are evaporated to a final known volume of water; (c) chemical functionalization with amphiphilic polymers, polyhydroxyl group, carboxylic acid and amino acid. The exceptional chemical features of fullerene produce great attention for real-time requirements to fabricate new material for medical applications [52]. Figure 4a shows the SEM image of hexagonal transparent fullerene nanosheets synthesized via liquid–liquid interfacial precipitation

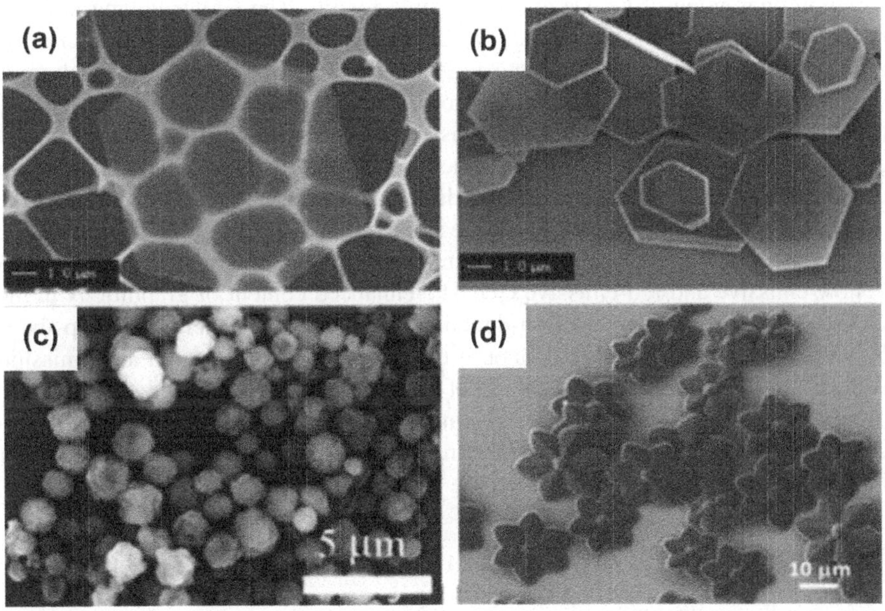

Fig. 4 SEM image of **a** fullerene nanosheet, **b** C_{60}/Fc nanosheet, **c** C_{60} konpeito-like nanocrystal, and **d** flower-shaped fullerene crystals (Reproduced with permission from Ref. [11])

(LLIP) method, using alcohol and CCl_4 [11]. A C_{60}/ferrocene (Fc) hybrid [11] was synthesized by the LLIP method and the corresponding SEM image is depicted in Fig. 4b. The effect of nonionic natural surfactants, diglycerol monolaurate ($C_{12}G_2$) and diglycerol monomyristate ($C_{14}G_{12}$) on the morphology of fullerene crystals and the crystallization of C_{60} and the used surfactants triggered a morphological change in fullerene from 1D to a zero-dimensional 'konpeito-like' fullerene crystals, as shown in Fig. 4c [11]. Fullerene flower, the most recent fullerene crystal, was produced via solution-phase crystallization [11], and the corresponding SEM image is presented in Fig. 4d.

4.5 Defects in Carbon Quantum Dots

Carbon dots (CDs) show the PL property, and they mention at least one dimension below 10 nm [17]. CDs contain polymeric accumulations or nitrogen/oxygen-containing functionalities. Primarily, the CDs comprise CQDs, polymer dots (PDs) and GQDs. GQDs are the small fragments of graphene, and electronic transportation is limited in all 3D dimensions. Graphene is a zero-bandgap semiconductor, and its exciton Bohr diameter is infinite. Therefore, any portion of graphene involvement is quarantine. Though the GQDs mention the graphene fragments with 20 nm below size, cutting of graphene sheet, the GQDs can be made-up. The edge effect and quantum quarantine of GQDs are responsible for PL. The GQDs having nonzero bandgap show PL after excitation [17]. The bandgap of GQDs can be altered by changing the surface chemistry and its size. PDs having π-conjugated polymers with small particle size show illumination. The PDs reveal multipurpose applications with fluorescence imaging. To considerably influence the wavelength of CD emission, heteroatom doping is performed in an operative way [53]. N doping, in the form of graphitic N, hydrazine, pyridinic and amino, is the utmost doping type that tempts PL redshift of CD. The largest content of N is present in blue-emissive and green-emissive CD in the form of amine/amides, whilst graphitic N governs in yellow-emissive and red-emissive CD. The enhanced content of graphitic N in the CD structure, which is an electron-donating element, could generate midgap states between LUMO and HOMO, thus ensuing in red-shifted absorption and emission (Fig. 5a) [53]. Furthermore, co-doping of heteroatom was established to control the optical behaviour of CD. N and S doping, as well as carboxyl group, synergistically decreases the energy gap of prepared CD, thus conducive to the red-shift of fluorescence, as shown in Fig. 5b.

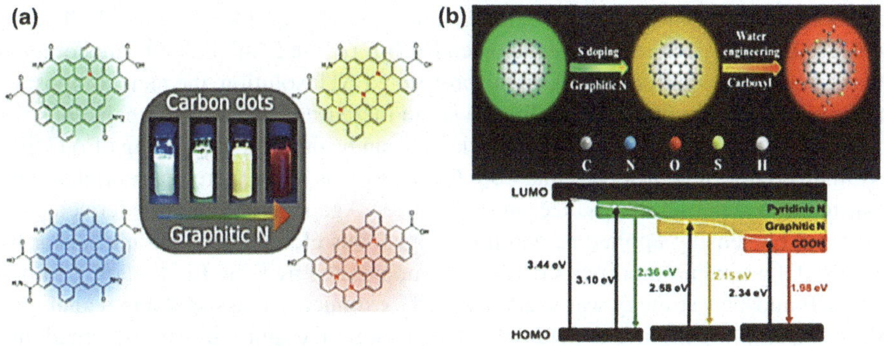

Fig. 5 **a** Schematic arrangements of red fluorescence in CDs triggered by graphitic nitrogen, and **b** Diagram of the possible (upper) structure models and (bottom) energy level .of the PL-tunable CDs (Reproduced with permission from Ref. [53])

5 Defected Carbon Nanostructures for Biomedical Application

5.1 Sensing Application

Recently, graphene and its analogues have been widely used for electrochemical sensing in biomedical sciences. As graphene has some unique electrical properties, it has drawn significant attention from the sensing communities and its analogues towards sensing applications [9]. For electrochemical sensing applications in biomedical purposes mainly some important features of the 2D material have been so far utilized. For the detection of circulating tumour cells and also for the purpose of detection glucose, graphene was used as a high-conducting substance [9]. This graphene sensor has also been used for the electrochemical detection of various drugs. In this regard, modified graphene, as well as glassy carbon electrodes (GCEs), have displayed better-quality improved sensitivity towards the detection of various drug molecules. Various nanoparticles also help in the proliferation of surface area, high conductivity as well as electro activity by virtue of the interaction of various functional molecules at the surface of them. Jun-jie Zhu reported that the electrocatalytic consequence of GO in Ru $(bpy)_3{}^{2+}$ oxidation can be detected at GO functionalized glassy carbon electrode in the absenteeism of co-reactants [54]. In this case, GO itself works as a co-reactant of the complex Ru $(bpy)_3{}^{2+}$. Thus, the electrochemiluminescence (ECL) biosensor may be fabricated. It has also been found from certain studies that the thiol group-ended ATP aptamer immobilized on the GO films were obtained by DNA hybridization. When the gold nanoparticles (AuNPs)/graphene oxide nanocomposites were functionalized with these aptamers by means of the creation of S–Au bond to generate an ever-known sandwich-like structure. ECL resonance energy transfer could arise between $Ru(bpy)_3^{2+}$ and AuNPs/GO nanocomposites. After the incubation of ECL in the preformed ATP solution, the nanocomposites come out from the electrode leading to some increase in the ECL signal. Thus the produced ECL aptasensor is very efficient and can be used for the finding of ATP [9]. From this report it is observed that both GO as well as AuNPs are appropriate materials to be used in ECL resonance energy transfer. Again, as a part of modern electrochemical biosensing applied research homogeneous electrochemical aptasensor by means of a graphene-modified GCE was demonstrated by Feng Li et al. [55]. In this case, a brand-new strategy was used, that is, T7 exonuclease-assisted target analogue recycling amplification. The detailed aptamer-target recognition is transformed into an ultrahigh-sensitive electrochemical output signal by means of this amplification approach. Such types of electrochemical aptasensor are also appreciated for the detection and measurement of biomolecules from a homogeneous solution without the immobilization of any kind of bio-probe on the electrode superficial area. For ultrahigh-sensitive detection and measurement of Hg^{2+} and MUC1 another GO/ECL-based sensor was introduced with an exonuclease I (Exo-I) stimulated target recycling amplification stratagem [9]. The signal strength of ECL for *N*-(aminobutyl)-*N*-(ethylisoluminol) (ABEI)-modified Ag nanoparticle-ornamented GO nanocomposites (GO–AgNPs–ABEI) were improved via ferrocene tagged ssDNA Fc-S1 when H_2O_2 was present in the media. By using these types of aptamers, associate ssDNA S2 and full thymine bases, ssDNA S3-modified gold nanoparticles etc. [9] were immobilized on the sensing surface by means of a hybridization reaction. As there is very tough and steady T–Hg^{2+}–T interaction, a plenty of Hg^{2+} was effectively caught on the surface of AuNPs-S2-S3 and efficiently inhibited the ECL reaction of ABEI. When the switch is in 'on' state for the signal, the mechanism was implemented by employing MUC1 as an aptamer-definite target to connect to the aptamer, which leads to a huge decrease in the concentration of captured Hg^{2+}. After that, exo is utilized to digest the aptamer bound with MUC1, which results in the discharge of MUC1 for recycling purposes which can again have a highly efficient measurable ECL signal.

The optical, structural and electronic properties are taken collectively to make SWCNTs an efficient material to produce nanostructures, especially those that are efficient for generating an array of sensing purposes [9]. For the glucose sensor, a stable dispersion of glucoseoxidase tagged with SWCNTs was involved. In these types of devices, small molecules like $Fe(CN)_6^{3-}$ are allowed to permeate through the surface coating and can attach to the surface through noncovalent bonding. This type of binding gives rise to quenching of emission. Nevertheless, the amount of H_2O_2 produced during the course of the reaction of glucose and concerned enzyme partly diminishes the surface tagged $Fe(CN)_6^{3-}$, which results in upsurged intensity of the emission [56]. This increase in emission then works as a sensitive reporter of the concentration of H_2O_2 present. An additional example of a transparent and stretchable electrochemical sensor for biomedical applications based on CNT was stated and described by Weihua Huang et al. in the year 2017 [57]. The sensor includes an attachment of SWCNT with some important conductive polymer to produce the composite films. These films are prepared via coating of distinct SWCNTs using poly(3,4-ethylenedioxythiophene) (PEDOT) followed by vacuum filtration and have the thickness in nanometres. PEDOT coating successfully removes the surface defects and also offers complimentary joint contacts between discrete SWCNTs. The conductivity, as well as the electrochemical properties of these composite films, is significantly improved and matched to only CNT films. PEDOT protects these junctions of SWCNT from the separation when it is stretched, which provides the sensor with high physio-mechanical strength and efficient electrochemical properties throughout the time of big distortion. This work has gathered and generated a highly stretchable sensor based on CNT. It is suggested that more real-time applications may be extended by the preparation of wearable and *in vivo* implantable electrochemical sensing devices. Figure 6 shows the schematic illustration of the SWCNT-based immune biosensor and its discriminating CD4 T cell capture [58].

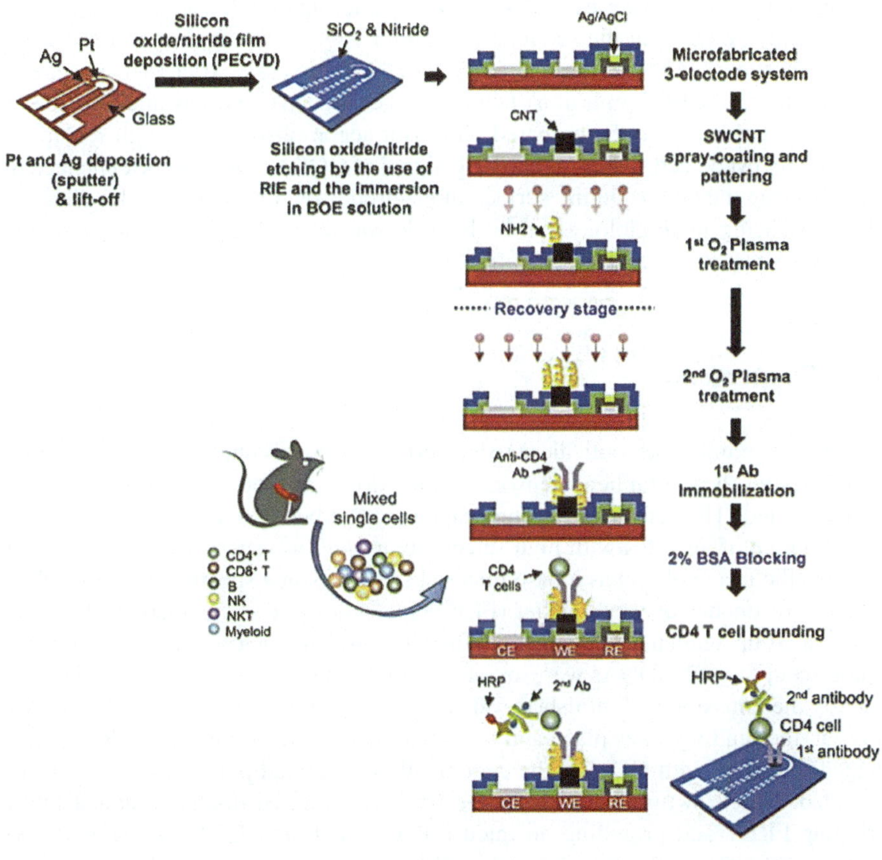

Fig. 6 Flowdiagram of the SWCNT immunosensor and its CD4 T cell capture (Reproduced with permission from Ref. [58])

Carbonaceous nanostructures have drawn great attention over the preceding years as imperative materials for sensing applications. CQDs is found to be a highly capable material for electrochemical biosensing by the number of reports. Recently, Yaqin Chai et al. prepared GQDs (2.3 nm) to construct an ECL biosensor derived from T7 exonuclease-assisted cyclic amplification and 3D DNA-mediated Ag enrichment for the analysis of miRNA [59]. Aminated 3,4,9,10-perylenetetracarboxylic acid (PTCA-NH_2) was loaded to GQDs by means of π–π stacking (GQDs/PTCA-NH_2), improving the application of GQDs in solid-state. Fe_3O_4–Au core-shell type nanocomposite was accepted as a probe to generate a new ECL signal tag for GQDs/PTCA-NH_2/Au@Fe_3O_4 nanocomposite. The hairpin probe functionalized on the electrode material was opened by helper DNA, which is then followed by accumulation of target to hybridize with the bare stem of the helper DNA. After that, T7 exonuclease was engaged to digest the DNA/RNA complex and activate the target recycling without demanding an exact recognition unit at the target sequence. Thus, it helps in the detection series of RNA/DNA by altering only the arrangement of the complementary DNA in the sequence. The ECL signal was again improved using Ag nanoparticles (AgNPs) derived from 3D DNA networks. After two subsequent amplifications, the ECL emission of GQDs was enhanced extremely and by virtue of that increment in signal the prepared biosensor accomplished very high sensitivity having 0.83 fM detection limits. The ECL signal of GQDs could also be enhanced by immobilizing the GQDs on the surface of electrodes owing to the smaller distance. This biosensing methodology can be long-drawn-out to discover a wide range of DNA/RNA targets for real-time applications.

5.2 Biomedicine

Recently, a number of optical sensors derived from graphene have been introduced. Using these graphene sensors various drugs and other biomolecules have been detected. The sensors based on graphene have been so far utilized to sense a variety range of analytes with high selectivity as well as sensitivity [9]. Xing et al. reported the usage of GO as a base material to sense some specific DNA by a fluorescence resonance energy transfer (FRET) technology. The members of this same research group again used this methodology to sense and measure the amount of K^+. Fluorescein-tagged ssDNAs were arrested onto the surface of GO sheets and consequently the emission is diminished. When the complementary strand of the DNA is present, the binding takes place and that restricts the interaction of the fluorescent-tagged sequence with GO. For the cationically conjugated polymeric systems PFP is added, which in turn interacts to the freshly generated double-stranded DNA, bringing FRET and providing an intense fluorescent signal. This sensing is very sensitive towards any single mismatches. When two or more mismatches are present in the sequence it remains bonded to the surface of GO. This is one of the best examples of a DNA sensor with 40 pM detection limit which possesses a fluorescence turn-on ratio of 7.60 when GO is present. Nanomaterials derived from graphene containing different degrees of oxidation were incorporated into tetronic tyramine hydrogels through the cross-linking process driven by certain enzymes [9]. The oxidation of graphene combined with amphipathic tetronic tyramine imperatively upgraded the amount of dispersibility of GO in aqueous media, resulting in a noteworthy improvement of composite hydrogels, i.e., tetronic–tyramine/GO. These types of nanocomposite hydrogel can be employed as injectable matrices for biomedical needs. Choong et al. prepared chemical vapour deposition (CVD)-aided methodology to produce 3D graphene foams (GFs). These were simultaneously spin-coated in the presence of certain polymeric materials to generate polymer-embedded 3D GFs which were extraordinarily conductive and flexible [60]. It opens up a new direction of upcoming next-generation uses of polymer-enriched 3D GFs to treat the problems associated with bone problems and also in further biomedical applications. Another set of experiments conveyed graphene platelet (GPL)-reinforced alumina (Al_2O_3) nanocomposites with excellent mechanical characteristics, and it also demonstrates best in vitro biocompatibility [61]. Mechanical features of Al_2O_3 medium are imperatively increased via addition of GPLs. The GPL/Al_2O_3 composites also possess very good cytocompatibility. Its outstanding mechanical and biological properties permit them to be used very extensively in various engineering and biological needs. In this regard, Qingjie Ma et al. reported a fluorescent/photoacoustic imaging-directed photothermal theranostics agent by embedding Au nanoparticles on the surface of the GO sheet. This improved the photo conversion efficacy and increased the photothermal tumour ablation effect of the existing materials [62]. The photothermal effect of GO/Au nanohybrid material was imperatively increased as compared to bare GO only. A number of studies inspire the usage of hybrid nanomaterials for image-assisted superior photothermal therapy in biomedicine applications, exclusively in cancer theranostics. The research group of Dong-Eum Kim has established a simplistic fluorometric device to detect miRNA by means of rolling loop amplification. For this purpose, GO and fluorescently tagged peptide nucleic acid were used [63]. The isothermally improved material having miRNA sequences is adsorbed onto the GO monolayer in a comparatively lesser amount and this weakens the quenching of fluorescence that was done by GO. Mandler et al. reported a method that was initiated by electrochemically triggered drug delivery interface based on a flexible electrode involving thin Au films, deposited onto Kapton coated with doxorubicin incorporated RGO thin films through electrophoretic deposition [64].

The belongings of CNT have been explored to generate different types of biosensors [9]. CNT can be further functionalized with effective biological sensing elements, like nucleic acids (DNA, RNA) and peptides. They can also be improved using appropriate groups proficient in distinguishing biomolecules like enzymes, antibodies etc. or regulating bioprocess. A noteworthy improvement was immobilization of glucose oxidase on the lateral walls of SWCNT to detect glucose molecules, which was first observed by Besteman et al. [65]. The semiconducting SWCNT with glucose oxidase coating was observed to be used for pH alterable biosensors and presented an increment in conductance upon addition of glucose, signifying the usage of them as a biosensor based on the enzymatic activity. However, another research group also used SWCNT but that is stabilized by sodium cholate and functionalized using derivatives of boronic acid to detect glucose selectively [9]. These biosensors allowed glucose detection limit ranges from 5 to 30 mM, which is biologically very important. Ahn et al. described a tag-free strategy accomplished by single protein–protein interaction detection spending an SWCNT fluorescent sensor [66]. Theranostics systems derived from SWCNT have found significant potential in dual modalities imaging-directed PTT/PDT together to treat tumours. CNTs functionalized with aptamer and horseradish peroxidase (HRP) were utilized as a probe to increase the impedimetric sensing performance of the aptamer/thrombin interaction. The HRP catalysed oxidation of 3,3-diaminobenzidine (DAB) in the presence of H_2O_2. The post electro-polymerization of insoluble precipitates produced from the enzymatic catalysis on the electrode supports was utilized as a signal enhancement route for the sensing process. Thrombin was sensed by aptamer 1 embedded on a GCE. The MWCNT-aptamer 2-HRP (MWCNT–apt2–HRP) device was connected to the complex (aptamer 1-thrombin) by means of the interaction with thrombin-aptamer 2. The post electro-polymerization of catalysed precipitates of DAB on the electrode potentially enhanced the transfer of electrons and creates a resistance barrier at the interface electrode material and the electrolyte solution. Cyclic voltammetry (CV) and electrochemical impedance spectroscopy were used to survey the subsequent stepwise production of the aptasensor and the proper detection of thrombin. Employing this method very low concentration of thrombin (~0.05 pM) could be measured. Herein, the reaction between the aptamer and thrombin may be significantly enhanced by means of MWCNT–apt2–HRP hybrids and post electro-polymerization of the residues generated by the enzymatically catalysed reaction within HRP and DAB. This type of material could be extensively used for the normal detection of various proteins with higher sensitivity and lower detection limits. Various CNT-based biosensors for the recognition of nucleic acids have also been developed and reported. The fluorescence emission of DNA-modified SWCNT has also been employed for concurrent detection of genotoxic analytes, chemotherapeutic alkylating drugs, reactive oxygen species as well as single-molecule detection of H_2O_2. Also, an easy and ultrasensitive miRNA electrochemical biosensor using MWCNT–polyamidoamine (PAMAM) dendrimer and methylene blue redox indicator is also developed. miRNA expression is accompanied by cancer beginning, tumour stages and response of tumour towards treatment. A GCE was functionalized with MWCNT–PAMAM, on which the oligonucleotide detention probes are immobilized. The functionalization of MWCNTs with PAMAM upholds the electrochemical properties of MWCNTs towards MB reduction but reduces the undesired adsorption of MB on the exterior of MWCNT with a 0.5 fM detection limit was found.

Nowadays CDs have been drawn potential attention towards the detection of biomolecules, a variety of drugs, hydrogen sulphide, miRNA, metal ions and glucose. To medical health, the expression levels of these biomolecules are very strictly associated. These have been sensed in various ways, along with fluorescence 'turn-off turn-on' approaches, and fluorescence quenching, which improved electrodes and target recognition. Yunsheng Xia demonstrated and reported the use of a fluorescence sensor [67]. This is an easy and operative approach that uses boronic acid-modified CDs for the detection of blood glucose levels. The boronic acid-modified particles were prepared via a single-step 'synthesis modification integration approach' by means of a phenyl boronic acid molecule as the sole precursor. The nanoparticles and CDs synthesized thus contain a high volume of boronic acid groups at the surface of CDs, which work as glucose recognition sites. In the presence of glucose, cross-linking of CDs takes place because of binding to receptor sites on different particles, which generates CDs emission quenching. This chemo sensor based on CDs is also cost-effective as it needs no costly enzymes and tedious surface functionalization procedures. This system has a detection limit of 1.5 mM. It has selectivity towards a variety of amino acids, a variety of carbohydrates like maltose, lactose, fructose and sucrose, and other important biomolecules. An additional example of the uses of CDs for cellular sensing was described by the Wu group [68]. They have produced a fluorescence resonance energy transfer (FRET) ratio metric CDs sensor for the detection of H_2S in water media and live cells. H_2S detection is very important as it is linked to diseases like arterial and pulmonary hypertension as well as Alzheimer's disease. For this purpose CDs work as the energy donor as well as the anchoring site for the probe. The sensor using CDs functionalized azo-naphthalimide probe is diminished to an amino-naphthalimide when H_2S is present. In absenteeism of H_2S, CDs exhibits blue emission. The generation of the amino-naphthalimide builds a suitable energy acceptor towards the CDs emission, which gives green FRET emission at 526_{nm}. The presence of the green emission was used to specify the occurrence of H_2S in HeLa cells as well as murine aneuploid fibrosarcoma cells. CDs are promising drug carriers due to their significant biocompatible and optical properties. Yanli Zhao and his group reported tumour extracellular microenvironment-sensitive drug nanocarriers based on cisplatin (IV) pro-drug incorporated charge exchangeable CDs (CDs–Pt(IV)@PEG-(PAH/DMMA)) for imaging-directed delivery of drug at the target site of interest [69]. An anionic polymer containing dimethyl maleic acid (PEG-(PAH/DMMA)) on the produced CDs–Pt(IV)@PEG-(PAH/DMMA) could undertake intriguing charge alteration to a cationic polymer in a slightly acidic extracellular microenvironment of tumour, leading to steady electrostatic repulsion which in turn release positive CDs–Pt (IV). Positively charged nanocarriers demonstrate high affinity towards negatively charged cell membrane of cancer cells. CDs increased the therapeutic effects of smart drug nanocarriers in vitro. Thus, these nanocarriers find potential clinical applications in the treatment of cancer. Shenqiang Zou and colleagues reported a novel theranostics gadolinium-doped CDs-based nanoplatform fabricated specifically for magnetic resonance imaging (MRI) of tumours and also radiotherapy of tumours [70]. Gadolinium-doped CDs work as an efficient passive tumour targeting agent (T1 contrast agents) *in vivo*, demonstrating high longitudinal relaxation efficiency, large circulation period, radio sensitization increments and complimentary biocompatibility. Recently, CNSs are being widely used in brain-targeted drug delivery [38]. The desirable size of CNSs allows blood–brain (BBB) barrier penetration and also permits easy loading or conjugation of drugs molecules. Therefore, CNSs are very helpful to deliver chemicals and drugs into the brain and can be used to treat neurodegenerative disorders and tumours in the brain [38]. Figure 7

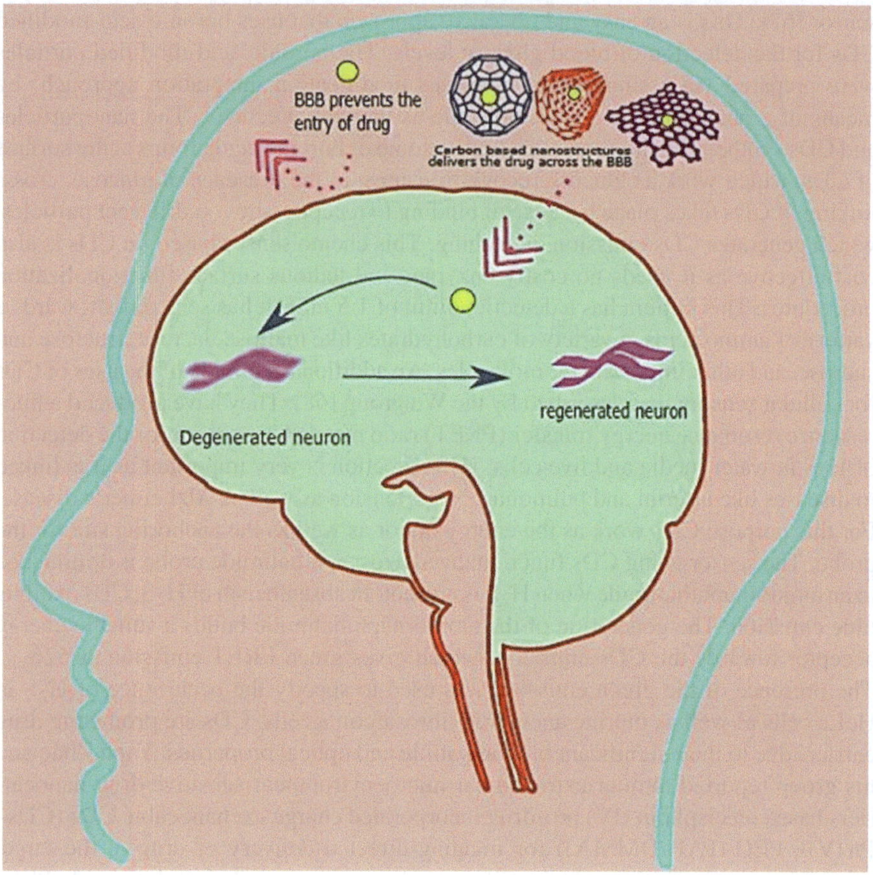

Fig. 7 Uses of CNSs for neurodegenerative disorders (Reproduced with permission from Ref. [38])

displays an illustrative demonstration of the uses of CNSs in brain-targeted drug delivery for neurodegenerative disorders.

5.3 Biological Imaging

Solution phase GO synthesis is easy; therefore, GO has maximum use for biomedical imaging propose as fluorescent graphene material. Sometimes back, Dai et al. reported their article using the intrinsic NIR emitting PEGylated GO for targeted cell imaging [71]. In their study, the very small-sized GO sheet was cracked into smaller fragments at the time of PEGylation of the certain functional groups (mainly carboxyl) available at the surfaces of the GO producing GO–PEG with ~20 nm size. Ultrahigh discernment towards the targeted fluorescence imaging in the cost of

NGO–PEG depicts individual degrees of expression of CD20 protein on cell lines and proposes that the fluorescent NGO may further be utilized in molecular phenotyping with adequate sensitivity. Moreover, the PL spectrum of GO is also used in the microscopic imaging of live cells in the NIR which exhibits a very minute amount of autofluorescence background effects. Logic gates of the imaging purpose of metal ions of various valancies were functional in the case of living cells. In this context Zhang et al. have prepared a 'smart' fluorescent device which responds to the change in concentration of Fe^{3+} ions using selective fluorescence quenching characteristics exhibited by GO [72]. Only GO cannot be utilized as a probe for molecular imaging for some membrane receptor proteins. But they can be used in intracellular functional microscopic imaging to detect the concentration of some specific metal ions. The bioimaging technology centred on graphene is also applicable for imaging many other small to large biomolecules present in living cells. Renowned scientist Li and his team first reported a nanosensor probe that produces direct fluorescence activation imaging of Cyt c. During the programmed cell death, i.e., cell apoptosis this Cyt c is released from mitochondria in the cells [73]. This approach depends on the spatially selective cytosolic release of fluorophore-labelled DNA aptamer on PEGylated graphene nanosheets-based nanosensors. The cytosolic delivery of the molecule Cyt c has the capability to separate the aptamer from graphene which in turn activates a fluorescence signal. This is the first time when the spatially selective immobilization of a nanosensor is fabricated for a 'turn-on' fluorescence imaging of intracellular translocation proceedings in live cells. In vitro, the sensor showed a large signal-to-background ratio. Also, fluorescence imaging based on hybridization chain reaction as well as the detection of intracellular telomerase concentration was produced using GO. The nanoflare probe contains gold nanoparticles modified with a dense shell of nucleic acid arrangements by the formation of Au–S bond. This nucleic acid arrangement is made up of many sequences like thiol-labelled sequence, telomerase primer sequence, as well as FAM-terminated reporter sequence. This hybridization chain reaction goes through the formation of two FAM-functionalized hairpin sequences that are adsorbed onto the surface of GO. This effort can sensitively recognize the activity of telomerase in live cells and also differentiate normal cells from cancer-affected ones. Recently, a receptor based on folate and cathepsin B activatable nanoprobes was developed with the intention of background-free microscopy and selective therapy in cancer cells [9]. This nanoprobe is fabricated by the noncovalent accumulation of phospholipid/poly (ethylene oxide) to functionalize folate and photosensitizer-tagged peptide onto the superficial areas of GO. After a certain time and after the selective internalization of the nanoprobe inside the lysosome of the cancer cells by means of folate receptor facilitated endocytosis, the peptide can be broken down to release the photosensitizer in the presence of cancer-related cathepsin B, which generates 18-fold fluorescence increment for cancer cell discrimination and detailed detection of intracellular cathepsin B. When irradiation is subjected, the released photosensitizer encourages the production of cytotoxic singlet oxygen to trigger photosensitive lysosomal cell death. After lysosomal damage, the irradiated photosensitizer disperses from the lysosome to the cytoplasm of the cell; this provides a visible methodology for in situ monitoring of therapeutic efficiency.

SWCNTs have a high potential for application in bioimaging owing to their outstanding features, like lower toxic effects, higher photostability, nonappearance of quenching and so on if tuned appropriately [9]. Improved SWCNTs are generally used for bioimaging [9]. Though SWCNTs are covalently attached to the organic fluorophores of smaller emission wavelengths, they are found to take the image of cells with the help of confocal fluorescence microscopy. A comprehensive study proving the imaging of cells using the fluorescence of SWCNT was accomplished by Weisman and his team. They fabricated SWCNT stabilized with pluronic and macrophages and magnificently imaged the spreading of fluorescent SWCNT inside the cells which were phagocytosed by the cells and has been permeated inside cytoplasm when incubated [74]. The diffusion of SWCNTs inside the cell has been observed as confined bright areas in the NIR-II images. This discovery delivers a great chance for detecting the biological interaction of SWCNT and biosystem by means of the direct recognition of the inherent fluorescence of SWCNT. This same research community has demonstrated that the fluorescence property of SWCNT can be utilized to map the sharing of intravenously injected SWCNT inside rabbit liver tissues *ex vivo* after the collection of the organs. The basic nonlinear photoluminescence (PL) property of chemically modified multi-walled nanotubes (m-MWNT) is developed by Festy and Al-Jamal [75]. Images of m-MWNT are taken by multiphoton photoluminescence and fluorescence imaging in immobile lung epithelial cancer cells and Kupffer cells in vitro. It may be too implanted in vivo in the subcutaneous areas in the tumours. NIR excitation-driven multiphoton microscopy in the target area affords an optimum sensitivity and resolution to track m-MWNTs inside the compartments inside the cells and enables the proper imaging of tumour as well as in vivo sentinel lymph node tracking. More vast potentials of m-MWNTs comprise engaging this strategy in imaging of the live cells in the presence of m-MWNTs in physiological environment to enable drug delivery helped by imaging. In many cases, X-ray fluorescence (XRF) may also be delivered using appropriate contrast agents (CAs). For this purpose CNTs can be functionalized with a variety of various inorganic compounds. Thus, they can be certainly used as 'CAs' if physiologically deficient components are needed to be monitored. These CAs are capable of highlighting exact organelles in multiplexed XRF mapping. Hence, they can be used as multipurpose means for bioimaging. A new therapeutic system made up of SWCNT was recognized for the delivery of Chlorin e6 and was reported by Lei Zhu et al. [76]. Generally, used photosensitizers exhibit very strong NIA emissions, which helps in collective synergistic photodynamic therapy (PDT) as well as photothermal therapy (PTT). The theranostics agents derived from SWCNT has significant potential in dual modalities imaging-directed PTT/PDT collective for the management of tumours. One more example is separation based on chromatographic technique centred on the chirality discriminating attraction between CNTs and a gel enclosing a surfactants mixture. Subsequently, the chirality of CNTs is measured via the chiral angle and diameter. This study also reports chiral angle specificity and selectivity in diameter. All these specificities lead to very good single chirality purification and have very low-scale (mg) output. Effective vascular micrographs in mice were taken using purified single chirality nanotubes (9, 4). These offer superior effects in bioimaging, especially in vascular imaging.

CDs, which is found to be nanosized graphene or GO with size less than 10 nm, have a lot of resemblance with GO in optical properties, especially in fluorescence [9]. CDs are considered to be very effective fluorescent tags [18]. CDs are also potentially photostable and photobleaching-resistant, so CD-tagged cell fluorescence imaging underneath prolonged illumination is possible. Moreover, one more advantage is that CDs do not contain any toxic components in the course of staining and imaging of live organisms. Also, CDs are highly compact fluorescent probes with exceptionally small footsteps. CDs have broadly been used in in vitro micrographs of cells. Sarkar et al. published an article on tagging Ehrlich ascites carcinoma cells with CDs as well as functioning microscopy of the tagged cells through gathering fluorescence properties of the CDs under both UV as well as visible light illuminations mainly blue [77]. This strategy is induced to synthesize these hydrophilic materials at the milligram level. The in vitro fluorescence characteristics of the CDs are helpful in cell imaging. These CDs penetrate inside the cells deprived of any additional modification. Hence, the inherent fluorescence of the CDs may be utilized to find their location inside the organisms with the help of a fluorescence microscope. CDs can be functionalized with some definite metal ions to achieve fluorescence emission modification which can be further used for biological imaging purposes. Lin et al. demonstrated FL excitation-sensitive CDs (also named as full-colour Cods and shortened as F-Cods), which display remarkably equivalent intensity of emission approximately including the whole visible spectrum range. However, there is a slight shift in the wavelength of excitation [78]. So, F-Cods may work as multicolour labelling components in the case of bioimaging. Tian and his research group described CDs and Cu^{2+} conjugate that progressively drops its fluorescence blue emission at ~500 nm when Cu^{2+} concentrations gradually enhance inside the organelles [79]. CdSe/ZnS core-shell/quantum dots nanohybrid (red-coloured fluorescence is inactive to the Cu^{2+}) may generate a dual emission fluorescent device that transforms its colour from blue (controlled using CDs) to red (controlled using CdSe QDs) depending on the addition of Cu^{2+} cation. Alike, the intracellular fluorescent sensors for the detection of Fe^{3+} as well as O^{2-} have also been fabricated. Again, CDs aptamer conjugates have been produced to target, find out and stain cancer cells while the normal and healthy cell lines remain uncoloured. One more advantage is that carbon can be utilized for cell imaging as well as for live animal imaging. In this consequence, Sun et al. first demonstrated in vivo fluorescence imaging of live mice employing the prepared CDs [80]. When CDs were subcutaneously injected below the dorsal skin the bright fluorescence of green (525 nm) as well as red (620 nm) colours can be evidently observed in the injected region under the excitations of 470 and 545 nm correspondingly. Intradermal migratory movement of these injected CDs towards axillary lymph nodes may easily remain detected after it was injected into the front margin by measuring the intensity of fluorescence emission of the CDs. These results recommend that the CDs exhibit very strong fluorescence in in vivo conditions, which might impose immense benefit in biological imaging. Also, carbon dot tagged cells can be utilized for super-resolution imaging and fluorescence microscopic imaging.

In recent times, Pompa et al. have reported that using stimulated emission depletion microscopy, the subdiffraction limit spreading of fluorescent CDs adjacent to the cell nucleus has 60–80 nm size [81]. CDs have also been observed to be photoswitchable. The excitation maxima of CDs-based devices are generally found to be in the blue region, which restricts the potential applications of them towards microscopic imaging in living organisms. Li et al. have reported CDs-centred fluorescent-based biosensor to detect active selenol present in live cells [82]. These CDs can radiate yellowish-green fluorescence (Y-G-CDs). The superficial area of these CDs is then modified using 2,4-dinitrobenzenesulfonyl chloride (DNS-Cl) (covalently) which produce 2,4-dinitrobenzene modified CDs (CD-DNS). This CD-DNS acts as nanosensors to detect the molecule selenol. When this material is treated with selenocysteine, the DNS entity of CD-DNS can be easily fragmented by selenolate. It involves a nucleophilic substitution reaction, which gives rise to the production of very intense fluorescent Y-G-CD. CDs-based fluorescent sensors for the detection of selenol were fabricated for the first time and served appropriately in bodily environments. This experiment depicts that the developed material CD-DNS can work as a beneficial medium for the fluorescence imaging of external and internal selenol that are present in live cells. NIR-emitting GQDs have also been fabricated for the reorganization of internal ascorbic acid in alive organism. GQDs exhibit good two-photon fluorescence characteristics with emission maxima at 660 nm. These NIR GQDs also demonstrated very less background (fluorescence) effects in the case of a live organism to provide more high resolution at the time of fluorescence imaging. This experiment was effectively employed for two-photon microscopy of internal ascorbic acid present in the deep tissue. This strategy could produce a noteworthy effect for the investigation of various biological samples both in vitro and in vivo conditions. In 2019, the research group of Hao developed second NIR-II-producing CDs using watermelon juice as a carbon source through the hydrothermal method (Fig. 8) [52]. These CDs exhibit emission spectra from 900 to 1200 nm with a very high quantum yield (0.4%). These CDs also possess biocompatibility and easy clearance by the renal route. Hence, they are efficiently used as contrast agents for NIR-II-driven fluorescence bioimaging.

5.4 Conclusion and Future Prospects

Defective carbon materials with unique characteristics have opened up a platform for biomedical application. Carbonaceous materials such as graphene, RGO, GO, CNT, CQDs etc. are intensely studied having the concern on biomedical research. Defects in the carbon nanostructure create some exceptional properties which improve their biocapacity, cytotoxicity, dispersibility etc. and these belongings make the defected carbon materials beneficial for usage in biomedicine, bioimaging, biosensing, gene delivery, tissue engineering etc. Thus, 2D materials have opened up an opportunity for next-generation biomedical applications. However, the proper investigations of these defective carbon materials in the biomedical field are in their early development

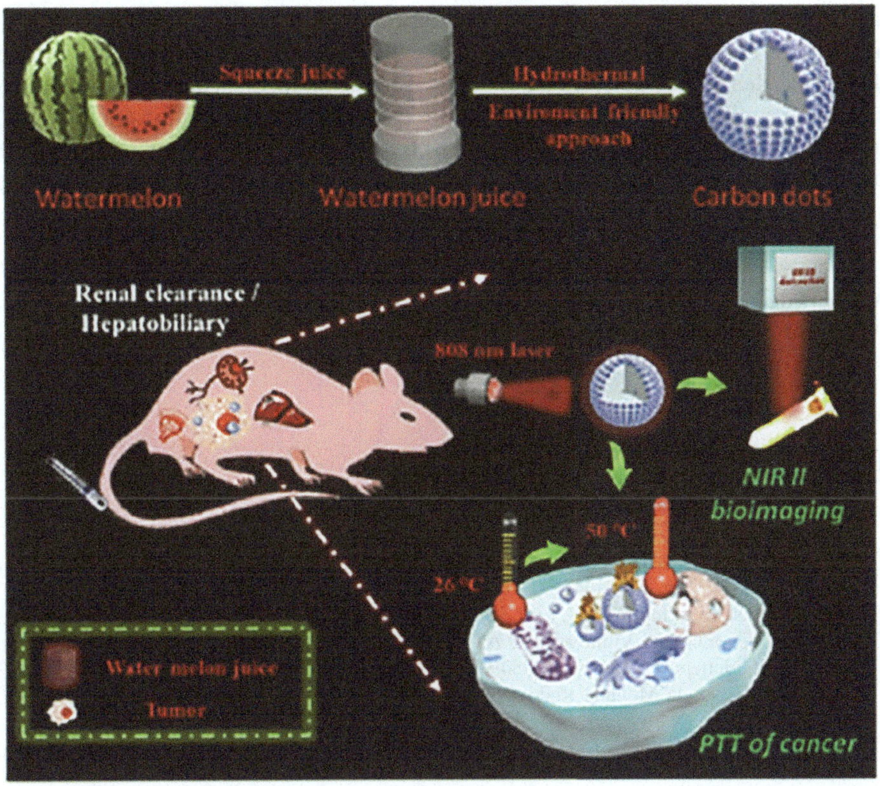

Fig. 8 Diagram of the fabrication of NIR-II-producing CDs for quick renal clearance NIR-II-guided bioimaging and PTT of cancer (Reproduced with permission from Ref. [52])

stage, and these materials still have a possibility in the broad biomedical field. For clinical testing, details of biosafety and biological evaluations of these materials are instantly required. The evaluations should be focused on the excretion, biodegradation, biodistribution and potential toxicity to specific organs, which may impact the prime development of real-time biomedical device. With the comprehensible distinction between the great promises of defective carbon materials and the incomplete studies and applications, more studies are essential to be carried out in this field to endorse its advancement furthermore.

The biomedical applications considered here only mark the commencement of a new era of applying these interesting materials to help fight challenging diseases and recover the life quality of human lives. It is obvious that the realizing of the potentiality of defective carbon materials to spread their integration in the biomedical application needs the accessibility of satisfactory quantities of safe and pure materials. Thus, some sustained efforts are required with respect to size-selective synthesis, separation and purification of CNSs for safety consideration to make these materials bright indeed in future. Appropriate methodology to synthesize CNSs in a controlled and standard way with suitable hydrophilicity, dispersity and surface functionalities are very difficult, which confines their application in vivo biological application and diagnostic/therapeutic performances. Moreover, the biomedical application of CNSs should be captivated on the biodegradation, excretion, biodistribution and potential toxicity to specific organs, which may inspire the initial development of real-time biomedical devices. Finally, CNSs can be smartly combined with other biocompatible systems, which can be additionally used for numerous therapeutics and diagnostics purposes.

Acknowledgements AR and CKM would like to acknowledge the IIT (ISM), Dhanbad, Jharkhand, India.

References

1. Reina, G., González-Domínguez, J.M., Criado, A., Vázquez, E., Bianco, A. and Prato, M., 2017. Promises, facts and challenges for graphene in biomedical applications. Chemical Society Reviews, 46(15), pp. 4400–4416.
2. Peng, J., Gao, W., Gupta, B.K., Liu, Z., Romero-Aburto, R., Ge, L., Song, L., Alemany, L.B., Zhan, X., Gao, G. and Vithayathil, S.A., 2012. Graphene quantum dots derived from carbon fibers. Nano letters, 12(2), pp. 844–849.
3. Hahm, M.G., Leela Mohana Reddy, A., Cole, D.P., Rivera, M., Vento, J.A., Nam, J., Jung, H.Y., Kim, Y.L., Narayanan, N.T., Hashim, D.P. and Galande, C., 2012. Carbon nanotube–nanocup hybrid structures for high power supercapacitor applications. Nano letters, 12(11), pp. 5616–5621.
4. Novoselov, K.S., Fal, V.I., Colombo, L., Gellert, P.R., Schwab, M.G. and Kim, K., 2012. A roadmap for graphene. nature, 490(7419), pp. 192–200.
5. Cao, X., Shi, Y., Shi, W., Lu, G., Huang, X., Yan, Q., Zhang, Q. and Zhang, H., 2011. Preparation of novel 3D graphene networks for supercapacitor applications. small, 7(22), pp. 3163–3168.
6. Li, H., He, X., Liu, Y., Huang, H., Lian, S., Lee, S.T. and Kang, Z., 2011. One-step ultrasonic synthesis of water-soluble carbon nanoparticles with excellent photoluminescent properties. Carbon, 49(2), pp. 605–609.
7. Cao, Q. and Rogers, J.A., 2009. Ultrathin films of single-walled carbon nanotubes for electronics and sensors: a review of fundamental and applied aspects. Advanced Materials, 21(1), pp. 29–53.
8. Shenderova, O.A., Zhirnov, V.V. and Brenner, D.W., 2002. Carbon nanostructures. Critical reviews in solid state and material sciences, 27(3-4), pp. 227–356.
9. Liu, H., Zhang, L., Yan, M. and Yu, J., 2017. Carbon nanostructures in biology and medicine. Journal of Materials Chemistry B, 5(32), pp. 6437–6450.
10. Krätschmer, W., Lamb, L.D., Fostiropoulos, K.H.D.R. and Huffman, D.R., 1990. Solid C 60: a new form of carbon. Nature, 347(6291), pp. 354–358.
11. Goodarzi, S., Da Ros, T., Conde, J., Sefat, F. and Mozafari, M., 2017. Fullerene: Biomedical engineers get to revisit an old friend. Materials Today, 20(8), pp. 460–480.
12. Sun, Y.P., Zhou, B., Lin, Y., Wang, W., Fernando, K.S., Pathak, P., Meziani, M.J., Harruff, B.A., Wang, X., Wang, H. and Luo, P.G., 2006. Quantum-sized carbon dots for bright and colorful photoluminescence. Journal of the American Chemical Society, 128(24), pp. 7756–7757.
13. Liu, H., Ye, T. and Mao, C., 2007. Fluorescent carbon nanoparticles derived from candle soot. Angewandte chemie, 119(34), pp. 6593–6595.

14. Gao, X., Ding, C., Zhu, A. and Tian, Y., 2014. Carbon-dot-based ratiometric fluorescent probe for imaging and biosensing of superoxide anion in live cells. Analytical chemistry, 86(14), pp. 7071–7078.
15. Lin, Z., Xue, W., Chen, H. and Lin, J.M., 2011. Peroxynitrous-acid-induced chemiluminescence of fluorescent carbon dots for nitrite sensing. Analytical chemistry, 83(21), pp. 8245–8251.
16. Wei, W., Xu, C., Ren, J., Xu, B. and Qu, X., 2012. Sensing metal ions with ion selectivity of a crown ether and fluorescence resonance energy transfer between carbon dots and graphene. Chemical Communications, 48(9), pp. 1284–1286.
17. Molaei, M.J., 2019. Carbon quantum dots and their biomedical and therapeutic applications: a review. RSC advances, 9(12), pp. 6460–6481.
18. Roy, A., Samanta, S., Singha, K., Maity, P., Kumari, N., Ghosh, A., Dhara, S. and Pal, S., 2020. Development of a Thermoresponsive Polymeric Composite Film Using Cross-Linked β-Cyclodextrin Embedded with Carbon Quantum Dots as a Transdermal Drug Carrier. ACS Applied Bio Materials, 3(5), pp. 3285–3293.
19. Polizu, S., Savadogo, O., Poulin, P. and Yahia, L.H., 2006. Applications of carbon nanotubes-based biomaterials in biomedical nanotechnology. Journal of nanoscience and nanotechnology, 6(7), pp. 1883–1904.
20. Allen, M.J., Tung, V.C. and Kaner, R.B., 2010. Honeycomb carbon: a review of graphene. Chemical reviews, 110(1), pp. 132–145.
21. Coleman, J.N., 2013. Liquid exfoliation of defect-free graphene. Accounts of chemical research, 46(1), pp. 14–22.
22. Paton, K.R., Varrla, E., Backes, C., Smith, R.J., Khan, U., O'Neill, A., Boland, C., Lotya, M., Istrate, O.M., King, P. and Higgins, T., 2014. Scalable production of large quantities of defect-free few-layer graphene by shear exfoliation in liquids. Nature materials, 13(6), pp. 624–630.
23. Chien, C.T., Li, S.S., Lai, W.J., Yeh, Y.C., Chen, H.A., Chen, I.S., Chen, L.C., Chen, K.H., Nemoto, T., Isoda, S. and Chen, M., 2012. Tunable photoluminescence from graphene oxide. Angewandte Chemie International Edition, 51(27), pp. 6662–6666.
24. Khan, M., Tahir, M.N., Adil, S.F., Khan, H.U., Siddiqui, M.R.H., Al-warthan, A.A. and Tremel, W., 2015. Graphene based metal and metal oxide nanocomposites: synthesis, properties and their applications. Journal of Materials Chemistry A, 3(37), pp. 18753–18808.
25. Chung, C., Kim, Y.K., Shin, D., Ryoo, S.R., Hong, B.H. and Min, D.H., 2013. Biomedical applications of graphene and graphene oxide. Accounts of chemical research, 46(10), pp. 2211–2224.
26. Vardharajula, S., Ali, S.Z., Tiwari, P.M., Eroğlu, E., Vig, K., Dennis, V.A. and Singh, S.R., 2012. Functionalized carbon nanotubes: biomedical applications. International journal of nanomedicine, 7, p. 5361.
27. Jain, S., Thakare, V.S., Das, M., Godugu, C., Jain, A.K., Mathur, R., Chuttani, K. and Mishra, A.K., 2011. Toxicity of multiwalled carbon nanotubes with end defects critically depends on their functionalization density. Chemical research in toxicology, 24(11), pp. 2028–2039.
28. Crinelli, R., Carloni, E., Menotta, M., Giacomini, E., Bianchi, M., Ambrosi, G., Giorgi, L. and Magnani, M., 2010. Oxidized ultrashort nanotubes as carbon scaffolds for the construction of cell-penetrating NF-κB decoy molecules. ACS nano, 4(5), pp. 2791–2803.
29. Dutta, D., Sundaram, S.K., Teeguarden, J.G., Riley, B.J., Fifield, L.S., Jacobs, J.M., Addleman, S.R., Kaysen, G.A., Moudgil, B.M. and Weber, T.J., 2007. Adsorbed proteins influence the biological activity and molecular targeting of nanomaterials. Toxicological Sciences, 100(1), pp. 303–315.
30. Cheung, W., Pontoriero, F., Taratula, O., Chen, A.M. and He, H., 2010. DNA and carbon nanotubes as medicine. Advanced drug delivery reviews, 62(6), pp. 633–649.
31. Liu, Z., Davis, C., Cai, W., He, L., Chen, X. and Dai, H., 2008. Circulation and long-term fate of functionalized, biocompatible single-walled carbon nanotubes in mice probed by Raman spectroscopy. Proceedings of the National Academy of Sciences, 105(5), pp. 1410–1415.
32. Liu, Z., Tabakman, S.M., Chen, Z. and Dai, H., 2009. Preparation of carbon nanotube bioconjugates for biomedical applications. Nature protocols, 4(9), pp. 1372–1381.

33. Liu, Z., Fan, A.C., Rakhra, K., Sherlock, S., Goodwin, A., Chen, X., Yang, Q., Felsher, D.W. and Dai, H., 2009. Supramolecular stacking of doxorubicin on carbon nanotubes for in vivo cancer therapy. Angewandte Chemie International Edition, 48(41), pp. 7668–7672.
34. Liao, X. and Zhang, X., 2011. Preparation, characterization and cytotoxicity of carbon nanotube–chitosan–phycocyanin complex. Nanotechnology, 23(3), p. 035101.
35. Liao, X., Zhang, B., Wang, X., Yan, H. and Zhang, X., 2011. Purification of C-phycocyanin from Spirulina platensis by single-step ion-exchange chromatography. Chromatographia, 73(3-4), pp.291-296.
36. Kim, D.G., Jang, M.J., Choi, C.Y., Kim, T.H., Jang, M.K. and Nah, J.W., 2007. Enhance of tumor targeting by receptor-mediated endocytosis using low molecular water-soluble chitosan nanoparticles loaded with anticancer agent. In Key Engineering Materials (Vol. 342, pp. 469–472). Trans Tech Publications Ltd.
37. Piret, J.P., Detriche, S., Vigneron, R., Vankoningsloo, S., Rolin, S., Mendoza, J.M., Masereel, B., Lucas, S., Delhalle, J., Luizi, F. and Saout, C., 2010. Dispersion of multi-walled carbon nanotubes in biocompatible dispersants. Journal of Nanoparticle Research, 12(1), pp. 75–82.
38. Henna, T.K., Raphey, V.R., Sankar, R., Shirin, V.A., Gangadharappa, H.V. and Pramod, K., 2020. Carbon nanostructures: the drug and the delivery system for brain disorders. International Journal of Pharmaceutics, p. 119701.
39. Wang, H., Maiyalagan, T. and Wang, X., 2012. Review on recent progress in nitrogen-doped graphene: synthesis, characterization, and its potential applications. Acs Catalysis, 2(5), pp. 781–794.
40. Reddy, A.L.M., Srivastava, A., Gowda, S.R., Gullapalli, H., Dubey, M. and Ajayan, P.M., 2010. Synthesis of nitrogen-doped graphene films for lithium battery application. ACS nano, 4(11), pp. 6337–6342.
41. Jin, Z., Yao, J., Kittrell, C. and Tour, J.M., 2011. Large-scale growth and characterizations of nitrogen-doped monolayer graphene sheets. Acs Nano, 5(5), pp. 4112-4117.
42. Deng, D., Pan, X., Yu, L., Cui, Y., Jiang, Y., Qi, J., Li, W.X., Fu, Q., Ma, X., Xue, Q. and Sun, G., 2011. Toward N-doped graphene via solvothermal synthesis. Chemistry of Materials, 23(5), pp. 1188–1193.
43. Panchakarla, L.S., Subrahmanyam, K.S., Saha, S.K., Govindaraj, A., Krishnamurthy, H.R., Waghmare, U.V. and Rao, C.N.R., 2009. Synthesis, structure, and properties of boron-and nitrogen-doped graphene. Advanced Materials, 21(46), pp. 4726–4730.
44. Ghosh, A., Late, D.J., Panchakarla, L.S., Govindaraj, A. and Rao, C.N.R., 2009. NO_2 and humidity sensing characteristics of few-layer graphenes. Journal of Experimental Nanoscience, 4(4), pp. 313–322.
45. Geng, D., Chen, Y., Chen, Y., Li, Y., Li, R., Sun, X., Ye, S. and Knights, S., 2011. High oxygen-reduction activity and durability of nitrogen-doped graphene. Energy & Environmental Science, 4(3), pp. 760–764.
46. Sheng, Z.H., Shao, L., Chen, J.J., Bao, W.J., Wang, F.B. and Xia, X.H., 2011. Catalyst-free synthesis of nitrogen-doped graphene via thermal annealing graphite oxide with melamine and its excellent electrocatalysis. ACS nano, 5(6), pp. 4350–4358.
47. Li, X., Wang, H., Robinson, J.T., Sanchez, H., Diankov, G. and Dai, H., 2009. Simultaneous nitrogen doping and reduction of graphene oxide. Journal of the American Chemical Society, 131(43), pp. 15939–15944.
48. Wang, Y., Shao, Y., Matson, D.W., Li, J. and Lin, Y., 2010. Nitrogen-doped graphene and its application in electrochemical biosensing. ACS nano, 4(4), pp. 1790–1798.
49. Banerjee, A.N., 2018. Graphene and its derivatives as biomedical materials: future prospects and challenges. Interface Focus, 8(3), p. 20170056.
50. Lee, S.U. and Han, Y.K., 2004. Structure and Stability of the Defect Fullerene Clusters of C_{60}: C_{59}, C_{58}, and C_{57}. The Journal of chemical physics, 121(8), pp. 3941–3942.
51. Hu, Y.H. and Ruckenstein, E., 2003. Ab initio quantum chemical calculations for fullerene cages with large holes. The Journal of chemical physics, 119(19), pp. 10073–10080.
52. Bakry, R., Vallant, R.M., Najam-ul-Haq, M., Rainer, M., Szabo, Z., Huck, C.W. and Bonn, G.K., 2007. Medicinal applications of fullerenes. International journal of nanomedicine, 2(4), p. 639.

53. Ding, H., Zhou, X.X., Wei, J.S., Li, X.B., Qin, B.T., Chen, X.B. and Xiong, H.M., 2020. Carbon dots with red/near-infrared emissions and their intrinsic merits for biomedical applications. Carbon.
54. Dong, Y.P., Zhou, Y., Wang, J. and Zhu, J.J., 2016. Electrogenerated chemiluminescence resonance energy transfer between Ru (bpy) 32+ electrogenerated chemiluminescence and gold nanoparticles/graphene oxide nanocomposites with graphene oxide as coreactant and its sensing application. Analytical chemistry, 88(10), pp. 5469–5475.
55. Ge, L., Wang, W., Sun, X., Hou, T. and Li, F., 2016. Affinity-mediated homogeneous electrochemical aptasensor on a graphene platform for ultrasensitive biomolecule detection via exonuclease-assisted target-analog recycling amplification. Analytical chemistry, 88(4), pp. 2212–2219.
56. Barone, P.W., Baik, S., Heller, D.A. and Strano, M.S., 2005. Near-infrared optical sensors based on single-walled carbon nanotubes. Nature materials, 4(1), pp. 86–92.
57. Jin, Z.H., Liu, Y.L., Chen, J.J., Cai, S.L., Xu, J.Q. and Huang, W.H., 2017. Conductive polymer-coated carbon nanotubes to construct stretchable and transparent electrochemical sensors. Analytical chemistry, 89(3), pp. 2032–2038.
58. Merum, S., Veluru, J.B. and Seeram, R., 2017. Functionalized carbon nanotubes in bio-world: applications, limitations and future directions. Materials Science and Engineering: B, 223, pp. 43–63.
59. Zhang, P., Zhuo, Y., Chang, Y., Yuan, R. and Chai, Y., 2015. Electrochemiluminescent graphene quantum dots as a sensing platform: a dual amplification for microRNA assay. Analytical chemistry, 87(20), pp. 10385–10391.
60. Wang, J.K., Xiong, G.M., Zhu, M., Özyilmaz, B., Castro Neto, A.H., Tan, N.S. and Choong, C., 2015. Polymer-enriched 3D graphene foams for biomedical applications. ACS applied materials & interfaces, 7(15), pp. 8275–8283.
61. Liu, J., Yang, Y., Hassanin, H., Jumbu, N., Deng, S., Zuo, Q. and Jiang, K., 2016. Graphene–alumina nanocomposites with improved mechanical properties for biomedical applications. ACS applied materials & interfaces, 8(4), pp. 2607–2616.
62. Gao, S., Zhang, L., Wang, G., Yang, K., Chen, M., Tian, R., Ma, Q. and Zhu, L., 2016. Hybrid graphene/Au activatable theranostic agent for multimodalities imaging guided enhanced photothermal therapy. Biomaterials, 79, pp. 36–45.
63. Hong, C., Baek, A., Hah, S.S., Jung, W. and Kim, D.E., 2016. Fluorometric detection of microRNA using isothermal gene amplification and graphene oxide. Analytical chemistry, 88(6), pp. 2999–3003.
64. He, L., Sarkar, S., Barras, A., Boukherroub, R., Szunerits, S. and Mandler, D., 2017. Electrochemically stimulated drug release from flexible electrodes coated electrophoretically with doxorubicin loaded reduced graphene oxide. Chemical Communications, 53(28), pp. 4022–4025.
65. Besteman, K., Lee, J.O., Wiertz, F.G., Heering, H.A. and Dekker, C., 2003. Enzyme-coated carbon nanotubes as single-molecule biosensors. Nano letters, 3(6), pp. 727–730.
66. Ahn, J.H., Kim, J.H., Reuel, N.F., Barone, P.W., Boghossian, A.A., Zhang, J., Yoon, H., Chang, A.C., Hilmer, A.J. and Strano, M.S., 2011. Label-free, single protein detection on a near-infrared fluorescent single-walled carbon nanotube/protein microarray fabricated by cell-free synthesis. Nano letters, 11(7), pp. 2743–2752.
67. Shen, P. and Xia, Y., 2014. Synthesis-modification integration: one-step fabrication of boronic acid functionalized carbon dots for fluorescent blood sugar sensing. Analytical chemistry, 86(11), pp. 5323–5329.
68. Yu, C., Li, X., Zeng, F., Zheng, F. and Wu, S., 2013. Carbon-dot-based ratiometric fluorescent sensor for detecting hydrogen sulfide in aqueous media and inside live cells. Chemical Communications, 49(4), pp. 403–405.
69. Feng, T., Ai, X., An, G., Yang, P. and Zhao, Y., 2016. Charge-convertible carbon dots for imaging-guided drug delivery with enhanced in vivo cancer therapeutic efficiency. ACS nano, 10(4), pp. 4410–4420.

70. Du, F., Zhang, L., Zhang, L., Zhang, M., Gong, A., Tan, Y., Miao, J., Gong, Y., Sun, M., Ju, H. and Wu, C., 2017. Engineered gadolinium-doped carbon dots for magnetic resonance imaging-guided radiotherapy of tumors. Biomaterials, 121, pp. 109–120.
71. Sun, X., Liu, Z., Welsher, K., Robinson, J.T., Goodwin, A., Zaric, S. and Dai, H., 2008. Nano-graphene oxide for cellular imaging and drug delivery. Nano research, 1(3), pp. 203–212.
72. Mei, Q., Jiang, C., Guan, G., Zhang, K., Liu, B., Liu, R. and Zhang, Z., 2012. Fluorescent graphene oxide logic gates for discrimination of iron (3+) and iron (2+) in living cells by imaging. Chemical Communications, 48(60), pp. 7468–7470.
73. Chen, T.T., Tian, X., Liu, C.L., Ge, J., Chu, X. and Li, Y., 2015. Fluorescence activation imaging of cytochrome c released from mitochondria using aptameric nanosensor. Journal of the American Chemical Society, 137(2), pp. 982–989.
74. Cherukuri, P., Bachilo, S.M., Litovsky, S.H. and Weisman, R.B., 2004. Near-infrared fluorescence microscopy of single-walled carbon nanotubes in phagocytic cells. Journal of the American Chemical Society, 126(48), pp. 15638–15639.
75. Rubio, N., Hirvonen, L.M., Chong, E.Z., Wang, J.T.W., Bourgognon, M., Kafa, H., Hassan, H.A.F.M., Al-Jamal, W.T., McCarthy, D., Hogstrand, C. and Festy, F., 2015. Multiphoton luminescence imaging of chemically functionalized multi-walled carbon nanotubes in cells and solid tumors. Chemical Communications, 51(45), pp. 9366–9369.
76. Xie, L., Wang, G., Zhou, H., Zhang, F., Guo, Z., Liu, C., Zhang, X. and Zhu, L., 2016. Functional long circulating single walled carbon nanotubes for fluorescent/photoacoustic imaging-guided enhanced phototherapy. Biomaterials, 103, pp. 219–228.
77. Ray, S.C., Saha, A., Jana, N.R. and Sarkar, R., 2009. Fluorescent carbon nanoparticles: synthesis, characterization, and bioimaging application. The Journal of Physical Chemistry C, 113(43), pp. 18546–18551.
78. Pan, L., Sun, S., Zhang, A., Jiang, K., Zhang, L., Dong, C., Huang, Q., Wu, A. and Lin, H., 2015. Truly fluorescent excitation-dependent carbon dots and their applications in multicolor cellular imaging and multidimensional sensing. Advanced materials, 27(47), pp. 7782–7787.
79. Qu, Q., Zhu, A., Shao, X., Shi, G. and Tian, Y., 2012. Development of a carbon quantum dots-based fluorescent Cu 2+ probe suitable for living cell imaging. Chemical communications, 48(44), pp. 5473–5475.
80. Yang, S.T., Cao, L., Luo, P.G., Lu, F., Wang, X., Wang, H., Meziani, M.J., Liu, Y., Qi, G. and Sun, Y.P., 2009. Carbon dots for optical imaging in vivo. Journal of the American Chemical Society, 131(32), pp. 11308–11309.
81. Leménager, G., De Luca, E., Sun, Y.P. and Pompa, P.P., 2014. Super-resolution fluorescence imaging of biocompatible carbon dots. Nanoscale, 6(15), pp. 8617–8623.
82. Wang, Q., Zhang, S., Zhong, Y., Yang, X.F., Li, Z. and Li, H., 2017. Preparation of yellow-green-emissive carbon dots and their application in constructing a fluorescent turn-on nanoprobe for imaging of selenol in living cells. Analytical chemistry, 89(3), pp. 1734–1741.

Hetero Atom Doped Carbon Nanomaterials for Biological Applications

Moganapriya Chinnasamy, Rajasekar Rathanasamy, Sathish Kumar Palaniappan, Surya Selvam, Gobinath Velu Kaliyannan, and Saravanakumar Jaganathan

Abstract A huge effort has been made in terms of synthesis, efficiency, mechanism and applications of carbon dots (CDs), as an emerging lighting material. CDs are used as non-harmful substitutes to replace conventional heavy metal quantum dots with high photostability, chemical stability, low cytotoxicity and high quantic yield. Graphene Quantum dots (GQDs) are the new Graphene-based nanomaterial, which have inspired comprehensive research in the fields of environmental, biological and other part of sciencedue to its low in vivo cytotoxicity, excellent stability, and durability. There are several GQD preparation methods available with special properties, such as absorption, PL, and electroluminescence, which can be accomplished with specific size tuning and practical adjustment methods by controlling the band difference. The current book chapter is focused on the heteroatom doping of such carbon nanostructures for biological applications.

Keywords Arbon nanomaterials · Hetero atom doping · Biological application · Nanostrucutres

M. Chinnasamy · R. Rathanasamy (✉) · S. Selvam
Department of Mechanical Engineering, Kongu Engineering College, Erode 638060, Tamil Nadu, India

S. K. Palaniappan
Department of Mining Engineering, Indian Institute of Technology Kharagpur, Kharagpur 721302, West Bengal, India

G. V. Kaliyannan
Department of Mechatronics Engineering, Kongu Engineering College, Erode 638060, Tamil Nadu, India

S. Jaganathan
Department of Engineering, Faculty of Science and Engineering, University of Hull, Hull HU6 7RX, UK

S. Sahoo et al. (eds.), *Defect Engineering of Carbon Nanostructures*, Advances in Material Research and Technology, https://doi.org/10.1007/978-3-030-94375-2_2

1 Introduction

Different allotropes of carbon have attracted significant interests from electronic components to bioimaging devices for their prospective applications. In the last 10 years, graphitic structural materials have been studied extensively, such as zero-dimensional spherical fullerene [1–4], DNs [2, 5–7], CODs [8–12], cylindrical 1D Carbon (CNT) [13–16], and 2D Graph Quantum Points (GQDs) [17, 18], Graphene [18–20]. Though DN, CD, and GQD reflect three identical quantum restricted fluorescent carbon materials, commonly utilized as biosensors and bioimagery bodies [21], spatial arrangements for carbon particles and atoms are different, resulting in distinct physical and chemical characteristics [22]. DNs are typically consists of a sp-hybridized core and a carbon layer. The CDs and GQDs are made, by contrast, mainly of oxygen, carbon, nitrogen and other heteroatomic medications [23, 24]. In fact, the CDs do not have perfectly crystal-like arrangements like GQDs [25–28]. Furthermore, the scale of light-emitting CDs is typically less than 10 nm [29–30], whereas light-emitting GQDs are up to 100 nm sideways [30–32].

2 Carbon Dots (CDs)

CDs are commonly considered as potential candidates in biosensing, imaging and other biologically-associated applications, because of their low cost, high quality, ample source, minimal cytotoxicity, outstanding chemical, and photo stabilization [24, 33–35]. CDs were first observed in the purification technique of carbon nanotubes produced in 2004 by Scrivens using arc discharge method [36]. In recent years, several precursor materials and synthetic methods, including electrochemical [26, 37–39], assisted routes [40–42], combustion/heating [43], acidic oxidation [44–46], hydrothermal [47–52], and plasma treatment have been established. These methods can be categorised in two types: upward and downward. The top-down approach refers to breaking large carbonates such as CNT [53], graphites [54], NDs [55] and industrial carbon activation. The sponsored route includes promoting the position of development of CDs by blocking the accumulation of nanoparticles through high therapy. A second artificial solution is precursor carbonisation, a simple and popular means of accessing CDs via a variety of methods of therapy such as hydrothermal, microwaveand heating. Candle was also employed to render fluorescent CDs as a precursor. The accumulated candle powder is refluent to oxide particulate matter by HNO_3 after combustion. After freezing, as-synthesized CDs are obtained through subsequent centrifugation or dialysis. Likewise, natural gas is employed to generate CDs via combustion process with a carbon source. Subsequently multiple precursors have been investigated to synthesise CDs including molecular compounds and natural sources. As molecular precursors for the attainment of luminescent CDs, Giannelis and colleagues utilized octadecylammonium citrate, sodium 11-aminoundecanoate, and ammonium citrate.

Tiny organic compounds were also commonly used as precursors. The CDs are typically enriched in functional surface groups comprising carboxy, amino, hydroxy, and others to encourage further changes to the operating properties of the optic. In several published reviews [56], the synthesis methods of CDs were systematically discussed [32].

2.1 *Synthesis of N-CDs*

The synthesis of nitrogen CDs (N-CDs) using H was reported by many researchers. Unatus fruit and marine ammonia have been used as the precursors for carbon and nitrogen, reepctively. Synthesized N-CDs were characterised by UV–Visible, fluorescence spectroscopy, high-resolution microscopy of transmission electrons (HR-TEMs), electron diffraction selection of areas (SAEDs), X-ray diffraction (XRD) and spectroscopy of the Fourier infrarot process. In addition, L-929 (Lymphoblastoid-929) and MCF-7 (Michigan Cancer Foundation-7) cells are tested for cytotoxicity and biocompatibility of N-CDs. Further, sodium borohydrate as a reducing agent was analysed for the catalytic activity of synthesised N-CDs in MB reduction [57]. Hydrothermal carbonization method was used to generate undatus extract. In the synthesis process, a mixture of approximately 29 mL extract and 1 mL of aqueous ammonia have been switched to 50 mL Teflon mounted stainless steel autoclave. The blend was held in a hot air oven at a temperature of 180 °C for 12 h. Autoclave was then refreshed to room temperature after the reaction. Further, the mixture was vigorously washed with DMSO to clear the unreacted organic moisture of H. The Undatus extract from the resulting darkened brown solution contained N-CDs. In addition, the prepared N-CDs were filtered with Whatmann 40 filter paper and ubsequently centrifuged for 30 min at 10,000 rpm to take the major particles out. At the end, it was collected and held at 4 °C for the characterization and applications [57].

The MTT assay L-929 (Mouse fibroblast cell line) and MCf-7 (Human Breast Adenocaricinoma cell line) were evaluated for in vitro cytotoxicity of the synthesised N-CDs. Microscopic images of L929 and MCF-7 treated cells with their percentage of cytotoxicity showed the N-CDs (different concentrations). The results showed that N-CDs have the capacity to anticipate MCF-7 without affecting cell viability. The corresponding images indicate that N-CDs affect the normal cells of L-929 less than the cancer cells of MCF-929 supported by the cell counts in the two cell wells. In cells with L-929, the cell viability was found to be more than 90% (Fig. 1a) and in cells with MCF-7 the cell viability in N-CDs was found to be 81% (Fig. 1b). The corresponding results also showed that N-CDs had no effect on L-929 cell viability. The N-CDs, on the other hand, have increased MCF-7 cell cytotoxicity. Prepared N-CDs induced more cell death in MCF-7 cells than L-929 cells which showed objective as well as selective activity of the prepared N-CDs. The prepared N-CDs may have followed passive diffusion mechanism on L-929 cells for cytotoxicity activity and endocytosis mechanism in anticancer activity on MCF-7 cells [57, 58]. In this regard, a precise N-CD formation is possible. Different concentrations of synthesised N-CDs for the

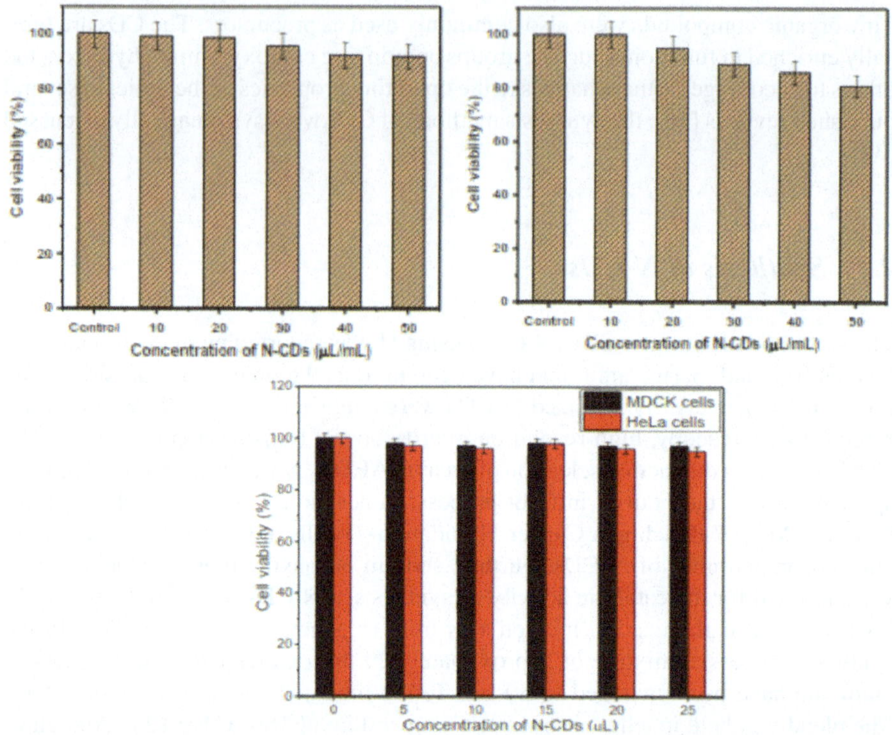

Fig. 1 **a** Cell viability (%) of L-929 cells after incubation of 24 h with various concentrations of synthesized N-CDs [57]. **b** Cell viability (%) of MCF-7 cells after incubation of 24 h with various concentrations of synthesized N-CDs [57] **c** Cytotoxicity of different concentrations of N-CDs on MDCK and HeLa cells. (Reproduced with permission from Ref. [60])

MDCK and HeLa cells were used for the CCK-8 cell viability test. Cells with a high level of concentration (25 µL) have been shown to have rather less cyto-toxicity for MDC (97 per cent) and hela cells, although the synthesised N-CDs have a very low cyto-toxicity and a strong bio-consistency for both MDCK and hela cells. This is shown in Fig. 1c. The N-CD fluoresce are ideal cells with limited cytotoxicity [59]. In the laser scanning confocal microscopy the cellular uptakes of the N-CDs were analysed and described in Fig. 1c and Fig. 2 respectively for MDCK and HeLa cells. Images show that N-CD incubated MDCK and HeLa cells are clearly visible in all thefigures. The overlay images display the N-CD's ability to infiltrate the MDCK and HeLa cell cell walls. Therefore, the synthesised N-CDs were employed successfully to stain cell imagery [60].

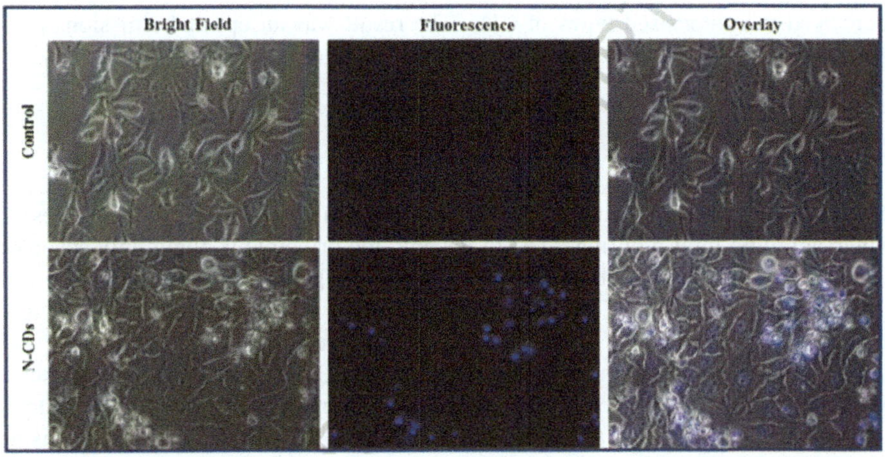

Fig. 2 Laser scanning confocal microscopy images of MDCK cells without N-CDs as a control and with N-CDs. (Reproduced with permission from Ref. [60])

2.2 *Characteristics of CDs for Bioapplications*

2.2.1 Photoluminescence and Absorbance

Though, huge effort has been made to examine the source of photoluminescence (PL), it quiet remains uncertain about the mechanism. The key emissive mechanism of CDs is currently called the quantitative effect and emission of defects. Generally, in the UV zone, CDs display heavy optical absorption with an extremity covering into the visible field. In the range of 280–360 nm, maximum of the one-step developed CDs display absorbance. Several surfaces passivations or modifications can be used to control the absorption band. CDs display typically intensity and wavelength excitement-dependant emission properties. The author stated that the passivized CDs of polyethyleneimine were colourful. Under longer wavelength excitation PL amplitude decreased exceptionally. This can result from the consistency of PEI passivated CDs in size and surface status. The excitement-free CDs have also been reported.

Agriculture-related emissions were associated with co-doped CDs (N, S-CDs) because N, S-CD's PL propensities are contingent on the external surface moderately than the morphology and S-CDs were consistent in the surface conditions of the N. Interrelated N-S-CDs have also been achieved in the previous study. Hu and Trinchi have recently recorded the new form of CDs whose wavelengths of the highest fluorescent emission are modified across the spectrum by reagent modification and synthesis conditions. While enormous work have been made to examine the source of photoluminescence, this tool is not yet clear in this research with many progresses in bio-applications ralted with the absorption and photoluminescence. The key emissive mechanism of CDs is currently called the quantitative effect and emission of defects. Still, in certain cases, a slight change in emission with different excitations were perceived in synthesised types of CDs. The result was an optimal emissions peak [32].

2.2.2 Quantum Yield

Sharp illumination and photo stability are dualisticchieffacts that must be addressed when applying light-emitting CDs. In general, the principle is commonly accepted that CDs achieve high photobleaching resistance [61]. Excellent photo stability was observed for uncluttered CDs, thermoreduced CD's, and N, S-CD's in the previous research. This supercilious optical feature attracts investigators to focus on CDs with greater QY growth. For luminescent nanomaterials, QY is a fundamental parameter. In initial tests, CDs made of candle dust, citric acid, nanofibered stapled graphite or short QY graphite are typically less than ten percent [44]. Diverse designs comprising surface modification, doping of element and salt inorganic doping were established to increase the QY of CDs. Surface passivation or alteration, due to its simple handling and high performance, is widely used to enhance CD quality. In this aspect, Sun et al. stated that non-luminescent CDs could endow with light, PEG1500N (diamine-terminated poly(ethylene glycol)). A thorough separation by an aqueous gel column provided the highest luminescent CD's with a solid 60% QYs. In a 1-stroke process, a combination of the low temperature carbonised CDs with ramified polyethylenemine have been made. Chi and his colleagues synthesised CDs with polyamine-functionalized, which possesses better QY. The QY of CDs has been improved through thermo-mining with fewer surface carboxyl groups in the previous research. Recently, CD QY was increased by elemental doping. A doping factor for nitrogen-doped CDs with huge QYs using different nitrogenic complexes has been commonly used [62]. The raw materials for CD preparation with better QYs by microwave technique was used by Jiang et al. Since certain nitrogen compounds can be used in place of mutually passivating surface as nitrogen source, doping elements and passivating surfaces of nitrogen components are difficult to distinguish. For example, Li and Yu et al.'s co-doped nitrogen, sulphur CDs with extremely high QY (73%) using hydrothermal methods. L-cysteine has been employed as passivator surface and a doped element. The N-QY, which includes nitrogen and sulphur, is the maximum (70%) of all luminescent CDs. Furthermore, Xu et al. synthesised the use of hydrothermal methods for doping sulphur with substantial fluorescence QY (67%) which showed that sulphur doping has led to successful QY enhancements. Boron is another doping element for improved CD and QY in addition to nitrogen and sulphur. The green and easy route for improving the quality of CDs through boron doping was stated by Burlinos and Zboril et al. and boron doped CDs attained considerably increased for non-linear photosensitive properties [11]. In the existence of inorganic ions, phosphorus was similarly doped into CDs to boost QY by microwave polyol pyrolysis [63]. A recent one-step approach to the synthesis of CDs with high quality has been documented by Yang et al. (about 80.3%). Glutathione and citric acid have been used as the precursors in the method [32].

2.2.3 Cytotoxicity

The essential requirement of luminescent nanomaterials is low for bioapplications. The cytotoxicity has been extensively studied in various published research articles [64]. Plentiful studies are being performed in order to investigate the in vivo properties and potentials of CDs. Previous studies have shown that in in vitro experiments, CDs are typically nontoxic [29]. The Sun Group has made a great effort to investigate the in vitros, in vivos and surface passivation agents for the biological compatibility and cytotoxicity of the surface-passivated CD's. The cytotoxicity of passive PEG1500N CDs was first tested through in vitro and vivo techniques. The cells PEG1500N-passivated CD and PEG1500N were incubating human breast cancer and colorectal adenocarcinoma. It clearly demonstrates that CDs are weaker than PEG1500N in their effect on the cell proliferation, mortality and viability. In vivo cytotoxicity and biodissemination of CDs were also studied by venous mice inoculation. In the mice, at the concentrations needed for PL bioimaging, no apparent toxic effects were noted on CDs. Some CDs were detected, but accumulation levels were very low in some of the most critical organs, like hepatitis, spleen and kidney that did not have any major toxicity. In addition, CDs were cleared in mice within approximately 24 h of renal excision, and after 28 days no clinical symptoms occurred. Moreover, four polymers were used to prepare CDs as surface passivation agents, and the cytotoxicity of CDs was found through derivative from passivation mole. Surface passivation compounds with nominal cytotoxicity are therefore ideally suited to grow highly compatible CDs to be used as bio-sensors and bio-imaging agents for bioapplications. Some researchers' studies have drawn similar conclusions. Tao et al. conducted a more than three-month in vivo cytotoxicity analysis in mices, which showed no death or a substantial decrease in body weight of treated mice [53]. Furthermore, in treated mice with injection CDs dosing (20 mg/kg), anydeceptive toxic properties have not been observed from blood and histological examinations of treated mice [32].

2.3 Bioapplications

2.3.1 Biosensors

CDs are capable of acting as either excellent electron donors or electron acceptors, along with their superior biocompatibility features. CDs may be used to detect ions, pH values, proteins, vitamins, enzymes and nucleic acid for intracellularly. Also, CDs from several raw materials have been employed to identify different ions, like Cu, Cr, Hg, and Ag. The surface function groups on CDs indicate distinctive affinities with various target ions, resulting by an electron or energy transmission process and high selectivity to other ions in the quenching of the PL strength. Qu et al. used, for example, dopamine for the synthesis of CDs as a raw material for Fe, Fe, Fe, K, Cl and H^{3+} Sensor with a strong 0.32 lM detection limit. Figure 3 demonstrates the basic process of CDs to detect Hg $^{2+}$ in aqueous solution. The interaction

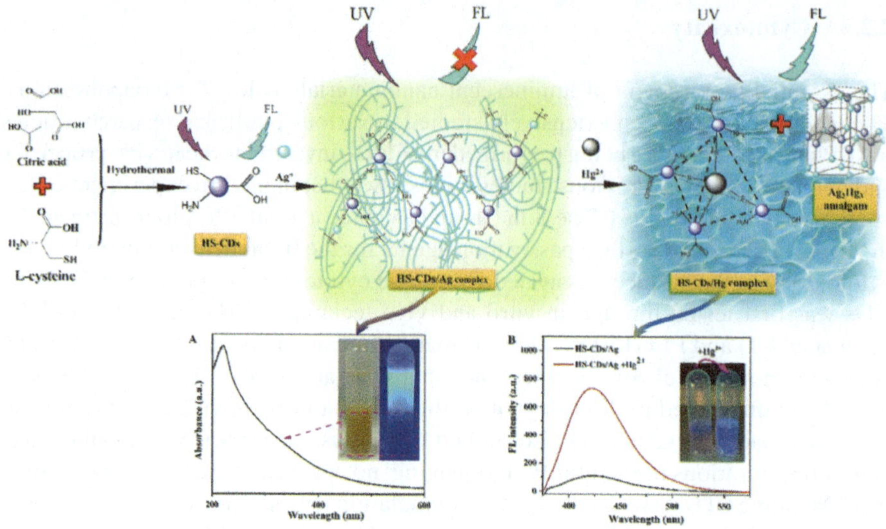

Fig. 3 Scheme of CDs to detect Hg 2+ in aqueous solution. (Reproduced with permission from Refs. [32, 65]).

between CDs and Hg^{2+} or other ions results in the quench of PL intensity. When the interaction breaks by external force, the PL is restored. The fluorescent turn-on system was recently created as well. A novel CDs (ODN) and optical graphene oxide sensor for Hg^{2+} were developed. Yang et al. demonstrated the fluorescence in DNO-CDs by fluorescence resonance energy transfer (FRET), which was efficiently extinguished by graphene oxide. The ultra-sensitive CD-based sensor was used in one-stage hydrothermal potato treatment to detect phosphate. CDs are also used for physiological assessment of living cells and tissues. Tian et al. synthesised aminomethylphenyltherpyridine (AE-TPy) CDs to measure the pH importance of changes in physiological condition. This PL sensor can be used to monitor pH values in a range between 6.0 and 8.5 with high sensitivity and selectivity. Furthermore, live cells and tumour tissue of mice were successfully added, which demonstrated further use of CD-based pH sensors both in vitro and in vivo. In recent years, the CDs have been used for detecting the intracellular pH value in living pathogenic fungal cells. CD-based sensors for nucleic acid have recently been developed. For example, Li et al. designed a CD-based DNA strategy that indicates that CDs are marked with ssDNA that quench PL from CDs. The PL was recovered when the ssDNA was separated from the cds, before the combination of objective DNA and the named ssDNA (dsDNA) was forming double stranded DNA (dsDNAs).

The interest of CD-based sensors to detect bio-molecules like vitamins and amino acids has been widespread. During our past studies, surface-functionalized CDs for target molecules were observed for Riboflavin and Vitamin B12. Due to the effect of the temperature variation and local variation of the probe concentration, the ratiometric sensing protocol is generally accepted. This is primarily due to the fact that the two intertwined PL signals are known as an index of detection. In theprevious study, CD-based ratio-metric sensors with high sensitivity (1.9 nM) were used to effectively detect riboflavin [62]. A new CD-based turn-on sensor for the detection of cysteine with great selectivity and sensitivity was recently developed. Liu and Zhang et al. also developed a nanosensor consisting of a CD and golden nanoparticle for multi-signal detection of cysteine. In addition, Yu et al. developed naphthalimide azide derivatives anchored CDs for detecting H2S, the lowest of fluorescent H2S sensors with a detection limit of 10 nM. The highly sensitive sensor was used in this process for the identification of CDs in biomedical areas not only in aqueous media and serum but also in live cells. A type of nitrogen rich was recently reported by Dong et al. [66]. The CDs could be used as double sensor platform for the electrochemical and fluorescent detection of 2,4,6-trinitrotolucióne (TNT) using a microwave-assisted pyrolysis method. The TNT amino interaction was used to improve the fluorescence sensor in this study, so as to effectively quench the brightness of amino-functional CDs through charge transference [67]. As analyzer of four heterocyclic aromatic amines, Algarra and coworkers investigated possible applications of CDs [66]. In order to measure concentration of hydroquinone in waste water, a fluorescent sensor based on CDs has also been used as depicted in Fig. 3 [32, 66].

2.3.2 Bioimaging

Although in vitro and in vivo bioimaging have been examined with semi-conductant quantum dots such as CdSe and other related core–shell nanoparticles, serious health and environmental problems restrict their bioapplications due to heavy metals. The optical imaging applications have been widely studied as an alternative to QDs for CDs with superior photostability and low cytotoxicity. Both in vitro and in vivo evaluations showed that, due to their visible stimulation and emission wavelength, high luminosity at individual dot level, CDs are excellent candidates in bioapplications. Many experiments were conducted using fluorocytes of different cell imagery cells from different cell lines including Caco-2 cells, Ehrlich ascites carcinome cells, HeLa cells, Escherichia coli cells etc. For example, for the labelling of bioimaging cell. Sun et al. have recorded the PEG1500N-passived and the PPEI-EI-passivated CDs [23]. Liu et al. produced CDs in the presence of TTDDA, which was synthesized using a one-step microwave-assisted glycerol pyrolysis. The hybrids of the HA/CDs showed a target specific delivery for the liver according to in vivo real-time biological surveys. The effect of PEG1500N-CZNS dots in the mouse body by a variety of injection methods was extensively studied by Sun's group [24, 65]. The results showed that CD's injected into mouse in different ways retained high in vivo fluorescence. CDs are very promising for bioimages and other related bioapplications due to their high biocompatibility and low cytotoxicity. The upward heavy PL and multicolor emissions are relative to downstream fluorescences. These CDs could be observed under a confocal mical microscope through multicolor PL in the cells. Divided CDs for the B16F1 cells and HEK 293 cells were developed by Hahn et al.

The microscopic images of B16F1 and HEK293 cells following incubation with CDs are shown in the confocal laser scanners Fig. 6 [56].

The effect of nitrogen doping ratio of CDs on bioimages in-vitro and in-vivo has recently been extensively studied by Wang and Leung et al. Fluorescence variant has many benefits, including noninvasive and deep NIR penetration in bioapplications. Several CDs can also recorded conversion PL properties [65]. Salinas-Castillo et al., for example, reported that PEI-CDs has excellent upgraded PL properties with a range of emissions from 308 to 550 nm with long-wavelength light excitation. These updated CDs have been used successfully for two photonic excitation in in vitro bio-imaging [23]. Moreover, Hahn et al. further developed the bioimaging application of CDs. CDs can be used in this report to bioimage target particular hyaluronic acid (HA) derivatives in real time. Moreover, Zhang, Kang, and Liu et al. have also researched in vivo NIR fluorescence imaging [32, 53].

3 Graphene Quantum Dots

Carbon-based nanomaterials have stimulated extensive research efforts over the recent decades and are known for their diverse morphologies and unique properties. As one-dimensional carbon nanotubes (CNTs) [68–73] and two-dimensional graphene [74, 75] were successively discovered, these have been applied in the fields of biotechnology, chemistry and the environment. Consequently, graphene quantum dots (GQD) were found recently in single layer, double layer and multiple layers (3 to <10) [76–78], as a graphite nanomaterial class with lateral dimension of less than 100 nm. They are superior to traditional semi-conductor QD's in terms of chemical inertity, ease of manufacture, photobleaching resistance, low cytotoxicity and excellent biocompatibility, making them very promising for captors, bioimaging, opto-electronic devices etc. [79, 80]. In addition, similar to graphhene, GQDs have excellent surface characteristics, large diameters, and finely grated grafts using the network or surface categories p-p conjugated and other physical features. In addition, their border carboxyl and hydroxyl groups allow them to demonstrate an excellent water solubility and suitability with a variety of organic, inorganic, polymeric or biological species to work successively [81–84]. The applications of GQDs in environmental and biological fields are subsequently identified. Furthermore, we speculate on some important issues for further research and future continuous growth [85].

As shown in Figs. 4 and 5, microwave irradiation is also needed for the synthesis of two colour GQDs using cleaving GO under acid conditions [86]. Simultaneous clearing and reducing processes were carried out without an external reduction agent using microwave therapy. The greenish yellow GQDs (gGQDs) are further reduced to 22.9% with $NaBH_4$ for light blue-blue luminescent GQDs (bGQDs). PL-compliance is due to the transition from the lowest unoccupied molecular orbital with a carbene-like triple soil state to the highest occupied molecular orbital as presented in Fig. 6.

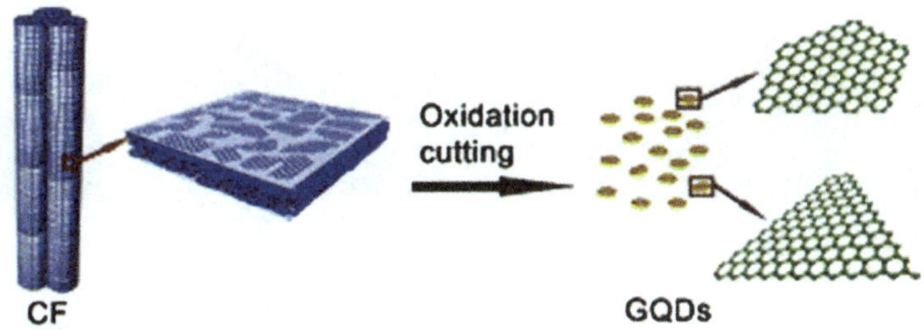

Fig. 4 Carbon-fiber (CF) oxidation cutting into graphene quantum dots (GQDs). (Reproduced with permission from Ref. [85])

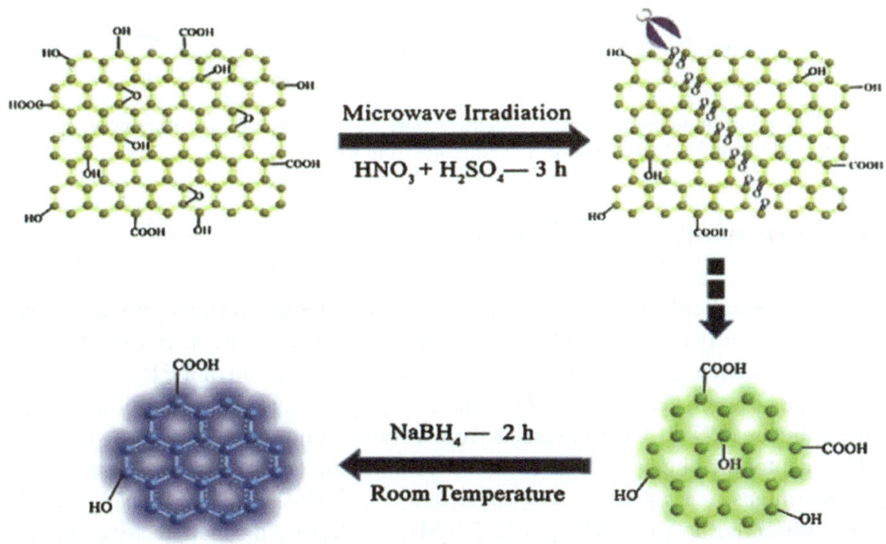

Fig. 5 The route for preparing greenish-yellow graphene quantum dots (gGQDs) and blue GQDs (bGQDs). (Reproduced with permission from Ref. [85])

Many groups have carried out the following methods for the synthesis of GQDs. The GQDs derived from different sugars, including glucose, sucrose and fructose, for production have been recorded in an effortless microwave-assisted process (Fig. 6) [87]. The heating time for the GQDs can be extended from 1 to 9 min in the range 1, 65–21 nm. Interestingly, most carbon-containing hydroxyl, H and O carboxyl or carbonyls, which can dehidrate in hydrothermal conditions are known as carbohydrates in 1:2:1 ratios, as H and O occur in the hydroxyl, carboxy and carbonyl classes [85].

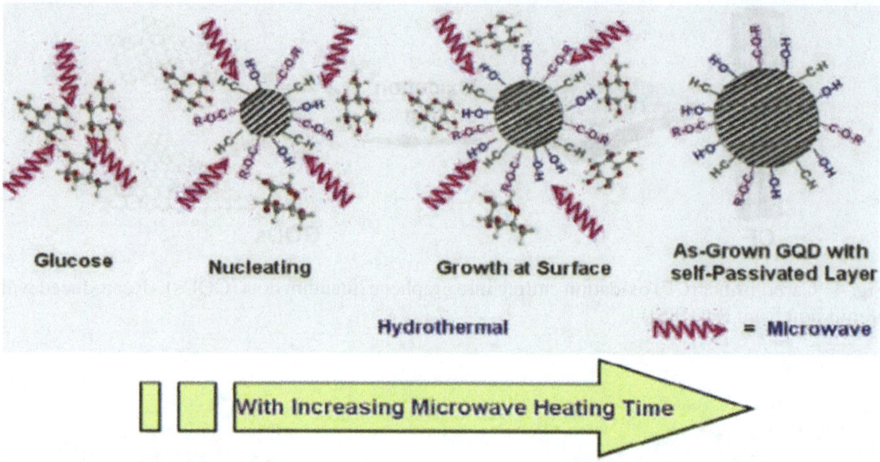

Fig. 6 The preparation of graphene quantum dots (GQDs) using the microwave-assisted hydrothermal method. (Reproduced with permission from Ref. [85])

3.1 Environmental Applications

Due to the excellent luminescent characteristics of GQD, various optical sensors have been recently constructed based on signal-off or signal-on processes. A GQD-based fluorescence sensing platform was developed with the aid of the P-P-p-package interaction between GQDs and TNT aromatical rings for the UVS detection of 2.4,6-trinitrotoluene (TNT). The fluorescence emission of donor GQD's from the radiative TNT acceptor was heavily suppressed by TNT on the GQD surface. Average fluorescence intensity attenuation in a linear range 4.95 10–1.82 10 4 1 TNT could be clearly observed up to 0.495 ppm (2.2 mM). By destroying their passive surface, free chlorine could substantially quench the GQD fluorescence signal (Fig. 7) [88]. The detection limit (LOD) for free chlorine was 0.05 lMin the range 0.05–10 lM, under ideal experimental conditions. The sensing device was also used in tap-water samples to quantify free residual chlorine and the findings agreed well with those obtained with the colorimetric process N–N-diethyl-penithylenediamine. These findings indicate that this sensing device could be used to track the quality of drinking water. As the Eu 3+ ion showed a greater affinity of phosphate oxygen-donor atoms (Pi) than carboxylate on the surface of the GQDs, an off–on PL test to detect Pi was performed [89]. The existence of Eu, which can quench the PL of GQDs via the transfer and/or electron transfer processes, might initially induce GQD aggregation. Then, due to the implementation of Pi, the aggregation of GQD was isolated. With a LOD of 0.1LM in the 0.5–190 lM Pi range, the sensing test showed a fast response in 5 min.

Similarly, for the test of ATP, the GQDs@GSH 'off-to-on' fluorescence mechanism was proposed because, at the end of the day, the GQDs@GSH fluorescence was

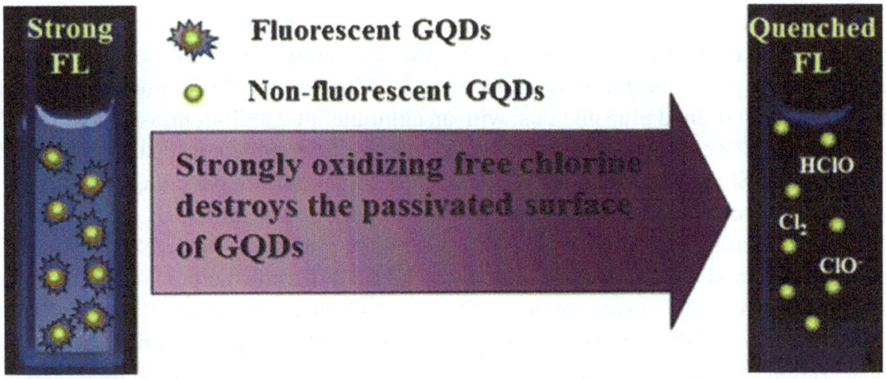

Fig. 7 Free chlorine destroying the passivated surface of graphene quantum dots (GQDs). (Reproduced with permission from Ref. [85])

switches off with growing concen—3+ Fe traction due to an efficient electron transference from the GSH at the GQDs@GSH to Fe [90] on the surface. In the presence of phosphate-containing molecules, the quenched fluorescence of GQDs@GSH was recovered due to its high affinity to Fe^{3+} by the Fe–O–P bonds. In the cell lysate and human blood serum this sensible assay was used to determine the concentration of ATP, and the findings are consistent with natural ATP levels of 2–10 mM in cells. In addition, the sensitive, selective Fe^{3+} detection sample with a LOD of 1 ppm was built based on the fluorescence quenching of GO nanosheets in light of the unique coordinate interaction between Fe and phenolic hydroxyl. Fluorescence quenching was the outcome of the reaction between the six member ring and Fe^{3+} [91] of the alpha-hydroxyquinoid. In the Tris–HCl buffer solution (0.05 M, pH 7.4) six ions (Ni) may also quench the ECL of gGQDs with 0.1 M K2S2O8 and 20 ppm gGQDs [86].

The result of coordinated reactions between the CD^{2+} and the coordination groups on the area of the gGQD could cause a 92% decrease in ECL intensity in gGQDs. In addition, GQDs showed high peroxidase-like activity. Cysteamine as a crosslinker was used in the installation of GQDs on an Au electrode [92]. With good stability and reusability, the resulting covalent assembled GQD/Au electrode was effective in H_2O_2 detection. In 10 s with low LODs down to 0.7 1 M in the range 0.002–8 mM was shown to be quick-amperometric in the GQD/Au electron. On account of its good electrocatalytic activityand high stability, the GQD/Au electrode was applied to detect HO in physiology and pathology.

3.2 *Bioimaging*

Semi-conductor QDs are largely constrained in their biological imagery due to the inner toxicity of the semiconductor QDs. GQDs are promising candidates for semiconductor QD for bio-imaging applications, offering physiological stability, low cytotoxicity and an eco-friendly design. The green GQD in the human breast cell lines T47D has been conducted with in vitro cell studies [93]. Fluorescent images clearly display the phase contrast image of T47D cells, the 4,6-diamino-2, phenyl indole (DAPI) stained blue nucleus, with an agglomerated high-contrast, green, gQD image around each nucleus, and the cell image overlaying the contrast phase, DAPI and gGQDs. Stem cells, neurosphere cells (NSCs), progenitor pancreas cells (PPCs) and cardiovascular progenitor cells (CPCs) were incubated in GQDs at 25 mg mL for 24 h at a concentration of 37C [94]. After three phosphate buffer solutions have been washed and fixed at room temperature with 4% paraformaldehyde for 20 min, confocal fluorescence images of the three stem cell forms (with an excited wavelength of 405 nm) clearly show cell morphology with the GQD's integrated cell due to strong observation of PL spots at the cytoplasmic region of these stem cells and their GQDs. This finding also confirmed GQDs penetrating easily but not entering the nucleus in the stem cells. In addition, there was no intensity alteration of the fluorescence in the microscope imagery, which suggests high photostability of the GQDs in the cells, after several rounds of repeated excitation. No substantial decrease in cell activity in 3-(4,5-dimethylthiazol-2-yl)-2,5-diphenyltetrazólic bromide (MTT) assay was seen by the addition of 400 mG GQDs to 150 mL culture medium (104 cells) [95]. (Fig. 8a). Confocal microscopic fluorescence images show a bright green area inside cells, indicating GQD's translocation through cell membrane (405 nm excitation) (Fig. 8c). A green–yellow colour could be observed when the wavelength for arousal shifted to 488 nm (Fig. 8d). The cell image under bright field is shown in Fig. 8b. These findings demonstrate that even when high cell concentrations are involved GQDs could be used for bio-imaging and other biomedical applications [85].

Because of the impressive array of many key advantages, carbon nanodots are superior to existing organic and inorganic fluorophores. While it remains in its infancy, a wide range of new fluorescence tags, based on these nanodots, offer unparalleled bioimaging possibilities and thus change the biomedical research landscape significantly [96].

4 Other Applications of C-Dots and GQDs

4.1 Cellular Imaging

Sun et al. who used PEG 1500 N passive c-dots for non-specifically stained Caco-2 cells first showed that C-dots could be used as fluorescent labels for cellular imaging [97]. Since then, c intracellular c-imagery in other cell forms of up-takes of C-dots has been shown, including E. coli, [98] honest ascites of carcinoma cells (EACs), [99]] Heli [100, 101] of cells, of Hepg2 cells, of [102] LLC-PK1, of cells [103] NIH-3T3 of fi broblast, of human lung cancer (A549) of the cell [104]. Passive uptake of hollow C-dots was also used to dye HEK 293 cells' cytoplasm [105]. GQDs have also been used for the labelling of a number of different types of cells, including T47D

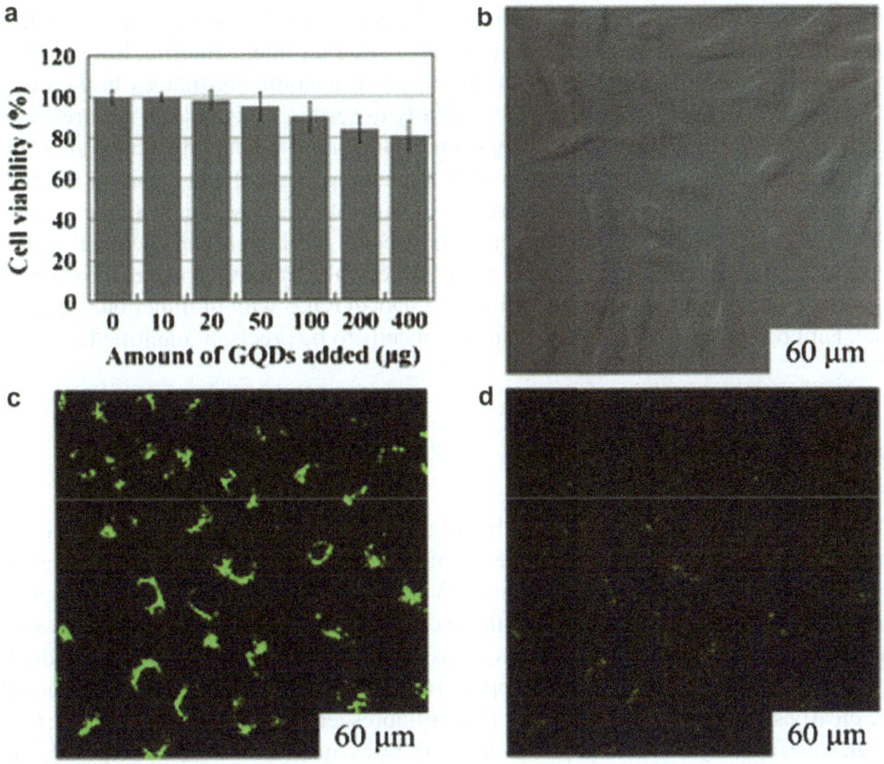

Fig. 8 Cellular toxicity and cellular imaging of graphene quantum dots (GQDs) **a** Effect of GQDs on viability of MG-63 cells; **b–d** are washed cells imaged underbright field, 405 nm and 488 nm excitations, respectively. (Reproduced with permission from Ref. [85])

[93, 106] HeLa, murine alveolar macrophage, [107] human liver carcinoma cells [108] and stem cells, including neurospheric cells, PPCs, cardiac progenitor cells and neural stem cells. For two-photon luminescencemicroscopy with an 800 nm excitation, potential for C-dots (5 nm, propionylethylenimine co-ethylenemine, or PPEI-EI passivated) were first explored in cells MCF-7. After 2 h of incubation at 37 °C, C-dots showed bright PL, both in cell membrane as well as in cytoplasm. C-dots were detected as temperatures based on cellular absorption, without internalisation at 4 °C. In the human cervical carcinoma cell (HeLa), under 800 nm excitation, the ability of dimethyl amine to conduct GQDs (To 3 Nm) in two-photon imagery has been investigated.

Pan et al. study showed that, after 10 min of continuous stimulation, GQDs (almost 3.0 nm, obtained by hydrothermal cutting of GA sheets) accumulated preferentially around the nucleus in HeLa cells and showed no apparent decrease in LP in intensity indicating long-term image viability. C-Dots (page 2–6 nm, synthesised from acidic carbon soot exfoliation) and GQDs (page 125 nm, synthesized by acidic black carbon exfoliation) have been shown to penetrate and mark the nucleus without using nuclear target motif to work. It is still uncertain why such an interesting nuclear position occurs. However, the distribution of these carbon nanodot substances in the cytoplasm is seen in most studies and their affinity to certain subcellular structures (e.g. endosomes) has not been studied rigorously GQDs are predicted to have substantial effect on their cellular penetration potential and intracellular position through charging status and functional classes. A comparative study, for example, shown that A549 cells take up photochemically reduced GQDs better than chemical GQDs because a photochemical decrease is more successful in removing negated carboxyll loading GQD parties. The combination of C-Dots/GQDs with biofunctional species (i.e. antibodies) enables molecular targets in cells to be precisely identified.

4.2 Optical Imaging in Vivo

Ideally, fluorescent samples should be bright, biocompatible and absorbed/emitted in the long, clear wavelength range of biological tissue for the in vivo imaging process. Yang et al. reported the first in vivo injection of C-dots in mice intradermally, intradermally or intravenously, with the use of the PEG-passivated C-dots (emission at 650 nm) and the ZnS-doped C-dots (emission at 510 nm) with good contrast, dual-color fluorescence imaging. In the scope of red excitement, NIR emitting C-dots could be clearly separated from the history of autofluorescence with a strong contrast. Cao et al. have reported that ZnS doped C-dots (QY by as much as 60%) are compared with CdSe and ZnS QD's well-established in vivo imaging. Yang et al. used PEG 1500 N for ex vivo imagery in mouse models, passivated C-dots. In sliced liver and spleen harvested with mouse six hours after intravenous injection, two-photon fluorescence imaging (green emission below 800 nm excitation) was obtained. Wu et al. injected extremely luminescent GQDs (QY = 54.5%) subcutaneously or intramuscularly into nude mice through L-glutamic acid pyrolysis (showing a blue to NIR emission of excitation-based emissions) and obtained strong in vivo fluorescence images. Intravenously injected polydopaminecoated GQDs in various organs of mices were demonstrated in Nurunnabi et al. and shown that their distribution depended on GQD size and polydopamine cover. Their distribution depended on the size of the mice. Dimethyl amine was synthesised in Liu et al.'s fascinating work with strong two-photon cross-section absorption (48,000 GM). These GQDs allow a broad depth of penetration in the imaging of tissues and show small photoblading and thermal effects with repeated laser irradiation of NIR femtosecond (800 nm) which promises long-term image of deep tissues [96].

4.3 Biosensing

CD/GQDs are sensitive to minute disturbances and are highly sensitive to PL, electronic or electrical properties. In addition, these nanodots can interact closely with biomolecules in a size similar to a biomolecule, resulting in an improved sensitivity and selectivity for detection.

4.3.1 Electrochemiluminescence (ECL) Sensors

ECL is a special and sensitive form of analysis. Because the species that emit light are produced in situ close to the surface of the electrode, ECL has an almost zero history and allows time and space to monitor reactions. For ultrasensitive ATP detection (up to 1.5 pm) the intensive and stable ECLs of the GQDs were used [109]. ATP molecules increase dose-dependent ECL signal because they allow additional ssDNA functional GQDs to be added to the ssDNA-functional electrode sample. A DNA sensor has also been established with a detection limit of 13 nM. In particular, GQD's ECL signal on the electrode is first tested for the ECL resonance energy transfer by conjugated gold nanoparticles (AuNPs), and ECL recoveries are recorded when the auNPs are released by hybridization with targets. The ratiometric C-doting ECL measurements were used for detection of Fe^{3+} ions of strong sensitivity (0.7 μM) and reproductivity [94].

4.4 Drug/Gene Delivery

The broad specific surface area and stable interaction of micrometre-size graph sheets with different molecules by stacking pressures, hydrophobic interaction, electrostatic attraction or physisorption has proven to possess a great drugs/gene loading capability. GQDs are smaller to allow fast cell intake and more biocompatible to reduce cytotoxic effects, while being inherited from these merits. The secure, efficient and noticeable distribution vector is anticipated for both GQDs and C-dot.

For cancer cell killing (1000 times more potent than an FDA-approved SN38 drug for clinical colon cancer treatment), the soluble NGO-MEG-SN38 complexes are highly effective. The efficacy of Doxorubicin physisorption into an anti-CD20 combination NGO-MEG by stacking the Raji B-cell lymphoma to be targeted has been demonstrated. GQDs passivated with PEG have proven to have a high capacity for drug loading (2.5 mg/mg anti-cancer medication doxorubicin pH 7.4), as predicted. To attain the precision of the targeting, GQDs for a target delivery for CD44-superexpressed tumours were anchored in hyaluronic acid moieties (HA). HA-GQDs will load up to 75% of the initially used doxorubicin (DOX), suggesting excellent agents to load hydrophobic drug molecules through the interaction of—β-molecules. DOX releases were rapidly 42% within the first six hours and near-completed releases at pH 5.5 were achieved within 48 h. This pH-dependent release is beneficial, because of the slightly acidic state in tumour tissues. In addition, in vivo or in vitro imaging HA-GQDs in balb/c female mice and A549 cells have confirmed the precise targeting of tumour tissues and cancer cells. Wang et al. have shown that DOX-accumulation can be effectively accelerated by GQDs (average potential of 30 nm) and DOX-divided activities can be markedly enhanced, thereby drastically enhancing the DOX cytotoxicity. Interestingly, MCF-7 drug resistant cancer cells are known to have DOX/GQD conjugates [96].

4.5 Anticancer Agents

The new in-vivo photosensitizer-based theranostic platform (Ce6) conjugated C-dot is planned for the use of NIR fluorescence imaging as photodynamic driven treatment to cure gastric cancer. Electrochemically prepared GQDs, which are possibly caused by defects in the GQD surface, were found to produce reactive oxygen species (ROS), like singlet oxygen (1 O 2) under the irradiation of blue light. Higher ROS levels due to up-to-date GQDs resulted in and subsequently significant oxidative stress. Human glioma U251 cells are apoptosis and autophagus. Photodynamic therapy may use this phenomenon. Ge et al. have most recently synthesised red light GQDs with a high 1–2 quantum yield over other photodynamic agents (1.3). With the in vivo use of these GQDs in mice, tumours were killed in a 17-day period of therapy. Zhou et al. found that supercoiled DNA was transformed to nicknamed DNA by GQD, Cu^{2+} by 90%. The high efficiency was due to the GQD's ability to interplay with DNA. Further systematic inquiries on this cleavage process have been carried out. They suggested that electro-rich GQDs pass electrons to the Cu^{2+} metal complex and that reactive oxygen species are formed once the complex is reduced. Oxidative DNA cleavage is attributed to the generated ROS. Such cleavage GQD helped DNA promises anti-cancer therapy applications [96].

4.6 Antibacterial and Antioxidant Activity

Mycoplasm is a common source and induces pneumonia, and other respiratory disorders in cell and clinical samples. Jiang et al. found that Mycoplasm inhibition could be rectified by amine-functional GQDs. The cytoprotection mechanism may be correlated with amine-GQD peroxidase-like behaviour. A small dose of GQD (10 μgmL^{-1}) is as effective as the commercial mycoplasma removal agent. Sun et al. applied GQDs and low-dose H_2O_2 to the band-aides for the disinfection of wounds with mice, taking advantages of both peroxidase-like activity and biocompatibility of GQD. GQDs transform H_2O_2 to $\cdot OH^-$radicals with strong antibacterial activity with low antibacterial activity and prevent adverse accumulation of H_2O_2. A wide variety of gram-negative and gramme-positive bacteria is impacted in this method. GQDs also have the photodynamic effect of killing Staphylococcus aureus and Escherichia coli, immune to methicillin. Growing ROS levels prevents dilation autoregulation in brain vasculature causing hypotension as a consequence of mild traumatic brain injury. A research on a rat model showed that, due to its antioxidant activity, PEG-functionalized hydrophilic pads (30–40 nm long, 2–3 nm wide, both graphical and oxidised) may be a solution. ROS can be annihilated in the CD graphical domains. It is assume that GQDs with sp^2 carbon nanodomains even have certain antioxidant characteristics too [96].

5 Conclusion

This chapter summarises an overview of CDs to introduce recent progress in bioapplications. Next, the physical and chemical properties including photoluminescence and cytotoxicity, which have been studied with respect to bioapplications. Bioapplications were then sysequently applied, such as biosensors and bioimaging agents. A huge effort was made in terms of synthesis, efficiency, mechanism and applications of CDs, as an emerging light material. CDs are used as non-harmful substitutes to replace conventional heavy metal QDs with high photostability and chemical stability, low cytotoxicity, and high quantic yield. As discussed in the chapter, GQDs are a new graphene-based nanomaterial that, due to its low cytotoxicity in vivot, excellent stability and durability, have inspired comprehensive research in the fields of environmental, biological and other science. There are several available GQD preparation methods with special properties, such as absorption, PL, and electroluminescence, which can be accomplished with specific size tuning and practical adjustment methods by controlling the band difference.

References

1. Guldi, D.M., Illescas, B.M., Atienza, C.M., Wielopolski, M., Martín, N.: Fullerene for organic electronics. Chemical Society Reviews 38(6), 1587–1597 (2009)
2. Barbot, A., Di Bin, C., Lucas, B., Ratier, B., Aldissi, M.: N-type doping and thermoelectric properties of co-sublimed cesium-carbonate-doped fullerene. Journal of Materials Science 48(7), 2785–2789 (2013).
3. Sall, M., Monnet, I., Moisy, F., Grygiel, C., Jublot-Leclerc, S., Della-Negra, S., Toulemonde, M., Balanzat, E.: Track formation in III-N semiconductors irradiated by swift heavy ions and fullerene and re-evaluation of the inelastic thermal spike model. Journal of Materials Science 50(15), 5214–5227 (2015).
4. Malgas, G.F., Motaung, D.E., Arendse, C.J.: Temperature-dependence on the optical properties and the phase separation of polymer-fullerene thin films. Journal of Materials Science 47(10), 4282–4289 (2012).

5. Laraoui, A., Meriles, C.A.: Approach to dark spin cooling in a diamond nanocrystal. ACS nano 7(4), 3403–3410 (2013).
6. Laraoui, A., Hodges, J.S., Meriles, C.A.: Nitrogen-vacancy-assisted magnetometry of paramagnetic centers in an individual diamond nanocrystal. Nano letters 12(7), 3477–3482 (2012).
7. Yu, M., George, C., Cao, Y., Wootton, D., Zhou, J.: Microstructure, corrosion, and mechanical properties of compression-molded zinc-nanodiamond composites. Journal of Materials Science 49(10), 3629–3641 (2014).
8. Hola, K., Bourlinos, A.B., Kozak, O., Berka, K., Siskova, K.M., Havrdova, M., Tucek, J., Safarova, K., Otyepka, M., Giannelis, E.P.: Photoluminescence effects of graphitic core size and surface functional groups in carbon dots: COO− induced red-shift emission. Carbon 70, 279–286 (2014).
9. Yu, X., Liu, J., Yu, Y., Zuo, S., Li, B.: Preparation and visible light photocatalytic activity of carbon quantum dots/TiO2 nanosheet composites. Carbon 68, 718–724 (2014).
10. Hu, L., Sun, Y., Li, S., Wang, X., Hu, K., Wang, L., Liang, X.-J., Wu, Y.: Multifunctional carbon dots with high quantum yield for imaging and gene delivery. Carbon 67, 508–513 (2014).
11. Li, C.-X., Yu, C., Wang, C.-F., Chen, S.: Facile plasma-induced fabrication of fluorescent carbon dots toward high-performance white LEDs. Journal of materials science 48(18), 6307–6311 (2013).
12. Hu, S., Tian, R., Dong, Y., Yang, J., Liu, J., Chang, Q.: Modulation and effects of surface groups on photoluminescence and photocatalytic activity of carbon dots. Nanoscale 5(23), 11665–11671 (2013).
13. De Volder, M.F., Tawfick, S.H., Baughman, R.H., Hart, A.J.: Carbon nanotubes: present and future commercial applications. science 339(6119), 535–539 (2013).
14. Wang, Z., Yin, L., Zhang, M., Zhou, G., Fei, H., Shi, H., Dai, H.: Synthesis and characterization of Ag 3 PO 4/multiwalled carbon nanotube composite photocatalyst with enhanced photocatalytic activity and stability under visible light. Journal of Materials Science 49(4), 1585–1593 (2014).
15. Rud, J., Lovell, L., Senn, J., Qiao, Q., Mcleskey, J.: Water soluble polymer/carbon nanotube bulk heterojunction solar cells. Journal of materials science 40(6), 1455–1458 (2005).
16. Tang, P., Zhang, R., Shi, R., Bin, Y.: Synergetic effects of carbon nanotubes and carbon fibers on electrical and self-heating properties of high-density polyethylene composites. Journal of materials science 50(4), 1565–1574 (2015).
17. Dinari, M., Momeni, M.M., Goudarzirad, M.: Dye-sensitized solar cells based on nanocomposite of polyaniline/graphene quantum dots. Journal of materials science 51(6), 2964–2971 (2016).
18. Liu, M., He, L., Liu, X., Liu, C., Luo, S.: Reduced graphene oxide and CdTe nanoparticles co-decorated TiO 2 nanotube array as a visible light photocatalyst. Journal of Materials Science 49(5), 2263–2269 (2014).
19. Wang, X., Pei, Y., Lu, M., Lu, X., Du, X.: Highly efficient adsorption of heavy metals from wastewaters by graphene oxide-ordered mesoporous silica materials. Journal of Materials Science 50(5), 2113–2121 (2015).
20. Su, S., Wang, J., Wei, J., Martínez-Zaguilán, R., Qiu, J., Wang, S.: Efficient photothermal therapy of brain cancer through porphyrin functionalized graphene oxide. New Journal of Chemistry 39(7), 5743–5749 (2015).
21. Wang, J., Qiu, J.: Luminescent graphene quantum dots: as emerging fluorescent materials for biological application. Science of Advanced Materials 7(10), 1979–1989 (2015).
22. Georgakilas, V., Perman, J.A., Tucek, J., Zboril, R.: Broad family of carbon nanoallotropes: classification, chemistry, and applications of fullerenes, carbon dots, nanotubes, graphene, nanodiamonds, and combined superstructures. Chemical reviews 115(11), 4744–4822 (2015).
23. Cao, L., Wang, X., Meziani, M.J., Lu, F., Wang, H., Luo, P.G., Lin, Y., Harruff, B.A., Veca, L.M., Murray, D.: Carbon dots for multiphoton bioimaging. Journal of the American Chemical Society 129(37), 11318–11319 (2007).

24. Yang, S.-T., Cao, L., Luo, P.G., Lu, F., Wang, X., Wang, H., Meziani, M.J., Liu, Y., Qi, G., Sun, Y.-P.: Carbon dots for optical imaging in vivo. Journal of the American Chemical Society 131(32), 11308–11309 (2009).
25. Tian, L., Ghosh, D., Chen, W., Pradhan, S., Chang, X., Chen, S.: Nanosized carbon particles from natural gas soot. Chemistry of materials 21(13), 2803–2809 (2009).
26. Zhao, Q.-L., Zhang, Z.-L., Huang, B.-H., Peng, J., Zhang, M., Pang, D.-W.: Facile preparation of low cytotoxicity fluorescent carbon nanocrystals by electrooxidation of graphite. Chemical Communications(41), 5116–5118 (2008).
27. Ray, S., Saha, A., Jana, N.R., Sarkar, R.: Fluorescent carbon nanoparticles: synthesis, characterization, and bioimaging application. The Journal of Physical Chemistry C 113(43), 18546–18551 (2009).
28. Zhou, J., Zhou, X., Li, R., Sun, X., Ding, Z., Cutler, J., Sham, T.-K.: Electronic structure and luminescence center of blue luminescent carbon nanocrystals. Chemical Physics Letters 474(4–6), 320–324 (2009).
29. Baker, S.N., Baker, G.A.: Luminescent carbon nanodots: emergent nanolights. Angewandte Chemie International Edition 49(38), 6726–6744 (2010).
30. Shen, J., Zhu, Y., Yang, X., Zong, J., Zhang, J., Li, C.: One-pot hydrothermal synthesis of graphene quantum dots surface-passivated by polyethylene glycol and their photoelectric conversion under near-infrared light. New Journal of Chemistry 36(1), 97–101 (2012).
31. Zhou, X., Zhang, Y., Wang, C., Wu, X., Yang, Y., Zheng, B., Wu, H., Guo, S., Zhang, J.: Photo-Fenton reaction of graphene oxide: a new strategy to prepare graphene quantum dots for DNA cleavage. ACS nano 6(8), 6592–6599 (2012).
32. Wang, J., Qiu, J.: A review of carbon dots in biological applications. Journal of Materials Science 51(10), 4728–4738 (2016).
33. Li, Q., Ohulchanskyy, T.Y., Liu, R., Koynov, K., Wu, D., Best, A., Kumar, R., Bonoiu, A., Prasad, P.N.: Photoluminescent carbon dots as biocompatible nanoprobes for targeting cancer cells in vitro. The Journal of Physical Chemistry C 114(28), 12062–12068 (2010).
34. Liu, C., Zhang, P., Zhai, X., Tian, F., Li, W., Yang, J., Liu, Y., Wang, H., Wang, W., Liu, W.: Nano-carrier for gene delivery and bioimaging based on carbon dots with PEI-passivation enhanced fluorescence. Biomaterials 33(13), 3604–3613 (2012).
35. Milosavljevic, V., Nguyen, H.V., Michalek, P., Moulick, A., Kopel, P., Kizek, R., Adam, V.: Synthesis of carbon quantum dots for DNA labeling and its electrochemical, fluorescent and electrophoretic characterization. Chemical Papers 69(1), 192–201 (2015).
36. Xu, X., Ray, R., Gu, Y., Ploehn, H.J., Gearheart, L., Raker, K., Scrivens, W.A.: Electrophoretic analysis and purification of fluorescent single-walled carbon nanotube fragments. Journal of the American Chemical Society 126(40), 12736–12737 (2004).
37. Zheng, L., Chi, Y., Dong, Y., Lin, J., Wang, B.: Electrochemiluminescence of water-soluble carbon nanocrystals released electrochemically from graphite. Journal of the American Chemical Society 131(13), 4564–4565 (2009).
38. Li, H., Ming, H., Liu, Y., Yu, H., He, X., Huang, H., Pan, K., Kang, Z., Lee, S.-T.: Fluorescent carbon nanoparticles: electrochemical synthesis and their pH sensitive photoluminescence properties. New Journal of Chemistry 35(11), 2666–2670 (2011).
39. Hou, Y., Lu, Q., Deng, J., Li, H., Zhang, Y.: One-pot electrochemical synthesis of functionalized fluorescent carbon dots and their selective sensing for mercury ion. Analytica Chimica Acta 866, 69–74 (2015).
40. Liu, R., Wu, D., Liu, S., Koynov, K., Knoll, W., Li, Q.: An aqueous route to multicolor photoluminescent carbon dots using silica spheres as carriers. Angewandte Chemie International Edition 48(25), 4598–4601 (2009).
41. Yang, Y., Wu, D., Han, S., Hu, P., Liu, R.: Bottom-up fabrication of photoluminescent carbon dots with uniform morphology via a soft-hard template approach. Chemical Communications 49(43), 4920–4922 (2013).
42. Li, S., Wang, L., Chusuei, C.C., Suarez, V.M., Blackwelder, P.L., Micic, M., Orbulescu, J., Leblanc, R.M.: Nontoxic carbon dots potently inhibit human insulin fibrillation. Chemistry of Materials 27(5), 1764–1771 (2015).

43. Kasibabu, B.S.B., D'souza, S.L., Jha, S., Singhal, R.K., Basu, H., Kailasa, S.K.: One-step synthesis of fluorescent carbon dots for imaging bacterial and fungal cells. Analytical Methods 7(6), 2373–2378 (2015).
44. Liu, H., Ye, T., Mao, C.: Fluorescent carbon nanoparticles derived from candle soot. Angewandte chemie 119(34), 6593–6595 (2007).
45. Peng, H., Travas-Sejdic, J.: Simple aqueous solution route to luminescent carbogenic dots from carbohydrates. Chemistry of Materials 21(23), 5563–5565 (2009).
46. Zhang, K., Deng, J., Xing, Y., Li, S., Gao, H.: Effect of microscale texture on cutting performance of WC/Co-based TiAlN coated tools under different lubrication conditions. Applied Surface Science 326, 107–118 (2015).
47. Dong, Y., Pang, H., Yang, H.B., Guo, C., Shao, J., Chi, Y., Li, C.M., Yu, T.: Carbon-based dots co-doped with nitrogen and sulfur for high quantum yield and excitation-independent emission. Angewandte Chemie International Edition 52(30), 7800–7804 (2013).
48. Wu, Z.L., Zhang, P., Gao, M.X., Liu, C.F., Wang, W., Leng, F., Huang, C.Z.: One-pot hydrothermal synthesis of highly luminescent nitrogen-doped amphoteric carbon dots for bioimaging from Bombyx mori silk-natural proteins. Journal of Materials Chemistry B 1(22), 2868–2873 (2013).
49. Zheng, X., Wang, H., Gong, Q., Zhang, L., Cui, G., Li, Q., Chen, L., Wu, F., Wang, S.: Highly luminescent carbon nanoparticles as yellow emission conversion phosphors. Materials Letters 143, 290–293 (2015).
50. Xu, Q., Pu, P., Zhao, J., Dong, C., Gao, C., Chen, Y., Chen, J., Liu, Y., Zhou, H.: Preparation of highly photoluminescent sulfur-doped carbon dots for Fe (III) detection. Journal of Materials Chemistry A 3(2), 542–546 (2015).
51. Xu, J., Zhou, Y., Cheng, G., Dong, M., Liu, S., Huang, C.: Carbon dots as a luminescence sensor for ultrasensitive detection of phosphate and their bioimaging properties. Luminescence 30(4), 411–415 (2015).
52. Wang, B., Tang, W., Lu, H., Huang, Z.: Hydrothermal synthesis of ionic liquid-capped carbon quantum dots with high thermal stability and anion responsiveness. Journal of Materials Science 50(16), 5411–5418 (2015).
53. Tao, H., Yang, K., Ma, Z., Wan, J., Zhang, Y., Kang, Z., Liu, Z.: In vivo NIR fluorescence imaging, biodistribution, and toxicology of photoluminescent carbon dots produced from carbon nanotubes and graphite. Small 8(2), 281–290 (2012).
54. Bottini, M., Balasubramanian, C., Dawson, M.I., Bergamaschi, A., Bellucci, S., Mustelin, T.: Isolation and characterization of fluorescent nanoparticles from pristine and oxidized electric arc-produced single-walled carbon nanotubes. The Journal of Physical Chemistry B 110(2), 831–836 (2006).
55. Wang, X., Qu, K., Xu, B., Ren, J., Qu, X.: Microwave assisted one-step green synthesis of cell-permeable multicolor photoluminescent carbon dots without surface passivation reagents. Journal of Materials Chemistry 21(8), 2445–2450 (2011).
56. Kargbo, O., Jin, Y., Ding, S.-N.: Recent advances in luminescent carbon dots. Current Analytical Chemistry 11(1), 4–21 (2015).
57. Arul, V., Edison, T.N.J.I., Lee, Y.R., Sethuraman, M.G.: Biological and catalytic applications of green synthesized fluorescent N-doped carbon dots using Hylocereus undatus. Journal of Photochemistry and Photobiology B: Biology 168, 142–148 (2017).
58. D'souza, S.L., Deshmukh, B., Bhamore, J.R., Rawat, K.A., Lenka, N., Kailasa, S.K.: Synthesis of fluorescent nitrogen-doped carbon dots from dried shrimps for cell imaging and boldine drug delivery system. RSC advances 6(15), 12169–12179 (2016).
59. Li, H., Li, F., Wang, G., Sun, H.: One-step synthesis of fluorescent carbon nanoparticles for degradation of naphthol green under visible light. Journal of luminescence 156, 36–40
60. Edison, T.N.J.I., Atchudan, R., Sethuraman, M.G., Shim, J.-J., Lee, Y.R.: Microwave assisted green synthesis of fluorescent N-doped carbon dots: Cytotoxicity and bio-imaging applications. Journal of Photochemistry and Photobiology B: Biology 161, 154–161 (2016).
61. Zhao, A., Chen, Z., Zhao, C., Gao, N., Ren, J., Qu, X.: Recent advances in bioapplications of C-dots. Carbon 85, 309–327 (2015).

62. Wang, J., Su, S., Wei, J., Bahgi, R., Hope-Weeks, L., Qiu, J., Wang, S.: Ratio-metric sensor to detect riboflavin via fluorescence resonance energy transfer with ultrahigh sensitivity. Physica E: Low-dimensional Systems and Nanostructures 72, 17–24 (2015).
63. Chen, L., Li, Y., Gu, W.: Synthesis of carbon dots by microwave pyrolysis of polyol in the presence of inorganic ions. Nanoscience and Nanotechnology Letters 7(1), 6–9 (2015).
64. Zhang, X., Wang, S., Zhu, C., Liu, M., Ji, Y., Feng, L., Tao, L., Wei, Y.: Carbon-dots derived from nanodiamond: Photoluminescence tunable nanoparticles for cell imaging. Journal of colloid and interface science 397, 39–44 (2013).
65. Li, H., He, X., Liu, Y., Huang, H., Lian, S., Lee, S.-T., Kang, Z.: One-step ultrasonic synthesis of water-soluble carbon nanoparticles with excellent photoluminescent properties. Carbon 49(2), 605–609 (2011).
66. Dong, Y., Su, M., Chen, P., Sun, H.: Chemiluminescence of carbon dots induced by diperiodato-nicklate (IV) in alkaline solution and its application to a quenchometric flow-injection assays of paracetamol, L-cysteine and glutathione. Microchimica Acta 182(5–6), 1071–1077 (2015).
67. Zhang, L., Han, Y., Zhu, J., Zhai, Y., Dong, S.: Simple and sensitive fluorescent and electro-chemical trinitrotoluene sensors based on aqueous carbon dots. Analytical chemistry 87(4), 2033–2036 (2015).
68. De Volder, M.F., Tawfick, S.H., Baughman, R.H., Hart, A.J.J.s.: Carbon nanotubes: present and future commercial applications. 339(6119), 535–539 (2013).
69. Park, S., Vosguerichian, M., Bao, Z.J.N.: A review of fabrication and applications of carbon nanotube film-based flexible electronics. 5(5), 1727–1752 (2013).
70. Terrones, M.J.A.R.O.M.R.: Science and technology of the twenty-first century: synthesis, properties, and applications of carbon nanotubes. 33(1), 419–501 (2003).
71. Nayak, G.C., Sahoo, S., Rajasekar, R., Das, C.K.C.P.A.A.S.M.: Novel approach for the selective dispersion of MWCNTs in the Nylon/SAN blend system. 43(8), 1242–1251 (2012).
72. Nayak, G.C., Rajasekar, R., Bose, S., Das, C.K.J.N.T.: Effect of MWNTs and SiC-coated MWNTs on properties of PEEK/LCP blend. (2009).
73. Rajasekar, R., Nayak, G.C., Malas, A., Das, C.K.M.D.: Development of compatibilized SBR and EPR nanocomposites containing dual filler system. 35, pp.878-885 (2012).
74. Geim, A.K.J.S.: Graphene: status and prospects. 324(5934), 1530–1534 (2009).
75. Huang, X., Yin, Z., Wu, S., Qi, X., He, Q., Zhang, Q., Yan, Q., Boey, F., Zhang, H.J.S.: Graphene-based materials: synthesis, characterization, properties, and applications. 7(14), 1876–1902 (2011).
76. Pan, D., Zhang, J., Li, Z., Wu, M.J.A.M.: Hydrothermal route for cutting graphene sheets into blue-luminescent graphene quantum dots. 22(6), 734–738 (2010).
77. Cheng, H., Zhao, Y., Fan, Y., Xie, X., Qu, L., Shi, G.J.A.N.: Graphene-quantum-dot assembled nanotubes: a new platform for efficient Raman enhancement. 6(3), 2237–2244 (2012).
78. Shen, J., Zhu, Y., Yang, X., Li, C.J.C.c.: Graphene quantum dots: emergent nanolights for bioimaging, sensors, catalysis and photovoltaic devices. 48(31), 3686–3699 (2012).
79. Zhang, Z., Zhang, J., Chen, N., Qu, L.J.E., Science, E.: Graphene quantum dots: an emerging material for energy-related applications and beyond. 5(10), 8869–8890 (2012).
80. Zhu, S., Tang, S., Zhang, J., Yang, B.J.C.C.: Control the size and surface chemistry of graphene for the rising fluorescent materials. 48(38), 4527–4539 (2012).
81. Tetsuka, H., Asahi, R., Nagoya, A., Okamoto, K., Tajima, I., Ohta, R., Okamoto, A.J.A.M.: Optically tunable amino-functionalized graphene quantum dots. 24(39), 5333–5338 (2012).
82. Luo, P., Ji, Z., Li, C., Shi, G.J.N.: Aryl-modified graphene quantum dots with enhanced photoluminescence and improved pH tolerance. 5(16), 7361–7367 (2013).
83. Shen, J., Zhu, Y., Chen, C., Yang, X., Li, C.J.C.C.: Facile preparation and upconversion luminescence of graphene quantum dots. 47(9), 2580–2582 (2011).
84. Shen, J., Zhu, Y., Yang, X., Zong, J., Zhang, J., Li, C.J.N.J.O.C.: One-pot hydrothermal synthesis of graphene quantum dots surface-passivated by polyethylene glycol and their photoelectric conversion under near-infrared light. 36(1), 97–101 (2012).

85. Lin, L., Rong, M., Luo, F., Chen, D., Wang, Y., Chen, X.: Luminescent graphene quantum dots as new fluorescent materials for environmental and biological applications. TrAC Trends in Analytical Chemistry 54, 83–102 (2014).
86. Li, L.L., Ji, J., Fei, R., Wang, C.Z., Lu, Q., Zhang, J.R., Jiang, L.P., Zhu, J.J.J.A.F.M.: A facile microwave avenue to electrochemiluminescent two-color graphene quantum dots. 22(14), 2971–2979 (2012).
87. Tang, L., Ji, R., Cao, X., Lin, J., Jiang, H., Li, X., Teng, K.S., Luk, C.M., Zeng, S., Hao, J.J.A.n.: Deep ultraviolet photoluminescence of water-soluble self-passivated graphene quantum dots. 6(6), 5102–5110 (2012).
88. Dong, Y., Li, G., Zhou, N., Wang, R., Chi, Y., Chen, G.J.A.c.: Graphene quantum dot as a green and facile sensor for free chlorine in drinking water. 84(19), 8378–8382 (2012).
89. Bai, J.M., Zhang, L., Liang, R.P., Qiu, J.D.J.C.A.E.J.: Graphene quantum dots combined with europium ions as photoluminescent probes for phosphate sensing. 19(12), 3822–3826 (2013).
90. Liu, J.-J., Zhang, X.-L., Cong, Z.-X., Chen, Z.-T., Yang, H.-H., Chen, G.-N.J.N.: Glutathione-functionalized graphene quantum dots as selective fluorescent probes for phosphate-containing metabolites. 5(5), 1810–1815 (2013).
91. Wang, D., Wang, L., Dong, X., Shi, Z., Jin, J.J.C.: Chemically tailoring graphene oxides into fluorescent nanosheets for Fe3+ ion detection. 50(6), 2147–2154 (2012).
92. Zhang, Y., Wu, C., Zhou, X., Wu, X., Yang, Y., Wu, H., Guo, S., Zhang, J.J.N.: Graphene quantum dots/gold electrode and its application in living cell H 2 O 2 detection. 5(5), 1816–1819 (2013).
93. Peng, J., Gao, W., Gupta, B.K., Liu, Z., Romero-Aburto, R., Ge, L., Song, L., Alemany, L.B., Zhan, X., Gao, G.J.N.l.: Graphene quantum dots derived from carbon fibers. 12(2), 844–849 (2012).
94. Zhang, M., Bai, L., Shang, W., Xie, W., Ma, H., Fu, Y., Fang, D., Sun, H., Fan, L., Han, M.J.J.O.M.C.: Facile synthesis of water-soluble, highly fluorescent graphene quantum dots as a robust biological label for stem cells. 22(15), 7461–7467 (2012).
95. Lu, J., Yang, J.-X., Wang, J., Lim, A., Wang, S., Loh, K.P.J.A.N.: One-pot synthesis of fluorescent carbon nanoribbons, nanoparticles, and graphene by the exfoliation of graphite in ionic liquids. 3(8), 2367–2375 (2009).
96. Zheng, X.T., Ananthanarayanan, A., Luo, K.Q., Chen, P.: Glowing graphene quantum dots and carbon dots: properties, syntheses, and biological applications. Small 11(14), 1620–1636 (2015).
97. Sun, Y.-P., Zhou, B., Lin, Y., Wang, W., Fernando, K.S., Pathak, P., Meziani, M.J., Harruff, B.A., Wang, X., Wang, H.J.J.o.t.A.C.S.: Quantum-sized carbon dots for bright and colorful photoluminescence. 128(24), 7756–7757 (2006).
98. Liu, R., Wu, D., Liu, S., Koynov, K., Knoll, W., Li, Q.J.A.C.I.E.: An aqueous route to multicolor photoluminescent carbon dots using silica spheres as carriers. 48(25), 4598–4601 (2009).
99. Ray, S., Saha, A., Jana, N.R., Sarkar, R.J.T.J.o.P.C.C.: Fluorescent carbon nanoparticles: synthesis, characterization, and bioimaging application. 113(43), 18546–18551 (2009).
100. Dong, Y., Pang, H., Yang, H.B., Guo, C., Shao, J., Chi, Y., Li, C.M., Yu, T.J.A.C.I.E.: Carbon-based dots co-doped with nitrogen and sulfur for high quantum yield and excitation-independent emission. 52(30), 7800–7804 (2013).
101. Li, Q., Ohulchanskyy, T.Y., Liu, R., Koynov, K., Wu, D., Best, A., Kumar, R., Bonoiu, A., Prasad, P.N.J.T.J.O.P.C.C.: Photoluminescent carbon dots as biocompatible nanoprobes for targeting cancer cells in vitro. 114(28), 12062–12068 (2010).
102. Xu, Y., Wu, M., Liu, Y., Feng, X.Z., Yin, X.B., He, X.W., Zhang, Y.K.J.C.A.E.J.: Nitrogen-doped carbon dots: a facile and general preparation method, photoluminescence investigation, and imaging applications. 19(7), 2276–2283 (2013).
103. Hsu, P.-C., Shih, Z.-Y., Lee, C.-H., Chang, H.-T.J.G.C.: Synthesis and analytical applications of photoluminescent carbon nanodots. 14(4), 917–920 (2012).
104. Wu, Z.L., Zhang, P., Gao, M.X., Liu, C.F., Wang, W., Leng, F., Huang, C.Z.J.J.o.M.C.B.: One-pot hydrothermal synthesis of highly luminescent nitrogen-doped amphoteric carbon dots for bioimaging from Bombyx mori silk-natural proteins. 1(22), 2868–2873 (2013).

105. Fang, Y., Guo, S., Li, D., Zhu, C., Ren, W., Dong, S., Wang, E.J.A.n.: Easy synthesis and imaging applications of cross-linked green fluorescent hollow carbon nanoparticles. 6(1), 400–409 (2012).
106. Pan, D., Guo, L., Zhang, J., Xi, C., Xue, Q., Huang, H., Li, J., Zhang, Z., Yu, W., Chen, Z.J.J.o.M.C.: Cutting sp 2 clusters in graphene sheets into colloidal graphene quantum dots with strong green fluorescence. 22(8), 3314–3318 (2012).
107. Wu, X., Tian, F., Wang, W., Chen, J., Wu, M., Zhao, J.X.J.J.o.M.C.C.: Fabrication of highly fluorescent graphene quantum dots using L-glutamic acid for in vitro/in vivo imaging and sensing. 1(31), 4676–4684 (2013).
108. Dong, Y., Chen, C., Zheng, X., Gao, L., Cui, Z., Yang, H., Guo, C., Chi, Y., Li, C.M.J.J.O.M.C.: One-step and high yield simultaneous preparation of single-and multi-layer graphene quantum dots from CX-72 carbon black. 22(18), 8764–8766 (2012).
109. Zhang, Z., Zhang, J., Chen, N., Qu, L.: Graphene quantum dots: an emerging material for energy-related applications and beyond. Energy & Environmental Science 5(10), 8869–8890 (2012).

Heteroatom Doping in Nanocarbon and Its Applications

Mohan Kumar Anand Raj, Rajasekar Rathanasamy, Sathish Kumar Palaniappan, GobinathVelu Kaliyannan, Moganapriya Chinnasamy, and Santhosh Sivaraj

Abstract Doped carbon nanomaterials have been applied in batteries, supercapacitors and fuel cells. The different carbon-based nanomaterials doped with heteroatoms like nitrogen, sulfur, boron, phosphorous and their mixers are used to replace platinum and other costlier materials for the application of Oxygen Reduction Reactions (ORR). In this chapter, adhering of the dopants, synthesis of heterodoped nanocarbons, nanocarbon material doping for the ORR and the mechanisms for ORR have been discussed in detail. Heteroatom-doped 3-dimensional carbon nanotube ORR and compound heteroatom co-doped 3-dimensional carbon nanotubes are also reviewed extensively. Additionally, different fullerenes, namely boron–nitrogen–carbon, boron–nitrogen and possible point defects that occurred in the corner of the crystallite graphite are also discussed.

Keywords Nanocarbon materials · N-doping in graphene · Oxygen Reduction Reactions · Heteroatom doping

1 Introduction

Carbon nanotubes and graphene have extraordinary and distinctive characterizations. The influence of the electronic nature of the Carbon Nanotubes (CNT) depends on the geometrical shape. Extensive work has been exerted to monitor the electronic characterization of a group of carbon nanotubes with substantial development [1–3]. Research on the refinement and parting of carbon nanotube samples have been investigated for last five years, and it includes an understanding of the properties compared to numerical prediction. There are many ways available to regulate the

M. K. A. Raj · R. Rathanasamy (✉) · M. Chinnasamy · S. Sivaraj
Department of Mechanical Engineering, Kongu Engineering College, Tamilnadu 638060, India

S. K. Palaniappan
Department of Mining Engineering, Indian Institute of Technology Kharagpur, West Bengal 721302, India

G. Kaliyannan
Department of Mechatronics Engineering, Kongu Engineering College, Tamilnadu 638060, India

S. Sahoo et al. (eds.), *Defect Engineering of Carbon Nanostructures*, Advances in Material Research and Technology, https://doi.org/10.1007/978-3-030-94375-2_3

optical and electronic characterization of carbon nanotubes and graphene. Although, the properties of graphite is not completely understood, this reduces its importance in various application. Different methods of structural functionalization can be used to control the characterization of graphene and nanotubes [4, 5].

Substitutional doping is different compared to other methods, such as donating electrons to the carbon nanotubes by electrical gating method or electrochemical method of charging and discharging. The two natural elements are nitrogen and boron, which are located closer in periodic table [5–8]. Nitrogen is most attracted compared to boron owing to its atomic size and more electrons equated to carbon.

Eventhough the nitrogen-doped multi-walled nanotubes widely exist, it is difficult to synthesize them at the beginning. However, the single-walled nanotube synthesis has been performed with lesser complications in recent years. Phosphorus is also one substitutional element. A few years back, researchers suggested phosphorous as a dopant for carbon-based systems theoretically. The experimental research work has been published recently on doping of phosphorus in carbon nanotubes and graphene [9–11].

1.1 Local Adhering of the Dopants

The local adhering of the dopants was responsible for achieving the required functionality and variation in the characterization of carbon materials. Various challenges arise for heteroatoms, for instance, in nanotubes, boron is likely to adhere in electronically and structurally direct-substitutional configuration, while nitrogen adheres with a different configuration. Meanwhile, nitrogen having additional electrons compared to carbon facilitates n-type doping to the host, when introducing the nitrogen atoms into the graphitic lattice. The numerical work showed that the replacement of carbon atom instead of nitrogen provides localized energy, which is above Fermi level [12, 13], Fig. 1a. This is due to the nitrogen atom holding 3 electrons in α-bonds and 1 in π-bond.

The electronic shapes of nitrogen atom are different from carbon atoms, hence, there is a possibility of defect formation called tripyridinic vacancy configuration. SP_2 bonds created by 3 nitrogen and 2 carbon atoms around a single vacancy [15] are shown in Fig. 1b. This is energetically advantageous due to no hanging bonds, even though other chances have also been assumed, Fig. 1. A plain replacement of Boron is the most promising bonding structure for this kind of bond. However, efforts to synthesize boron-doped materials have described the creation of boron nano-fields along the nanotube configuration [16, 17]. Only a small concentration of boron atoms are capable for the creation at low accept condition in the band gap.

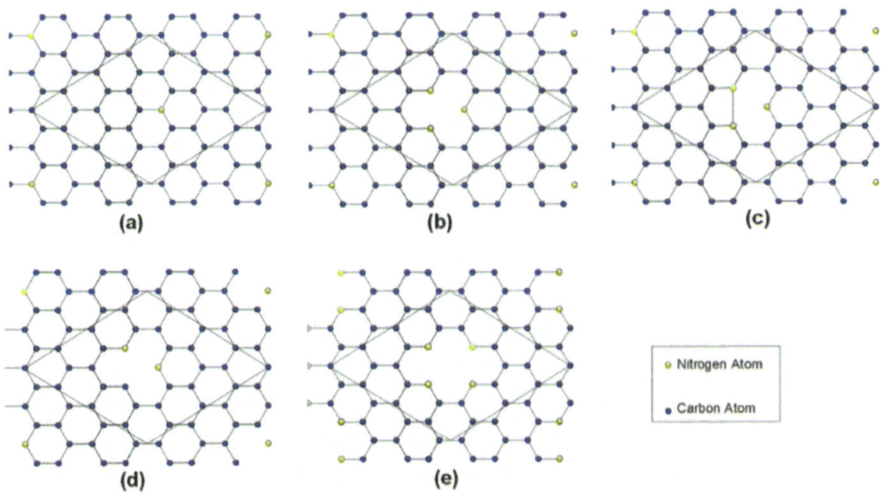

Fig. 1 Nitrogen dopant patterns in a typical graphene element cell indicated by the line **a** alternative, **b** tripyridinic void, **c** mono-pyridinic void, **d** dual-pyridinic void and **e** multiple pyridinic void (Reproduced with permission from Ref. copyright [14])

2 Synthesis of Heteroatom-Doped Nanocarbons

The synthesis of heteronanocarbons has been classified into two types, namely low- and high-temperature approaches; it is based on vaporization and decomposition of carbons. Chemical vapor deposition is one of the good techniques for the synthesis of doped nanotubes and graphene, which is established on pyrolysis of gases [5]. High-temperature approach is also known as the physical method, which is used to break the molecules by using arc or laser as a power source. Separate guidelines are available for making pristine materials in most of the chemical vapor deposition approaches. In a thermochemical substitution reaction, it is reported that the post-synthesis treatment of the pristine materials will affect the doping [14, 18]. It is clear that the nitrogen replacement on both adjacent places is also absorbed in graphene and the boron replacement corresponds to Fig. 1a–e.

Until modern years, it has been stated that the ratio of dopant atom and carbon atom were proportionate to each other in precursors and it does not analyze fully about the concentration of dopants in actual product. A very minimal amount of dopant has been used in the product due to the challenges of low doping concentration. The measurement of the amount of dopant present in Single-Walled Carbon Nanotubes (SWCN) by using indirect measurement is still reported in case studies. Various techniques were used to include heteroatoms into the structure. Recent research articles based on Nitrogen- (N) or Boron (B)-doped Multi-Walled Carbon Nanotube (MWCN) materials are listed in Table1. Researchers also reported the influence of graphene and nitrogen or boron-doped carbon nanotubes [19, 20]. Yi and Beernholc proposed the doping of CNTs with nitrogen or boron numerically [6]. Synthesis

Table 1 Doped Nanocarbon synthesis found from various literature

Method	Materials	Precursors	References
Laser ablation	Boron Carbon Nitride—MWCN	B	[21]
	Boron—SWCN	B	[37]
	Nitrogen—SWCN	N_2	[25]
Arc discharge	Boron Carbon Nitride—MWCN	BN	[38]
	Boron—SWCN	B	[15]
	Boron—MWCN	BN	[39]
	Nitrogen—SWCN	Melamine	[40]
	Boron–Graphene	Pyridine, B_2H_6, B	[41]
	Nitrogen–Graphene	NH_3	[42]
Chemical vapor deposition	Nitrogen—MWCN	Ni phthalocyanine	[22]
	Boron–Graphene	Phenylboronic acid	[43]
	Nitrogen–Graphene	NH_3	[42]
	Boron—SWCN	Triisopropyl borate	[44]
	Nitrogen—SWCN	Acetonitrile	
Thermochemical	Boron Carbon Nitride—SWCN	B_2O_3, N_2	[45]
	Boron–Graphene	B_2O_3	[46]
Solvothermal	Boron–Graphene	BBr_3	[47]
	Nitrogen–Graphene	Li_3N	[48]
	Nitrogen—MWCN	Cyanuric chloride	[49]
Ion bombardment	Nitrogen—SWCN	N_2^+	[50]
	Nitrogen—MWCN	N_2^+	[51]

of nitrogen and MWCNs were explored extensively in the year 1997 [21–24]. The effective synthesis was attained later using the laser ablation technique [25]. In addition to this, many researchers reported the synthesis of nitrogen on SWCNs using different chemical vapor deposition methods [26–36].

3 Nanocarbon Material Doping for the ORR

ORR at the cathode is a basic method and important part in metal and air batteries and fuel cells. However, a few drawbacks of the electrochemical efficiency have been still confronted by the lack of reliability of conventional platinum-based ORR and high cost. The heteroatom-doped nanomaterials with viable activity will enhance the

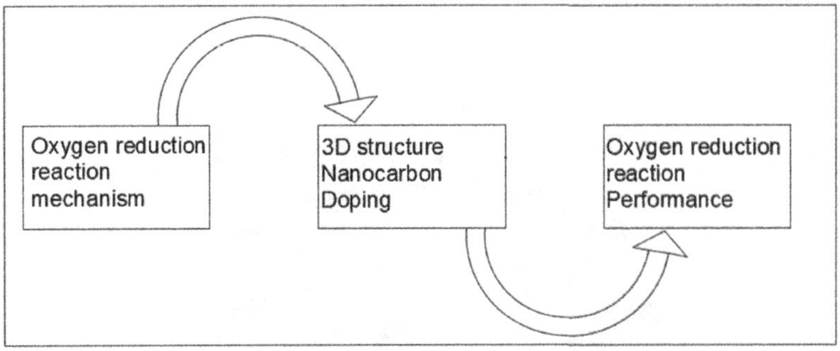

Fig. 2 The correlation of ORR mechanism, catalyst design ORR performance

reliability and have increasing interest toward an alternative for precious metal catalysts. Especially, the 3-dimensional porous design seems to be required for attaining high catalytic ORR movement by giving high definite surfaces and volume exposed for mass transportation to the electrocatalysts.

Many researchers have studied the heteroatom-doped nanocarbon materials and their application in the oxygen reduction reactions in last few years [52–55]. However, special importance has not been given to these heteroatom-doped nanocarbon materials on ORR application. In this chapter, recently recognized mechanisms for ORR have been described briefly as depicted in Fig. 2.

The recent balanced model of various 3-dimensional doped carbon nanomaterials, namely 3-dimensional graphene, 3-dimensional carbon nanotube nano-architectures, porous carbon and their mixtures has been discussed in detail, and also focuses on the recently enhanced ORR performance. The configuration-based ORR efficiency of 3-dimensional doped nanocarbon materials has been discussed.

3.1 *Mechanisms for ORR*

ORR is a basic reaction and the main controlling factor of performance for fuel cells and metal and air batteries [56, 57]. In common, ORR involves several electrochemical reactions and can produce water (H_2O) ($4H^+ + O_2 + 4e^- \rightarrow 2H_2O$, in acidic medium) and OH^- ($4H^+ + O_2 + 4e^- \rightarrow 4OH^-$, in alkaline medium) directly by proceeding in either a four-electron path as the last products or a low performance 2 step, 2 electron trail with the production of H_2O_2 ($O_2 + 2H^+ + 2e^- \rightarrow H_2O_2$, $H_2O_2 + 2H^+ + 2e^- \rightarrow 2H_2O$, in acidic medium) or HO_2^- ($O_2 + H_2O + 2e^- \rightarrow HO_2^- + OH^-$, $HO_2^- + H_2O + 2e^- \rightarrow 3OH^-$) as the middle product [54, 58]. Figure 3 shows a diagrammatic representation of oxygen or hydrogen fuel cell based on proton exchange membrane fuel cell H_2 and O_2 or air uninterruptedly pass in the negative and positive terminals, respectively. The hydrogen particles (fuel molecule) have oxidized at the negative terminal ($H_2 \rightarrow 2H^+ + 2e^-$). In this method,

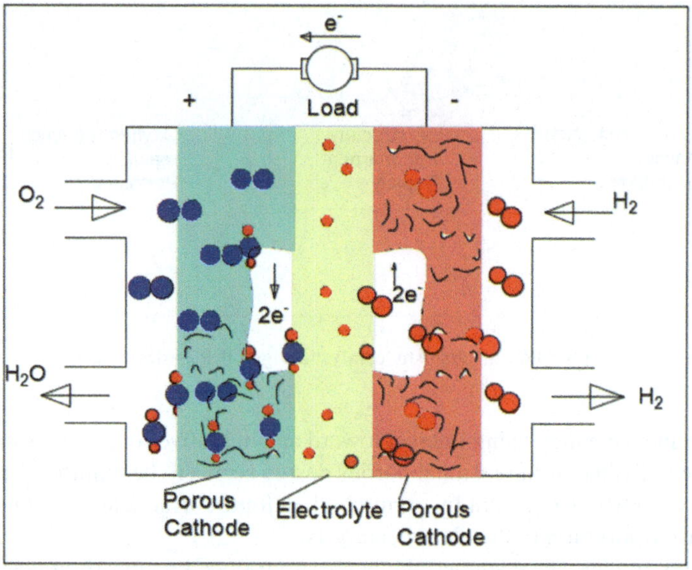

Fig. 3 Schematic diagram of a hydrogen fuel or oxygen cell and its responses based on the Proton exchange membrane fuel cell (Reproduced with permission from Ref. copyright [59])

the electron stream out of the negative terminal gives electrical energy, although protons are drawn out crosswise the electrolyte sheath toward the negative terminal and respond with absorbed O_2 to the formation of H_2O ($4H^+ + O_2 + 4e^- \rightarrow 2H_2O$). The 4-electron decline technique is the most promising method for ORR due to its performance benefit and the neglecting of H_2O_2 intermediate types that can affect the sheath and ionomer [59].

ORR at the negative terminal suffers from difficult electron relocations in a steep decline path which are depicted in Fig. 4. The O–O bond of adsorbed O_2 breaks down into 2–O* middle which is to be declined into OH^- and water as the finishing product in acidic and alkaline states correspondingly. However, for a fractional drop to take place, oxygen is first consumed onto the substance exterior and compared to the consumed oxygen pairs. Two protons to form HOOH* in between before the O–O bond is sliced and it is prominent to a high production of H_2O_2 or HO_2 through a dual electron path, which must be eliminated throughout the ORR process as represented in Fig. 4b [60, 61]. ORR is essentially numerous orders of magnitude lesser compared to the oxidation reaction, and substances are essential to inferior initiation obstacles of the slow ORR during an acidic state. Up to now, some materials only provide suitable activity and stability throughout the ORR in an acidic medium, such as heteroatom-doped carbon nanomaterials and the changeover metal-based nanoparticle operationalized nanocarbon materials [62, 63]. However, ORR activity has been very favorable in alkaline medium; it is also applicable to non-precious metal, namely doped nanocarbon materials and metal oxides [59].

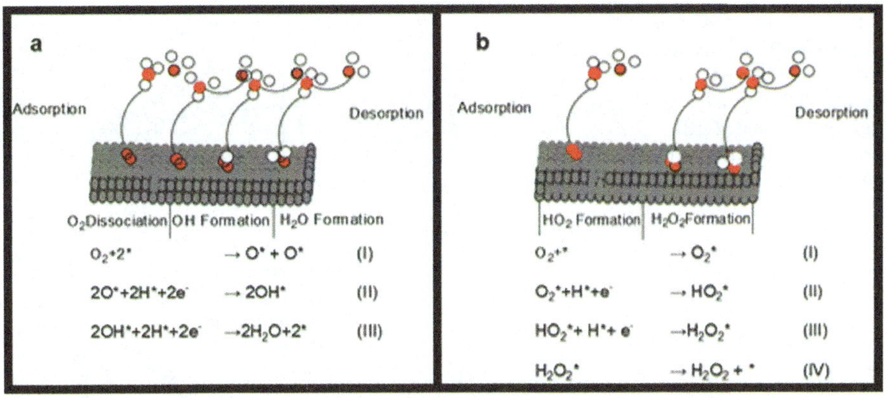

Fig. 4 Projected mechanism representations of **a** complete reduction and **b** fractional reduction (Reproduced with permission from Ref. copyright [64])

Carbon materials, particularly nanocarbon materials, have been used in manufacturing electrodes for energy harvesting. Common advantages of carbon-based energy storage devices are electrical conductivity; ion absorption area has larger and very good chemical corrosion resistance [65–68]. Additionally, flexible surface chemistry, ample structural variations, carbon nanomaterials combined to ORR at cost and act as active substances for fuel cells and metal air batteries [69, 70]. The structural optimization of carbon nanomaterials is one of the predominant method to adequately expose and activate th material as catalytic substances. It is also noted that the different carbon nanomaterials such as 0, 1 and 2-dimensional exposing on active surface have limited applications, which is unsuitable for electron and mass transfer during the ORR process [65, 71]. However, 3-dimensional nanocarbon and porous shaped materials have high surface area and exposed active surfaces, providing good electrocatalytic efficiency. Essentially, the 3-dimensional nanomaterials play a major role in oxygen diffusion and ion transfer [71]. In the past years, carbon nanomaterial-based non-metal electrocatalysts doped with heteroatom, namely nitrogen, sulfur, boron, phosphorous and their mixtures, have appeared as predominant to substitute platinum and other costlier metals for extremely capable ORR [69, 64].

The electron donor properties and electronic structure can vary in the sp^2 frame of graphitic carbon; it has been ensured by experimental and theoretical calculations. The results imply that the oxygen absorption and enhancement of the oxygen reduction using electro carbon have been performed by the breaking neutrality of electrons of sp^2 carbon; it will improve the electrocatalytic movement of ORR [58, 72, 73]. However, platinum-based substances show enhanced ORR activities compared with other carbon-based non-metal substances with good stability and cost effectiveness. Enhanced evidence shape that effective ORR carbon nanomaterial substances must be preferred to have ample reachable active sites for executing reaction, good electrical conductivity and appropriate porous structure for the transport of mass. The occurrence of acceptable responsive spots in mixture with novel structural design

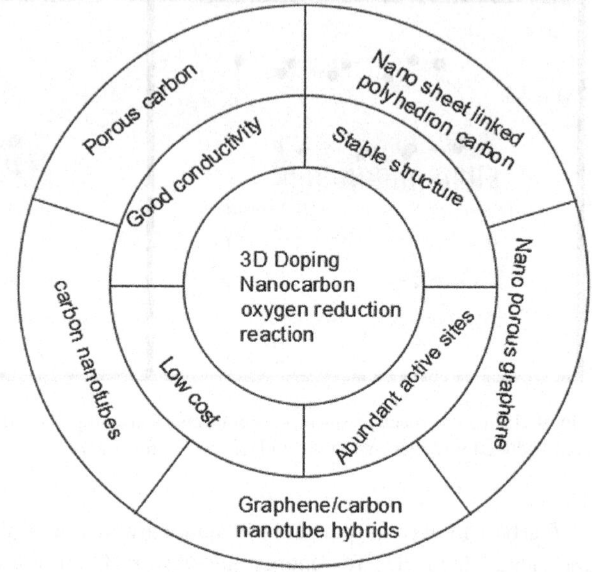

Fig. 5 Benefits of non-metal three-dimentional doped carbon nanomaterials and their applications as high-performance ORR electrocatalysts (Reproduced with permission from Ref. copyright [77])

produces smart non-metal substances. To attain the above-mentioned advantages, numerous heteroatom-doped 3-dimensional carbon nanomaterials, namely nitrogen-doped carbon nanotube aerogels [74], nitrogen-doped graphene with nanopores [75], nitrogen- and phosphorous-doped porous carbon and carbon nanotube or graphene mixure materials have been established as high-efficiency ORR substances as shown in Fig. 5 [75, 76].

3.2 *Heteroatom-Doped 3-dimensional Carbon Nanotubes ORR*

As distinctive 1-dimensional sp^2 crossed nanocarbon materials, carbon nanotubes can be assumed as 2-dimensional graphene plates rolled up into nano-dimensional tubes [78, 79]. Due to their excellent physicochemical and structural characteristics such as huge surface area, excellent mechanical characterization, good conductivity of electron transfer and good chemical property have inspired continuous attention in the nanotechnology field, especifically in energy and atmospheric areas [71, 80]. When heteroatoms are suitably doped into the ORR carbon matrix, improved ORR performance can be attained. Doped carbon nanotubes function in improving electron transfer, mass transfer of oxygen, deterioration opposition and water elimination of substances, prominent to enhanced catalytic activity and strength [81, 82]. Essentially, placing the tiny cylindrical nanotubes composed into an important 3-dimensional outline with interconnection and balanced circulation and controlling the porous structure can give samples showing active spots and steady electron and mass transport skeleton [37].

4 Compound Heteroatom Co-Doped 3-dimensional Carbon Nanotubes

In addition to the mono heteroatom-doped CNT, co-doped carbon nanotubes with various heteroatoms were examined to depict much better electrocatalyst ORR performance due to the combined outcome between various heteroatoms. Vertically allied multi-walled carbon nanotube arrays co-doped with phosphorous atoms and nitrogen atoms were first produced by an injection-aided chemical vapor deposition method. Due to the synergetic outcome rising from co-doping carbon nanotubes with both phosphorous and nitrogen, the obtained phosphorous, nitrogen co-doped multi-walled carbon nanotubes arrays essentially depict extraordinary electrocatalytic action toward ORR compared to the conventional platinum or carbon electrode and essentially better than that of carbon nanotubes doped by phosphorous and nitrogen only. Consequently, another nitrogen, phosphorous dual-doped carbon nanotube array was produced by one-pot synthetic approach with an aminophosphonic acid resin as the nitrogen, phosphorus and carbon sources [77]. Related with conventional bamboo-shaped nitrogen carbon nanotubes and nitrogen-phosphorous based carbon nanotubes, the huge hollow channels gives plenty of active sites in inner walls throughout oxygen ORR [83–85]. The advantages of metal-free 3D doped carbon nanomaterials and their applications as high-performance ORR electrocatalysts are depicted in Fig. 5.

The energy storage method has been classified into two types, namely surface charge storage and bulk charge storage also known as electric dual coating capacitance and pseudocapacitance, correspondingly. Materials having carbon atom, namely activated carbons [86], graphene [87], carbon nanotubes [88, 89], carbide derivative carbons [90] and carbon fibers [91] are the very commonly used electrodes in EDLC. The storage of electric charges using EDLC and electrostatic capacitors are similar, though, in the case of the EDLC the positive and negative ions are passed between two different electrical charge layers from carbon electrodes and electrolyte ions, respectively [92, 93] as shown in Fig. 6. Exact capacitance of a capacitor can be calculated using Eq. (1)

$$\mathrm{C} = \epsilon_0 \epsilon_\mathrm{r} \frac{\mathrm{A}}{\mathrm{d}} \tag{1}$$

EDLC continues definite capacitance 6–9 orders of magnitude greater when compared with traditional capacitors [94]. Subsequently, charge separation 'd' is much lesser during the development of an electric double layer, and the definite surface area 'A' of an active material is much greater (up to 3000 $\mathrm{m^2g^{-1}}$) [95–98] when related with electrostatic capacitors. Electrical energy stored in EDLC is

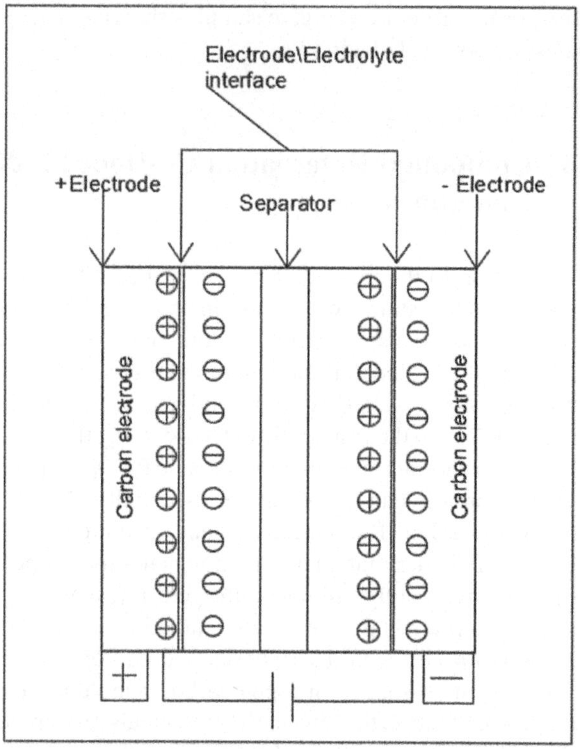

Fig. 6 Diagrammatic representation of the electric charge storage technique of an electrical dual layer capacitor (EDLC) (Reproduced with permission from Ref. copyright [93])

entirely in physical form due to the charging time being very low. EDLC is suitable for high power applications along with high durability [99–102]. Energy storage in pseudocapacitor is very fast and the electron transfer happed in the interface of electrolyte and electrode; the storage mechanism is totally reversible faradic charge transfer [103, 104] as shown in Fig. 7. In pseudocapacitors, energy density has higher then EDLC due to the higher specific capacitance, which is much influenced by the higher energy density.

Although many advantages, a few drawbacks are lower efficiency and cycle life compared to EDLC due to the charge deposited within the active materials. The reason for lower cycle efficiency is the adverse effect within the active material [105, 106]. Commonly used metal oxides and conducting polymers for the electrodes of pseudocapacitorsare manganese, iron ruthenium, nickel, polyaniline and polyacetylene [89, 107–109].

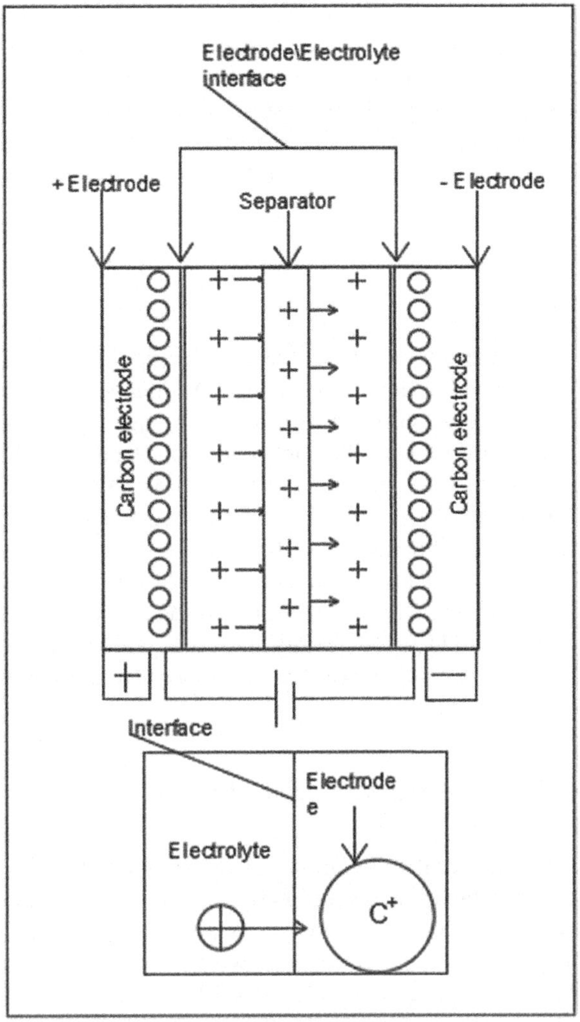

Fig. 7 Schematic diagram of the electric charge storage mechanism of a pseudocapacitor (Reproduced with permission from Ref. copyright [103])

4.1 Advantages of Electrical-Based Capacitor

The electrochemical capacitors maintain high energy densities, however, they produce lesser energy density compared to other batteries and fuel cell, and in order to meet high energy applications, the capacitors need higher energy capabilities. To enhance the energy capabilities of the electrochemical capacitors, researchers carried out different research works [110, 111].

$$E = \frac{1}{2}\mathrm{CV}^2 \tag{2}$$

where E—Energy density, C—capacitance and V—operating voltage.

The kind of electrolyte usage in energy storage is influenced by operation voltage. Energy density (Eq. (2)) of the electrochemical cell has been enhanced by increasing either operating voltage or specific capacitance. However, power density (P) of the electrochemical capacitor is to be reduced when using electrolytes having higher working voltages (ΔV) and is inversely proportional to resistance (R) shown in Eq. 3.

$$P = \frac{1}{2}\frac{(\Delta V)^2}{R} \tag{3}$$

Another method to increase the energy density of an electrochemical capacitor is enhancing specific capacitance. Adding conducting polymers or heteroatoms like phosphorous, sulfur, nitrogen, boron and metal oxides into a pseudocapacitor is to increase the specific capacitance of electrochemical cell [112]. Because of low conductivity and being costlier, metal oxides are used in limited applications [113]. The capacitive performance of the heteroatoms is higher because of the pseudo-capacitive influence through reversible Faradic reaction [114]. For supercapacitor application, many researchers have carried out research based on the functional materials, namely phosphorous, nitrogen, boron and sulfur in the field of energy harvesting [115–118]. For the past few years, research on nitrogen (heteroatom) has been carried out, and research on other heteroatoms have been considered recently.

4.2 *Endohedral Fullerene*

An endohedral fullerene normally exists in a caged appearance containing metal atoms, which has been reported by many researchers. However, metallofullerenes have been rarely studied. The nanoparticles of different metal dichalcogenide are displayed in the form of hollow cage structures, namely nanotubes, fullerene and nested fullerene structures [119–121]. The metal dichalcogenide structure consists of a metal layer present between the upper and lower chalcogen layers. For the understanding of the cage formation, it is required to be compared with carbon fullerenes. Figure 8 shows the schematic diagram of WS_2 crystallites and graphites. In controlled atmospheric conditions, graphite is fully stable, and the carbon atoms are arranged like a honeycomb structure. The carbon atom is joined with three adjacent atoms via planar sp^2 bonds. The shifting of the π electrons within the plane of the graphene sheets, lead to the metallic behavior to the perpendicular to the c axis. Most of the carbon atoms have been not attached with more than two atoms located in the periphery of the graphite nanoclusters.

The structure of IF-MoS_2 is different from carbon fullerenes and the point defect located in IF is represented in Fig. 9a. The rhomboidal point defect has been arranged on the corner of the IF which formed the rectangular projection. Euler's theory states

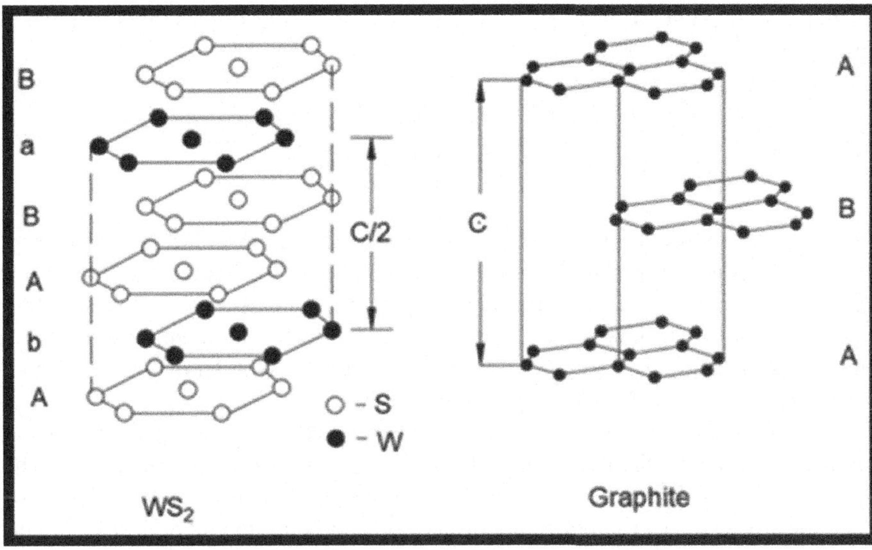

Fig. 8 Schematic diagram of the WS_2 nano-crystalites and graphite (Reproduced with permission from Ref. copyright [121])

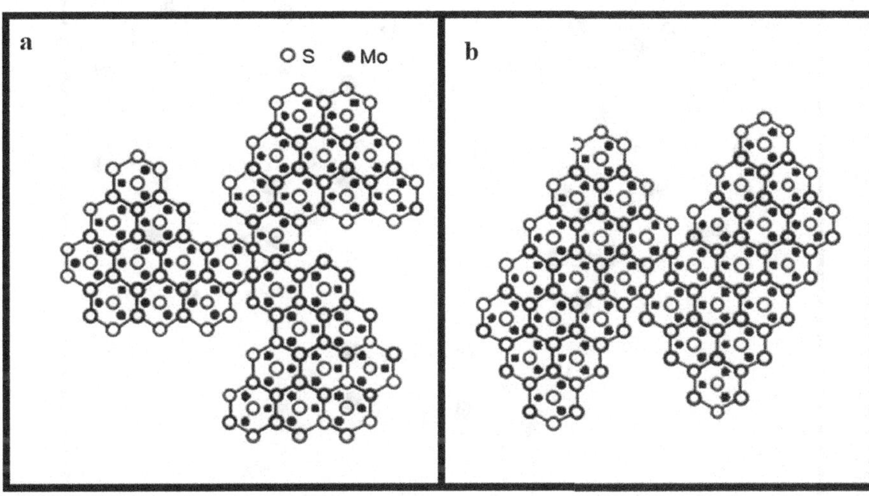

Fig. 9 Schematic diagram of the possible point defects in corner of IF-MoS: **a** triangular, **b** rhomboidal (Reproduced with permission from Ref. copyright [121])

that the closed polyhedron formed by the six rhomboidal point defects shown in Fig. 9b.

4.3 Fullerene Formed from B–C–N Phases

Many researchers reported the relocation of the carbon atom in the C_{60} network by boron and nitrogen atoms. It formed different combinational structures as shown in Fig. 10. However, the boron and nitrogen atoms paired unlikely in the initial stage due to destabilizing the fullerene cage and forming chemical hindrance. The following stable clusters, namely $C_{70}N_2$, $C_{59}N_6$, $C_{59}N_4$ and $C_{59}N_2$ formed in solution, which are all microscopic levels of evidence reported by researchers.

The relative stability of B–C–N has been compared with heterofullerenes using a number of theoretical calculations which was reported by a few researchers. This chapter describes the rejoining of the B–N in a way of hemispherical fullerenes, hence the weakness of the bonds has been removed, and the reaction of bonds has been improved.

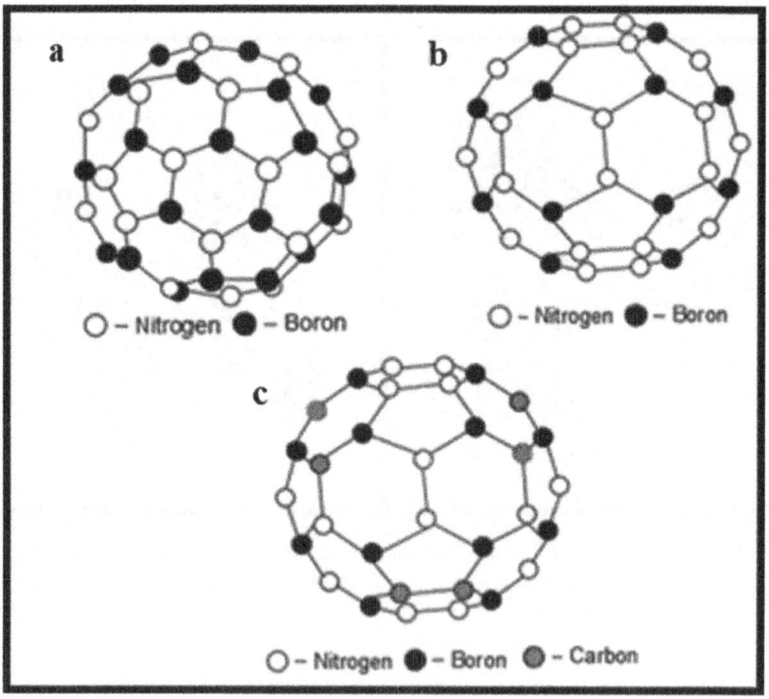

Fig. 10 Schematic diagram of BoronvNitrogen–Carbon nanoclusters **a** $B_{30}N_{30}$, **b** $B_{24}N_{36}$, **c** $C_{12}B_{24}N_{24}$ (Reproduced with permission from Ref. copyright [122])

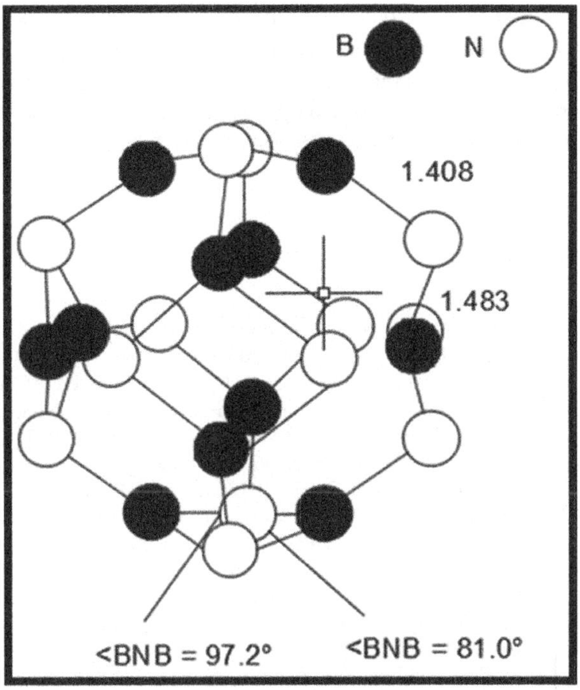

Fig. 11 Schematic diagram of determined structure $B_{12}N_{12}$ with truncated octahedron structure (Reproduced with permission from Ref. copyright [123])

4.4 B–N Fullerenes

The current calculations [122] display that the B_{30}–N_{30} and B_{24}–N_{36} cluster fullerene structures and adjacent boron–boron and nitrogen–nitrogen in Fig. 10a, b are metastable compared to bulk analogies. However, this metastability can be compared with C_{60} which changes into graphite or diamond by applying 150 $_{\text{k bar}}$ pressure. Hence, not able to produce large quantity of B_{30}–N_{30} and B_{24}–N_{36}. The fullerene structure has not clear when consider odd number of atoms. However, the presence of the B_{36}–N_{24} has been ensured by the laser ablation experiment [123]. Assuming the stability of boron–nitrogen compared with the boron–boron or nitrogen–nitrogen bonds, the boron nitride equivalent of C_{60} has been recommended to be $B_{12}N_{12}$, which comprises 6- and 4-membered rings [123] as shown in Fig. 11. In 4-cage rings, only peripheral confirmation has been made for the fullerene-like cages. Recently, the presence of concentric shells of B–N structures have been found.

5 Conclusion

Different doped nanocarbon synthesis methods, materials and precursors have been listed. Electric charge storage mechanism of the electric dual-layer capacitor and pseudocapacitor have been discussed with suitable schematic diagrams. Advantages of electrochemical capacitors on the energy capabilities, and the relation between the specific capacitance energy density have been discussed. Additionally, different fullerenes and graphite, namely endohedral fullerene, fullerene formed from B–C–N phases and B–N fullerenes have been discussed.

References

1. Arnold, M.S., et al., Sorting carbon nanotubes by electronic structure using density differentiation. Nature nanotechnology, 2006. 1(1): p. 60–65.
2. Miyata, Y., et al., Diameter analysis of rebundled single-wall carbon nanotubes using X-ray diffraction: verification of chirality assignment based on optical spectra. The Journal of Physical Chemistry C, 2008. 112(41): p. 15997–16001.
3. Krupke, R., et al., Separation of metallic from semiconducting single-walled carbon nanotubes. Science, 2003. 301(5631): p. 344–347.
4. Krupke, R., et al., Separation of metallic from semiconducting single-walled carbon nanotubes. Science, 2003. 301(5631): p. 344–347.
5. Ayala, P., et al., The physical and chemical properties of heteronanotubes. Reviews of modern physics, 2010. 82(2): p. 1843.
6. Yi, J.-Y. and J. Bernholc, Atomic structure and doping of microtubules. Physical Review B, 1993. 47(3): p. 1708.
7. Ewels, C. and M. Glerup, Nitrogen doping in carbon nanotubes. Journal of nanoscience and nanotechnology, 2005. 5(9): p. 1345–1363.
8. Ayala, P., et al., The doping of carbon nanotubes with nitrogen and their potential applications. Carbon, 2010. 48(3): p. 575–586.
9. Cruz-Silva, E., et al., Heterodoped nanotubes: theory, synthesis, and characterization of phosphorus– nitrogen doped multiwalled carbon nanotubes. ACS nano, 2008. 2(3): p. 441–448.
10. Cruz-Silva, E., et al., Electronic transport and mechanical properties of phosphorus-and phosphorus– nitrogen-doped carbon nanotubes. ACS nano, 2009. 3(7): p. 1913–1921.
11. Some, S., et al., Highly air-stable phosphorus-doped n-type graphene field-effect transistors. Advanced materials, 2012. 24(40): p. 5481–5486.
12. Zheng, B., P. Hermet, and L. Henrard, Scanning tunneling microscopy simulations of nitrogen- and boron-doped graphene and single-walled carbon nanotubes. ACS nano, 2010. 4(7): p. 4165–4173.
13. Czerw, R., et al., Identification of electron donor states in N-doped carbon nanotubes. Nano Letters, 2001. 1(9): p. 457–460.
14. Golberg, D., et al., MoO3-promoted synthesis of multi-walled BN nanotubes from C nanotube templates. Chemical Physics Letters, 2000. 323(1–2): p. 185–191.
15. Carroll, D., et al., Effects of nanodomain formation on the electronic structure of doped carbon nanotubes. Physical Review Letters, 1998. 81(11): p. 2332.
16. Quandt, A., et al., Boron doped graphene nanostructures. physica status solidi (b), 2008. 245(10): p. 2077–2081.
17. Borowiak-Palen, E., et al., Efficient production of B-substituted single-wall carbon nanotubes. Chemical physics letters, 2003. 378(5–6): p. 516–520.
18. Terrones, H., et al., The role of defects and doping in 2D graphene sheets and 1D nanoribbons. Reports on Progress in Physics, 2012. 75(6): p. 062501.
19. Chen, T., et al., Semiconductor to metal transition by tuning the location of N 2 AA in armchair graphene nanoribbons. Journal of Applied Physics, 2014. 115(5): p. 053707.

20. Stephan, O., et al., Doping graphitic and carbon nanotube structures with boron and nitrogen. Science, 1994. 266(5191): p. 1683–1685.
21. Yudasaka, M., et al., Nitrogen-containing carbon nanotube growth from Ni phthalocyanine by chemical vapor deposition. Carbon, 1997. 35(2): p. 195–201.
22. Koziol, K., B.O. Boskovic, and N. Yahya, Synthesis of carbon nanostructures by CVD method, in Carbon and Oxide Nanostructures. 2010, Springer. p. 23–49.
23. Terrones, M., et al., Efficient route to large arrays of CN x nanofibers by pyrolysis of ferrocene/melamine mixtures. Applied Physics Letters, 1999. 75(25): p. 3932–3934.
24. Lin, H., et al., Combined STM/STS, TEM/EELS investigation of CNx-SWNTs. physica status solidi (b), 2008. 245(10): p. 1986–1989.
25. Min, Y.-S., et al., Growth and characterization of nitrogen-doped single-walled carbon nanotubes by water-plasma chemical vapour deposition. Nanotechnology, 2007. 18(28): p. 285601.
26. Villalpando-Paez, F., et al., Synthesis and characterization of long strands of nitrogen-doped single-walled carbon nanotubes. Chemical Physics Letters, 2006. 424(4–6): p. 345–352.
27. Ayala, P., et al., Chemical vapor deposition of functionalized single-walled carbon nanotubes with defined nitrogen doping. physica status solidi (b), 2007. 244(11): p. 4051–4055.
28. Ayala, P., et al., Influence of the catalyst hydrogen pretreatment on the growth of vertically aligned nitrogen-doped carbon nanotubes. Chemistry of Materials, 2007. 19(25): p. 6131–6137.
29. Ayala, P., et al., Tailoring N-doped single and double wall carbon nanotubes from a nondiluted carbon/nitrogen feedstock. The Journal of Physical Chemistry C, 2007. 111(7): p. 2879–2884.
30. Elias, A., et al., Spectroscopic characterization of N-doped single-walled carbon nanotube strands: An x-ray photoelectron spectroscopy and Raman study. Journal of nanoscience and nanotechnology, 2010. 10(6): p. 3959–3964
31. Ibrahim, E., et al., Synthesis, characterization, and electrical properties of nitrogen-doped single-walled carbon nanotubes with different nitrogen content. Diamond and related materials, 2010. 19(10): p. 1199–1206.
32. Liu, Y., et al., Nitrogen-Doped Single-Walled Carbon Nanotubes Grown on Substrates: Evidence for Framework Doping and Their Enhanced Properties. Advanced Functional Materials, 2011. 21(5): p. 986–992.
33. Susi, T., et al., High quality SWCNT synthesis in the presence of NH3 using a vertical flow aerosol reactor. physica status solidi (b), 2009. 246(11-12): p. 2507–2510.
34. Pint, C.L., et al., Supergrowth of nitrogen-doped single-walled carbon nanotube arrays: active species, dopant characterization, and doped/undopedheterojunctions. ACS nano, 2011. 5(9): p. 6925–6934
35. Koós, A.A., et al., N-SWCNTs production by aerosol-assisted CVD method. Chemical Physics Letters, 2012. 538: p. 108–111.
36. Gai, P.L., et al., Structural systematics in boron-doped single wall carbon nanotubes. Journal of Materials Chemistry, 2004. 14(4): p. 669–675.
37. Terrones, M., et al., The role of boron nitride in graphite plasma arcs. Fullerene science and technology, 1998. 6(5): p. 787–800.
38. Charlier, J.-C., et al., Enhanced electron field emission in B-doped carbon nanotubes. Nano Letters, 2002. 2(11): p. 1191–1195.
39. Glerup, M., et al., Synthesis of N-doped SWNT using the arc-discharge procedure. Chemical Physics Letters, 2004. 387(1–3): p. 193–197.
40. Panchakarla, L., et al., Synthesis, structure, and properties of boron-and nitrogen-doped graphene. Advanced Materials, 2009. 21(46): p. 4726–4730.
41. Wei, D., et al., Synthesis of N-doped graphene by chemical vapor deposition and its electrical properties. Nano letters, 2009. 9(5): p. 1752–1758.
42. Wang, H., et al., Synthesis of boron-doped graphene monolayers using the sole solid feedstock by chemical vapor deposition. small, 2013. 9(8): p. 1316–1320.
43. Ayala, P., et al., A one step approach to B-doped single-walled carbon nanotubes. Journal of Materials Chemistry, 2008. 18(46): p. 5676–5681.

44. Golberg, D., et al., Large-scale synthesis and HRTEM analysis of single-walled B-and N-doped carbon nanotube bundles. Carbon, 2000. 38(14): p. 2017–2027.
45. Sheng, Z.-H., et al., Catalyst-free synthesis of nitrogen-doped graphene via thermal annealing graphite oxide with melamine and its excellent electrocatalysis. ACS nano, 2011. 5(6): p. 4350–4358.
46. Lin, H., et al., Many-body effects in electronic bandgaps of carbon nanotubes measured by scanning tunnelling spectroscopy. Nature materials, 2010. 9(3): p. 235–238.
47. Deng, D., et al., Toward N-doped graphene via solvothermal synthesis. Chemistry of Materials, 2011. 23(5): p. 1188–1193.
48. Cao, C., et al., Synthesis of carbon nitride nanotubes via a catalytic-assembly solvothermal route. Chemistry of materials, 2004. 16(25): p. 5213–5215.
49. Morant, C., et al., XPS characterization of nitrogen-doped carbon nanotubes. physica status solidi (a), 2006. 203(6): p. 1069–1075.
50. Xu, F., et al., Nitrogen ion implantation in single wall carbon nanotubes. Surface science, 2007. 601(13): p. 2819–2822.
51. Zhang, J. and L. Dai, Heteroatom-doped graphitic carbon catalysts for efficient electrocatalysis of oxygen reduction reaction. ACS Catalysis, 2015. 5(12): p. 7244–7253.
52. Yang, Z., et al., Recent progress in doped carbon nanomaterials as effective cathode catalysts for fuel cell oxygen reduction reaction. Journal of Power Sources, 2013. 236: p. 238–249.
53. Liu, J., et al., Recent advances in heteroatom-doped metal-free electrocatalysts for highly efficient oxygen reduction reaction. Electrocatalysis, 2015. 6(2): p. 132–147.
54. Ma, R., et al., Recent advances in heteroatom-doped graphene materials as efficient electrocatalysts towards the oxygen reduction reaction. Nano Adv, 2016. 1: p. 50–61.
55. Cheng, F. and J. Chen, Metal-air batteries: from oxygen reduction electrochemistry to cathode catalysts. Chemical Society Reviews, 2012. 41(6): p. 2172–2192.
56. Suntivich, J., et al., Design principles for oxygen-reduction activity on perovskite oxide catalysts for fuel cells and metal-air batteries. Nature chemistry, 2011. 3(7): p. 546–550.
57. Zhang, L. and Z. Xia, Mechanisms of oxygen reduction reaction on nitrogen-doped graphene for fuel cells. The Journal of Physical Chemistry C, 2011. 115(22): p. 11170–11176.
58. Liew, K.B., et al., Non-Pt catalyst as oxygen reduction reaction in microbial fuel cells: A review. international journal of hydrogen energy, 2014. 39(10): p. 4870–4883.
59. Gu, W., et al., Recent Advancements in Transition Metal-Nitrogen-Carbon Catalysts for Oxygen Reduction Reaction. Electroanalysis, 2018. 30(7): p. 1217–1228.
60. Stacy, J., et al., The recent progress and future of oxygen reduction reaction catalysis: A review. Renewable and Sustainable Energy Reviews, 2017. 69: p. 401–414.
61. Wang, Y.C., et al., S-doping of an Fe/N/C ORR catalyst for polymer electrolyte membrane fuel cells with high power density. AngewandteChemie, 2015. 127(34): p. 10045–10048.
62. Shui, J., et al., N-doped carbon nanomaterials are durable catalysts for oxygen reduction reaction in acidic fuel cells. Science advances, 2015. 1(1): p. e1400129.
63. He, W., et al., Structural effects of a carbon matrix in non-precious metal O 2-reduction electrocatalysts. Chemical Society Reviews, 2016. 45(9): p. 2396–2409.
64. Liu, X. and L. Dai, Carbon-based metal-free catalysts. Nature Reviews Materials, 2016. 1(11): p. 1–12.
65. Xiong, D., et al., Controllable oxygenic functional groups of metal-free cathodes for high performance lithium ion batteries. Journal of Materials Chemistry A, 2015. 3(21): p. 11376–11386.
66. Xiong, D., et al., Scalable synthesis of functionalized graphene as cathodes in Li-ion electrochemical energy storage devices. Applied Energy, 2016. 175: p. 512–521.
67. Li, C., et al., Three dimensional graphene networks for supercapacitor electrode materials. New Carbon Materials, 2015. 30(3): p. 193–206.
68. Zhou, M., H.-L. Wang, and S. Guo, Towards high-efficiency nanoelectrocatalysts for oxygen reduction through engineering advanced carbon nanomaterials. Chemical Society Reviews, 2016. 45(5): p. 1273–1307.

69. Tu, Y., D. Deng, and X. Bao, Nanocarbons and their hybrids as catalysts for non-aqueous lithium-oxygen batteries. Journal of energy chemistry, 2016. 25(6): p. 957–966.
70. Lin, Z., et al., Carbon nanotube sponges, aerogels, and hierarchical composites: Synthesis, properties, and energy applications. Advanced Energy Materials, 2016. 6(17): p. 1600554.
71. Daems, N., et al., Metal-free doped carbon materials as electrocatalysts for the oxygen reduction reaction. Journal of Materials Chemistry A, 2014. 2(12): p. 4085–4110.
72. Zheng, Y., et al., Nanostructured metal-free electrochemical catalysts for highly efficient oxygen reduction. Small, 2012. 8(23): p. 3550–3566.
73. Du, R., et al., Nitrogen-Doped Carbon Nanotube Aerogels for High-Performance ORR Catalysts. Small, 2015. 11(32): p. 3903–3908.
74. Ito, Y., et al., Bicontinuousnanoporous N-doped graphene for the oxygen reduction reaction. Advanced Materials, 2014. 26(24): p. 4145–4150.
75. Tian, G.L., et al., Nitrogen-doped graphene/carbon nanotube hybrids: in situ formation on bifunctional catalysts and their superior electrocatalytic activity for oxygen evolution/reduction reaction. Small, 2014. 10(11): p. 2251–2259.
76. Li, J.C., P.X. Hou, and C. Liu, Heteroatom-Doped Carbon Nanotube and Graphene-Based Electrocatalysts for Oxygen Reduction Reaction. Small, 2017. 13(45): p. 1702002.
77. Su, X.-L., et al., Three-dimensional porous activated carbon derived from loofah sponge biomass for supercapacitor applications. Applied Surface Science, 2018. 436: p. 327–336.
78. Avouris, P. and C. Dimitrakopoulos, Graphene: synthesis and applications. Materials today, 2012. 15(3): p. 86–97.
79. Xiao, X., et al., Freestanding mesoporous VN/CNT hybrid electrodes for flexible all-solid-state supercapacitors. Advanced Materials, 2013. 25(36): p. 5091–5097.
80. Yang, L., et al., Boron-doped carbon nanotubes as metal-free electrocatalysts for the oxygen reduction reaction. AngewandteChemie International Edition, 2011. 50(31): p. 7132–7135.
81. Yu, D., Y. Xue, and L. Dai, Vertically aligned carbon nanotube arrays co-doped with phosphorus and nitrogen as efficient metal-free electrocatalysts for oxygen reduction. The Journal of Physical Chemistry Letters, 2012. 3(19): p. 2863–2870.
82. Zhu, J., et al., One-pot synthesis of a nitrogen and phosphorus-dual-doped carbon nanotube array as a highly effective electrocatalyst for the oxygen reduction reaction. Journal of Materials Chemistry A, 2014. 2(37): p. 15448–15453.
83. Rajasekar, R., Nayak, G.C., Malas, A., Das, C.K.M.D., Development of compatibilized SBR and EPR nanocomposites containing dual filler system. 2012. 35, pp.878–885.
84. Nayak, G.C., Rajasekar, R., Bose, S., Das, C.K.J.N.T., Effect of MWNTs and SiC-coated MWNTs on properties of PEEK/LCP blend. 2009.
85. Nayak, G.C., Sahoo, S., Rajasekar, R., Das, C.K.C.P.A.A.S.M., Novel approach for the selective dispersion of MWCNTs in the Nylon/SAN blend system, 2012. 43(8): p. 1242–1251.
86. Li, X. and L. Zhi, Graphene hybridization for energy storage applications. Chemical Society Reviews, 2018. 47(9): p. 3189–3216.
87. Rey-Raap, N., et al., Influence of multiwalled carbon nanotubes as additives in biomass-derived carbons for supercapacitor applications. ACS applied materials & interfaces, 2019. 11(6): p. 6066–6077.
88. Shi, W., et al., High-performance capacitive deionization via manganese oxide-coated, vertically aligned carbon nanotubes. Environmental Science & Technology Letters, 2018. 5(11): p. 692–700.
89. Yan, P., et al., Ultrahigh-power supercapacitors based on highly conductive graphenenanosheet/nanometer-sized carbide-derived carbon frameworks. Nanotechnology, 2018. 29(25): p. 255403.
90. Vijayakumar, M., et al., Activated carbon fibres as high performance supercapacitor electrodes with commercial level mass loading. Carbon, 2018. 140: p. 465–476.
91. Sharma, K., A. Arora, and S.K. Tripathi, Review of supercapacitors: Materials and devices. Journal of Energy Storage, 2019. 21: p. 801–825.
92. Zhao, X., et al., Fabrication, characteristics and applications of carbon materials with different morphologies and porous structures produced from wood liquefaction: a review. Chemical Engineering Journal, 2019. 364: p. 226–243.

93. Jayalakshmi, M. and K. Balasubramanian, Simple capacitors to supercapacitors-an overview. Int. J. Electrochem. Sci, 2008. 3(11): p. 1196–1217.
94. Yu, D., et al., KOH activation of wax gourd-derived carbon materials with high porosity and heteroatom content for aqueous or all-solid-state supercapacitors. Journal of colloid and interface science, 2019. 537: p. 569–578.
95. Wang, J.-N., et al., RIPK1 inhibitor Cpd-71 attenuates renal dysfunction in cisplatin-treated mice via attenuating necroptosis, inflammation and oxidative stress. Clinical Science, 2019. 133(14): p. 1609–1627.
96. Zhang, L., et al., Ion-matching porous carbons with ultra-high surface area and superior energy storage performance for supercapacitors. Journal of Materials Chemistry A, 2019. 7(15): p. 9163–9172.
97. Wang, N., et al., Incomplete phase separation strategy to synthesize P/N co-doped porous carbon with interconnected structure for asymmetric supercapacitors with ultra-high power density. ElectrochimicaActa, 2019. 298: p. 717–725.
98. Jin, Z.Y., et al., Ionic Liquid-Assisted Synthesis of Microporous Carbon Nanosheets for Use in High Rate and Long Cycle Life Supercapacitors. Advanced Materials, 2014. 26(22): p. 3700–3705.
99. Huang, T., et al., Tri-high designed graphene electrodes for long cycle-life supercapacitors with high mass loading. Energy Storage Materials, 2019. 17: p. 349–357.
100. Sobhani-Nasab, A., et al., Sonochemical synthesis of terbium tungstate for developing high power supercapacitors with enhanced energy densities. Ultrasonicssonochemistry, 2018. 45: p. 189–196.
101. Liu, X., et al., Ni-doped cobalt-cobalt nitride heterostructure arrays for high-power supercapacitors. ACS Energy Letters, 2018. 3(10): p. 2462–2469.
102. Wang, G., et al., Beyond Activated Carbon: Graphite-Cathode-Derived Li-Ion Pseudocapacitors with High Energy and High Power Densities. Advanced Materials, 2019. 31(14): p. 1807712.
103. Jiang, Y. and J. Liu, Definitions of pseudocapacitive materials: a brief review. energy& environmental materials, 2019. 2(1): p. 30–37.
104. Wang, Y., et al., Electrochemical double layer capacitors containing carbon black additives for improved capacitance and cycle life. Carbon, 2018. 133: p. 1–5.
105. Wang, D., et al., High performance electrode materials for electric double-layer capacitors based on biomass-derived activated carbons. ElectrochimicaActa, 2015. 173: p. 377–384.
106. Prataap, R.V., et al., Effect of electrodeposition modes on ruthenium oxide electrodes for supercapacitors. Current Applied Physics, 2018. 18(10): p. 1143–1148.
107. Zeng, Y., et al., Iron-based supercapacitor electrodes: advances and challenges. Advanced Energy Materials, 2016. 6(24): p. 1601053.
108. Zheng, Y.-z., H.-y. Ding, and M.-l. Zhang, Preparation and electrochemical properties of nickel oxide as a supercapacitor electrode material.Materials Research Bulletin, 2009. 44(2): p. 403–407.
109. Lu, M., Supercapacitors: materials, systems, and applications. 2013: John Wiley & Sons.
110. Naoi, K., et al., Second generation 'nanohybridsupercapacitor': evolution of capacitive energy storage devices. Energy & Environmental Science, 2012. 5(11): p. 9363–9373.
111. Olabi, A. and M. Abdelkareem, Energy Storage Systems Towards 2050. 2020, Elsevier.
112. Abbas, Q., M. Mirzaeian, and A.A. Ogwu, Electrochemical performance of controlled porosity resorcinol/formaldehyde based carbons as electrode materials for supercapacitor applications. International Journal of Hydrogen Energy, 2017. 42(40): p. 25588–25597.
113. Abbas, Q., et al., Effect of physical activation/surface functional groups on wettability and electrochemical performance of carbon/activated carbon aerogels based electrode materials for electrochemical capacitors. International Journal of Hydrogen Energy, 2020. 45(25): p. 13586–13595.
114. Zhou, J., et al., Nitrogen-doped highly dense but porous carbon microspheres with ultrahigh volumetric capacitance and rate capability for supercapacitors. Journal of Materials Chemistry A, 2019. 7(2): p. 476–485.

115. Wang, D., et al., In situ double-template fabrication of boron-doped 3D hierarchical porous carbon network as anode materials for Li-and Na-ion batteries. Applied Surface Science, 2019. 464: p. 422–428.
116. Patiño, J., et al., Phosphorus-doped carbon-carbon nanotube hierarchical monoliths as true three-dimensional electrodes in supercapacitor cells. Journal of Materials Chemistry A, 2016. 4(4): p. 1251–1263.
117. Wu, Z.-S., et al., Bottom-up fabrication of sulfur-doped graphene films derived from sulfur-annulated nanographene for ultrahigh volumetric capacitance micro-supercapacitors. Journal of the American Chemical Society, 2017. 139(12): p. 4506–4512.
118. Hershfinkel M, Gheber L, Volterra V, Hutchison J, Margulis L, Tenne R (1994) Nested polyhedra of MX2 (M= W, Mo; X= S, Se) probed by high-resolution electron microscopy and scanning tunneling microscopy. Journal of the American Chemical Society 116 (5):1914–1917.
119. Tenne R (1995) Doped and heteroatom-containing fullerene-like structures and nanotubes. Advanced Materials 7 (12):965–995.
120. Margulis L, Salitra G, Tenne R, Talianker M (1993) Nested fullerene-like structures. Nature 365 (6442):113–114.
121. Bardo RD, Stanton CT, Jones WH (1995) Predicted Structures of Precursors B4N6H8, B4N8H8, and B4N8H6 and the B24N36 Analog of C60. Inorganic Chemistry 34 (5):1271–1272.
122. Pan L-L, Li J, Wang L-S (2008) Low-lying isomers of the B 9– boron cluster: The planar molecular wheel versus three-dimensional structures. The Journal of chemical physics 129 (2):024302.
123. Majedi S, Rauf HG, Boustanbakhsh M (2019) DFT study on sensing possibility of the pristine and Al-and Ga-embeded B12N12 nanostructures toward hydrazine and hydrogen peroxide and their analogues. Chemical Review and Letters 2 (4):176
124. Choi, H.J., et al., Defects, quasibound states, and quantum conductance in metallic carbon nanotubes. Physical Review Letters, 2000. 84(13): p. 2917

Doping of Carbon Nanostructures for Energy Application

Gobinath Velu Kaliyannan, Rajasekar Rathanasamy, Raja Gunasekaran, Manju Sri Anbupalani, Moganapriya Chinnasamy, and Sathish Kumar Palaniappan

Abstract Carbon nanomaterials are seen as potential alternative for high energy storage devices, because of their admirable mechanical properties, maximum thermal, and electrical conductivity. The carbon nanostructures such as fullerenes, graphene combined with numerous other materials have been identified in variety of applications. In the electrochemical energy storage application, the carbon nanostructures are found to be very encouraging. Their usage in high-frequency applications is limited due to their poor frequency response. These limitations could be overcome by the doping of carbon nanostructures by N, B, P/N, BIN, and Si. The electronic and transport properties can also be altered by doping with nitrogen or boron. This concise review objective is to address on the recent development of carbon nanostructures for energy storage. The advancements in testing and design of doped carbon nanostructures for energy storage, especially super capacitors, fuel cells, and lithium-ion batteries have been discussed with extensive examples.

Keywords Doping · Carbon nanostructures · Supercapacitor · Fuel cell · Li-ion batteries · Electrochemical energy

---

G. V. Kaliyannan
Mechatronics Engineering, Kongu Engineering College, Erode 638060, Tamil Nadu, India

R. Rathanasamy (✉) · M. Chinnasamy
Mechanical Engineering, Kongu Engineering College, Erode 638060, Tamil Nadu, India

R. Gunasekaran
Mechanical Engineering, Velalar College of Engineering and Technology, Erode 638012, Tamil Nadu, India

M. S. Anbupalani
Chemical Engineering, Kongu Engineering College, Erode 638060, Tamil Nadu, India

S. K. Palaniappan
Department of Mining Engineering, Indian Institute of Technology Kharagpur, Kharagpur 721302 West Bengal,, India

S. Sahoo et al. (eds.), *Defect Engineering of Carbon Nanostructures*, Advances in Material Research and Technology, https://doi.org/10.1007/978-3-030-94375-2_4

1 Introduction

On-world growing demand of inorganic energy sources and its environmental effects made the researchers to search for the alternate and viable renewable energy sources including wind, solar, hydropower, etc., [1, 2].

The current energy crisis can be sort out by developing dynamic energy storage device, while there is an increasing cost of energy and more usage of electricity by the people [3, 4]. The efficiency of the system can be increased by accumulating the energy when they are available in excess and it is released during the demand [3]. The currently available different energy storage systems including storage of compressed air, thermal energy, hydro energy, and like that electrochemical energy storage systems such as super capacitors, batteries, electrochemical capacitors (ECs), and fuel cells are also available [5, 6]. The electrode materials in these energy storage devices can be replaced by using carbon based materials as they are less cost and easily available [7]. As seen from the past 20 years, many researchers working on carbon nanostructures to improve its scalability and processability [8].

The carbon nanotube incorporation with the electrode material possesses major advantage such as improved electrical and mechanical properties. The fabrication of devices is economically cheap because of the precursor material used in CNT synthesis is less cost [9]. The large specific surface area and their improved electrical and thermal behavior makes the CNTs and their composites for electrochemical energy storage systems (EESS) [10, 11–13]. The CNTs are first discovered by Iijima in 1991 [14] which has been created interest in nanostructured carbon. Carbon nanotubes are cylindrical in shape and it is a macromolecule having radius of few nanometers which is about 10,000 times smaller than the human hair. Generally, CNTs are a type of elongated fullerene, as their diameters are very less than 1 mm and lengths of several mm. CNTs may be either single- or multi-walled which are produced from sheets of graphene as accessible in the Fig. 1.

Based on the number of cylinders present and their preparation, carbon nanotubes can be classified variously. According to elementary geometry, the diameter of single-walled carbon nanotubes (SWCNT) varies from 0.4 to 3 nm. By undulation graphene sheets, they can be developed. Over the chiral vector (n, m), the arrangement of carbon nanotubes can be defined by tube axis point of reference which is concerned with the hexagonal structure. Zigzag ($m = 0$) or armchair ($n = m$) are the types of nanotubes. Various cylinders whose diameter is about 100 nm are arranged in concentric arrays and they are present in multi-walled carbon nanotubes. DWCNT has concentric cylinders which is an exclusive classification of multi-walled carbon nanotubes [16].

The CNTs are produced by various methods including arc discharge method, laser ablation, and chemical vapor deposition (CVD) method [17, 18]. Carbon nanotubes can be processed in several techniques. CVD predates the improvement of CNT. According to this method, either microwave plasma CVD or thermal CVD improves the growth of carbon nanotubes. In the production of carbon nanotubes under industrial manufacturing, the most important technique is thermal CVD process. CVD

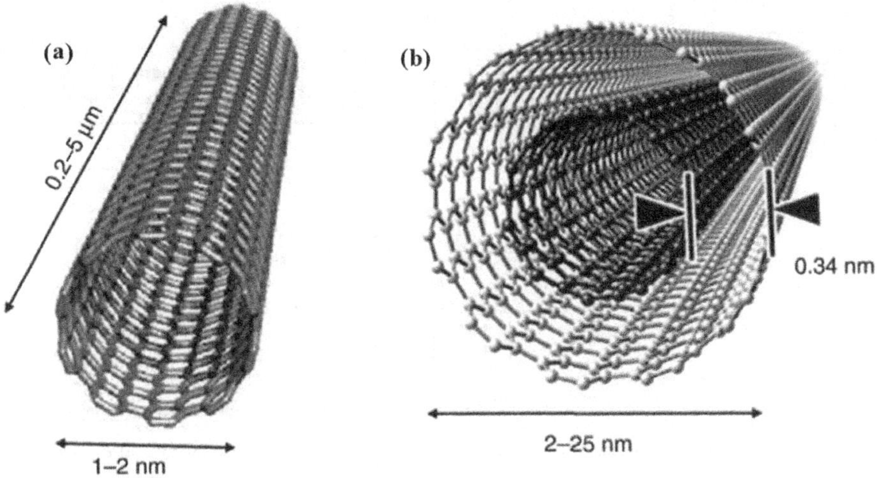

Fig. 1 Structural illustration of different types of carbon nanotubes and carbon structures: **a** Single-Walled CNT & **b** Multi-Walled CNT (Reproduced with permission from Ref. [15])

was the catalytic disintegration of carbon monoxide or hydrocarbon with transition metal which is acting as a catalyst. To probe carbon nanotubes for product purity and high-scale production, the most convenient technique is CVD technique.

Electric arc discharge produced under inert atmospheric gas of either argon or helium in between two electrodes made of graphite, is provoked. For manufacturing carbon nanotubes, the most broadly used process is electric arc discharge process. This electric arc discharge process probes the fullerene molecule and thus carbon nanotubes are exposed. By transformation technique, one of the graphite rods are removed during another rod is fixed. At maximum temperature about ~4000 K, the sublimation of carbon exists between the two rods. In a water-cooled reaction chamber, probe was executed. Initially, the chamber is displaced, after that it is replete with inert atmosphere (Ar or He 660 mbar). To reduce the distance, the anode shift toward the cathode. The distance must be lesser than 1 mm, to organize plasma and cross the current (100 A) via electrodes [19]. Laser ablation is the other strong process to develop carbon nanotubes. High impurities of about 30–40% are present in carbon nanotubes which are developed by thermal process. Even carbon nanotubes has maximum yield rate, when it is probed by laser ablation process. In the gas phase, fullerene cluster is produced by this process. Carbon is vaporized when a laser pulse is directed from solid graphite disk surface into maximum flow density of argon or helium. A long tube of quartz is fixed in a temperature-controlled furnace. The graphite target is kept at the mid of quartz tube. After moving the sealed tube, temperature is maximized to 1200 °C in the furnace. The tube is replete with inert gas and through the spherical lens a scanning laser is directed to the target material. The laser beam examines the target surface for vaporization and to manage flatten homogeneous surface. Laser vaporization generates the carbon

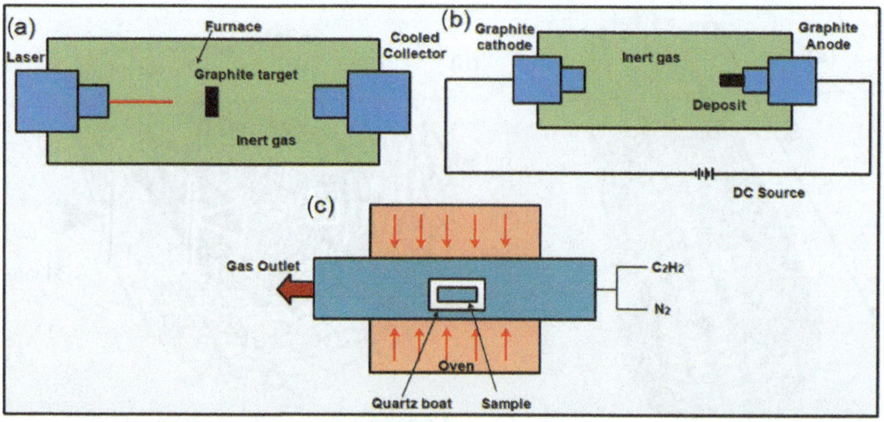

Fig. 2 Carbon nanotubes preparation methods **a** laser ablation, **b** arc discharge, and **c** CVD

species. From the maximum temperature zone, those carbon species are cleaned by inert gas and the tightened water-cooled copper collector collects those carbon species. Two consecutive laser pulses are implied to decrease the quantity of carbon collected. Laser removes the bigger particles at first and then breaks it. After that, those broken particles are delivered into the growing nanotube structure [20]. The schematic diagram of these three methods was represented in Fig. 2.

The distinct hybridization states such as sp, sp^2, and sp^3 have different chemical and physical properties as the covalent bonds are formed by coupling the carbon atoms [21]. The sp^2-hybridized CNT, specifically one-dimensional (1D) CNTs plays a vital role in energy applications as they possess high electrical conductivity. The CNTs finds application in various fields such as supercapacitors, sensors, photovoltaics devices, biomaterials, fuel cells, and field emission transistors due to their excellent electrical, mechanical, thermal, and optical properties [22]. The 1D CNTs are generally sheets of graphene are small strips that are rolled to produce single-walled nanotubes. These graphene sheets are rolled in many ways and they are denoted by pair of indices (n, m). It helps to define both chirality and diameter of single-walled CNTs [23]. The structure of the nanomaterial and interfacial reactions with other surrounding material contribute to the chemical and physical properties. The merits and demerits vary based on the different materials. Even though the CNTs improve the energy density of electrochemical capacitors, their applications are very limited due to the high production cost. The energy density is increased because of their immense electrical property and distinct tubular structure [24].

The thin film CNT electrodes are prepared by various methods such as wire wound rod coating [25], electrophoretic deposition [26] vacuum filtration [27]. CNTs can also be incorporated with polymers that are conducting called as polypyrrole (PPy) and polyaniline (PANI) composites are developed [28]. It is found that there is better charge propagation while incorporating CNTs with composites as they have increased mesoporosity and conducting properties. Large effort have taken by the researchers to develop distinct morphology CNT composites and also to increase the electrochemical performances [29]. Multi-walled carbon nanotubes with nanoparticles like bimetallic Pt–Pd are prepared by the pyrolysis method without using reducing agent [10]. It is seen from many literatures that there are many structural defects in multi-walled carbon nanotubes while studied in transmission electron microscopy (TEM) and also no ideal morphology. The doping of CNT by the intercalation reaction with electron acceptors and electron donor are very effective in enhancing electrical property [30]. Single-walled CNTs can be doped with electron acceptors or electron donors and they can be characterized by any of the following methods like Raman spectroscopy, optical absorption spectroscopy, X-ray diffraction (XRD), electron energy loss spectroscopy (EELS), transmission electronic microscopy (TEM), electron spin resonance (ESR), etc.

As seen the two-dimensional (2D) graphene is the least known synthetic carbon allotropes. Single-layer graphene are initially developed by the mechanical exfoliation of graphite in 2004 using Scotch Tape [31]. Carbon nanostructure materials such as graphene, grapheme oxide are mostly used as the support material in different types of absorbent to eradicate desulfurization [32]. It is found that the reduced graphene oxide (rGO) are used in various applications such as nanoelectronic devices, biosensors, and field effect transistors [33, 34] because of their electrochemical, functional, and electronic properties.

In recent times, zero-dimensional (0D) fullerenes have attracted much because of its low-cost solar energy devices with high potential. For processing on electrode surfaces, the solar cells with fullerenes or electron donor–acceptor by spin coating has been emerged [35]. The exclusive molecular structures and the effective properties of fullerene make them as a family of carbon precursors. The sub-nanometer molecules of fullerene consist of π-electrons, which are considered as the perfect building blocks for the construction of supramolecular structures. This can be further converted into mesoporous carbon material with high quality and π-systems with sp2-carbon frames [36, 37]. The carbon materials derived from fullerenes are highly stable and increased electrical properties which made them an enormous potential material in energy devices. As seen, Shrestha et al. investigated the development of mesoporous carbon from C70 microtubes [38], C60 nanotubes, C60 nanorods [37], C70 cubes [39], microbelts [40], for supercapacitors.

In addition, Vinu et al. [41] prepared porous carbon by using the C60-nanorods as precursor material and employed them as an electrode material for super capacitors. These indications prove that the porous carbon derived from fullerene has potential applications in energy storage device. This encourages the researchers to tailor the electrochemical and structural properties. Precisely, the efficient regulation of porosity by the 2D porous carbon may develop an effective electrode material in the field of energy storage by incorporating the merits of 2D morphological features that results in fast transfer of electrons, satisfies the electrolyte interaction, and reduces the defects at atomic level.

The oxygen reduction reaction (ORR) can be boosted by doping the carbon nanostructures with heteroatom as the electrode material which is also helpful in tailoring the electronic properties and structural stability [42]. The heteroatoms such as N, P,

O, and S and the transition metal atoms such as Fe, Co, and Ni doped with carbon nanostructures has been engaged as the replacement for the metal electrocatalysts in ORR [43, 44]. As similar to nitrogen doped carbon nanotubes the other heteroatom doped such as phosphorus and boron doped have been also increases the oxygen reduction reactions electrocatalytic property than the undoped CNTs [45, 46]. The phosphorus and boron have the same effect as that of nitrogen-doped CNT on the ORR. Anyhow boron- and nitrogen-doped carbon nanotubes possess distinct mechanisms. Now a days carbons doped with nitrogen has been used widely as it possess high capacitance because of the n-type and it is attained even if the material has a small surface area [47]. In few cases, it is seen that there has been rise in capacitance up to threefold [48].

The different types of functionalities of carbon materials may be affected by the precursor material that is chosen. Numerous methods have been adopted for doping nitrogen containing groups with compounds and also includes pretreatment with melamine, polyacrylonitrile, urea, and aldehyde resins [49–51]. It is reported that the nitrogen-doped CNT possess lower surface area up to 400 m^2/g [52]. This is considered to be very less than both pure multi-walled CNTs and single-walled CNTs have surface area larger than 830 m^2/g and 1315 m^2/g, respectively. It states that the large part of the total capacitance is occupied by pseudo-capacitance [53, 54]. Zhang et al. [55] reported that multi-walled CNTs doped with nitrogen using CVD method results high capacitance up to 44.3 F/g, that is two times higher than the undoped multi-walled CNTs.

Lee et al. [56] reported that nitrogen doped on CNTs that has vertical alignment enhance the wettability and capacitance will be improved to a certain limit due to the increase in electron donation by nitrogen. Both doped and undoped carbon nanotubes are formed directly on the substrate of stainless steel by CVD methods [56]. The conductivity of the CNTs are reduced due to excess nitrogen doping and it also inhibit charge storage. Boron is also considered as the subsequent material for doping carbon nanotubes because it possess p-type behavior and it also encourages the growth of CNT and maximizes the nanotubes oxidation temperature [57]. Anyhow the development of boron-doped carbon nanotubes were not fully established for using as a supercapacitors electrode material [52]. Report by Shiraishi et al. [42] reveals that there is increased in capacitance of B-doped multi-walled CNT from 6.5–6.8 $\mu F/cm^2$per surface area. CVD method is used for the synthesis of electrode. Wang et al. investigated B-doped carbon with either acid electrolyte (1 M H_2SO_4) or alkaline electrolyte (6 M KOH) increases the capacitance by 1.5–1.6 times at the interface [58].

Oxygen reduction reaction (ORR) can also be enhanced by using the carbon nanostructures extracted from metal organic frameworks [59]. Anyhow many drawbacks like poor electronic conductivity and there will be microstructural disorder and self-aggregation restricts the formation of catalysts at the time of pyrolysis process. It is seen that the HOMO–LUMO gap is maximized by doping boron adjacent to the metals, which helps to stabilize the electronic properties and improves spin multiplicity at their ground level. This makes the spreading of spin density to the overall system [60]. Palaniselvam et al. [61] investigated that the carbon doped with active

FeN of three types such as FeNC-68, FeNC-69, and FeNC-70. The doping of FeNC-70 results improved oxygen reduction reaction among the other two FeN-doped carbons and it has high content of nitrogen. As seen the FeNC-70 possess a structure of biomodal and it is observed in HR-TEM analysis that it is comprised of meso-porous shell and spherical macroscopic pore that is about 200 nm. Based on many researches [46, 47] n-doped CNTs it is demonstrated that efforts has been taken to attain maximum oxygen reduction reaction and also different processing techniques to develop N-doped CNTs [62, 63]. Not only nitrogen-doped CNTs other materials of carbon that has nitrogen like N-doped graphene [64, 65] and (NOMGAs) N-doped graphitic arrays [66] depicts maximum oxygen reduction reaction.

Figure 3 depicts the SEM, EDAX, and STEM images of nitrogen doped CNT [67]. It is clear that the tubes length is in few mm and diameter of nm. From the EDS spectrum, it is observed that there is rise in peaks of C and N, whereas the O peak is reduced which implies the higher C and N percentages. From the STEM analysis, it is found that the outer diameter is 120–140 nm and the inner diameter is 50–70 nm and the treatment at high temperatures pretends the outer surface nitrogen-doped CNTs as amorphous. These smaller diameters will enhance the different electronic properties like easy transportation of electrolytic ions and the surface area possess high electrochemical performance makes them superior for supercapacitors.

On the conductive substrate of nickel, doped CNTs have been precisely developed and it express a porous and open structure that is useful for quick ion kinetics and transport. Many reviews have been published on doping of carbon nanostructures anyhow only less are fixate on applications of energy storage [68–70]. Considering electrical properties carbon nanotubes applications are highly depends on parity,

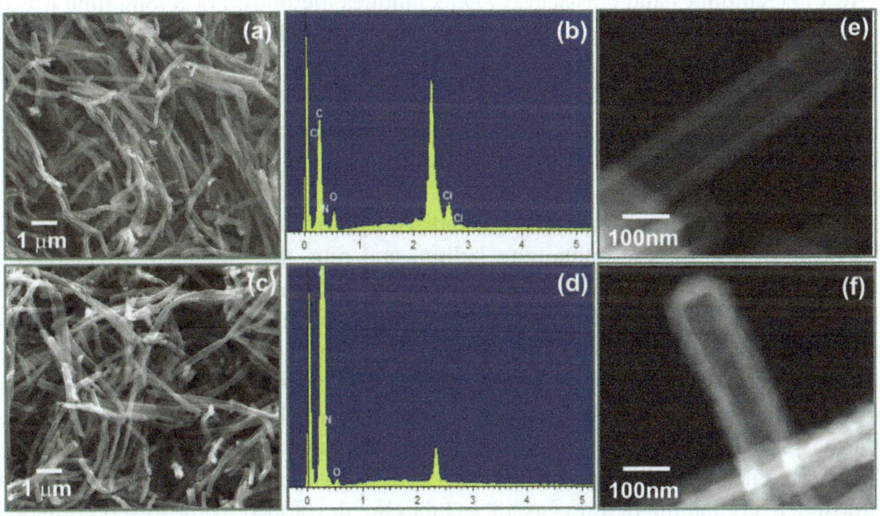

Fig. 3 FE-SEM, EDAX, and STEM images of nitrogen-doped CNTs (Reproduced with permission from Ref. [67])

helicity, and diameter [71]. By doping carbon nanotubes with nitrogen and boron makes CNTs as p-type and n-type behavior, respectively. B-doped CNTs tends to revealed increasing emission of electron field at edges of carbon nanotubes due to the presence of B-atoms [72]. Even though there is tube chirality-nitrogen doped carbon nanotubes depicts n-type behavior [73]. This paper briefly discuss about the current advancements in energy storing applications involving doped carbon nanotubes. This review reveals the methods of doping carbon nanostructures and their applications on energy storage in three separate areas such as super capacitors, fuel cells, and lithium-ion batteries (LIBs).

2 Doping Methods of Carbon Nanostructures

An electrical arc-discharge method initially used for doping and the reactions was done by Rb and K on multi-walled nanotubes [74]. By using the two-bulb method, Rb and K were doped in the vapor phase which leads to the MC8 saturation composition ($M = R_b/K$). Lithium the light alkali metal also dopes MWNTs [75]. By using the measurement of XRD under high pressure, the reactions of MWNTs in the ratio 2Li:C gave an intercalation compound. An aerosol-assisted chemical vapor deposition (AACVD) process contains carbon nanostructures with several dopants, and on the insulating quartz substrates they were really produced for the extension of CNT arrays [76, 77]. Zhao Jun Han et al. illustrated that ethylbenzene ($C_6H_5CH_2CH_3$) contains 5 wt% purified ferrocene ($Fe(C_5H_5)_2$) produces the pristine carbon nanostructure [78]. MWNTs are doped with nitrogen and it was incorporated by both AACVD techniques and classical CVD techniques methods. For a long time it was pretended that incorporating nitrogen doped multi-walled nanotubes has been a troublesome task, and nearly low doping concentration was handled in the early works [79]. Currently, using AACVD techniques by various literatures demonstrated that there is a chance of synthesizing important concentrations of nitrogen into MWNTs which is around 20 atom%. The different synthesis methods for doped CNTs are shown in Table 1.

Droppa et al. reported that by using arc-discharge method produces nitrogen doped CNTs in an atmosphere containing helium and nitrogen. Intellectual calculation proposes that the bending and the exaggeration of the tubes are due to the nitrogen present in pentagonal defects [84]. In nitrogen plasma atmosphere, nitrogen atoms partially replaces carbon atoms and, for incorporating N-CNTs, this method is used often [81]. Rather than this technique, temperature synthesis techniques like laser ablation and arc-discharge techniques can also be used [85]. The high temperature is used in all these cases to evaporate the carbon source, normally graphite [86]. During initial stages, doping of MWNTs was done by arc-discharging process. In an atmosphere containing helium gas, these tubes are developed by the evaporation of a graphite rod with its catalyst through an electric arc [87]. Laser ablation is one of the high-temperature growth method, in which a pulsed laser ablates a catalyst consist of graphite target, and the flowing gas nitrogen immediately removes the flowing

Table 1 Doped CNTs synthesis methods

S.No	Synthesis method	Parameters of synthesis	Dopant and their content (at %)	References
1	Thermal decomposition	CH_4+NH+Ar mixture and CH_4+pyridine+Ar mixture over a $Mo_{0.1}$ $Fe_{0.9}$ $Mg_{13}O$	Nitrogen (1 atom %)	[80]
2	Laser ablation	B_2O_3 in a flowing nitrogen atmosphere at 1503–1773 K over 30–240 min	B: 10% N: 2%	[81]
3	Chemical vapor deposition	Atmosphere of Ar/H_2 (100 sccm/40 sccm) at a time of 30 min for deposition at 800 °C	Nitrogen (2.27%)	[55]
4	Aerosol-Assisted Chemical Vapor Deposition (AACVD) process	Precursor solution of 5wt% purified ferrocene ($Fe(C_5H_5)_2$) in ethylbenzene ($C_6H_5CH_2CH_3$)	Nitrogen and boron	[78]
5	Electric arc discharge method	Buffer gas with He/N2 (volume ratio: 9:1) at a total pressure of 530–550 Torr and a current of 100 A	B: 1.20% N: 2.19%	[82]
6	Plasma treatment	Adding ammonia gas at flow rate of 180 sccm	Nitrogen	[83]

gas. By using arc-discharge technique some studies have tested the probability of nitrogen-doped SWNTs. In an arc-discharge observation, Droppa et al. created N-doped SWNTs under nitrogen–helium atmosphere by evaporating a catalyst consist of graphite rod [84].

This arc-discharge technique was also used by Glerup et at. for the creation of hetero-SWNTs [88]. In a helium atmosphere, they employed composite anodes rich in nitrogen precursors successfully to synthesis nitrogen into the tubes at 1 atom%. Due to the usage of dedicated EEL spectrometer and high nitrogen content, it was easy to examine the nanotubes directly. Suenaga et al. applied magnetron sputtering method successfully for doping, by attaining high nitrogen concentrations of 15–30% at 350 °C [79].

3 Doping of Carbon Nanostructures for Super Capacitors

Super capacitor is a progressive energy storage device whose magnificent characteristics are maximum power density, high life span, safe operation, and lightweight

[89]. Super capacitors perform a vital role in hybrid electrical vehicles, renewable energy utilizations, cell phones implantable biomedical appliances, compact laptops, suitable power sources, and in the improvement of compact electronics [90]. By having robust rate capacity, correlating to lithium-ion battery (LIB), quick charging and discharging asses, much higher power delivery (>10 kW/kg) for lesser times, empowers fast or ultrafast charging or discharging (in seconds) and maximum power density, so these super capacitors had obtained powerful consideration among the different energy storage systems. Most of the super capacitor electrodes has the distressing pore structure, by which a huge surface area is attained can restrict ion transportation and so it is ion kinetics [91]. By forming the electrode materials with reduced tortuosity and an open arrangement, the frequency response is improved and those limitations are reduced [92, 93].To develop the frequency response, vertical alignment, and ordered mesoporosity are determined on novel CNTs, such as porous carbon, graphene, and carbon nanostructures (CNTs) [92, 94–96].

The carbonized zIF-8 particles cause open channels and pores to provide the carbon composite substance with a high-speed ion transport pathways and a maximum specific surface area. On gathering super capacitor electrodes, the resultant composite accomplishes high range of cycling and maximum capacitance [97]. An efficient way to raise the capability of carbon substance as a substrate for electrodes in supercapacitors was doping, and to improve further charge storage, the dopants convert carbon layered with inert graphene into an active material which is electrochemical in nature.

The doping of nitrogen increases the capacitance of CNTs which is clearly observed form the previous studies [98, 99]. As seen many nitrogen-based materials with various structures are produced and used as alternative for electrode materials in supercapacitors which includes, porous nitrogen doped carbons, 1D nitrogen-doped CNTs, nitrogen-doped nanofibers, 2D nitrogen-doped graphenes, and also 3D nanostructures with variable dimensions. The production of B-doped carbon nanostructures is not fully developed for supercapacitor electrode substance [52]. The investigations by Shiraishi et al. proved that doping of boron increases the capacitive behavior up to 6.8 $\mu F/cm^2$ in 0.5 M LiBF4/PC [100]. By using CVD, the electrodes are incorporated repeatedly. In 2008, Wang et al. [58] described that, in B-doped carbon, interfacial capacitance increases up to 1.5–1.6 times than carbon with no boron but consist of acid electrolyte (1 M H_2SO_4) and alkaline electrolyte (6 M KOH).

For the application of super capacitors, Deng et al. [101] has studied various preparation processes of N-doped carbon. Due to the addition of nitrogen-enabled functionalities to the pseudo-capacitance, wettability of the electrodes develops by expanding active surface area available to the electrolyte and enlarging the substance's conductivity, nitrogen carbon-based substances are much interesting. N-based functionalities such as pyridinic-, quaternary-/graphic-N, pyrrolic-, and nitrogen oxides with pyridine are able to improve the electrochemical performance including conductivity, specific capacitance, cycling stability, etc.

To improve capacitance of supercapacitors, CNTs with various dopants such as P/N, B/N, Si, B, and N are introduced by Han et al. [78]. The minimum relaxation time constant of ~77 μs and the fast frequency range of about 13,200 Hz at −45°

phase angle was achieved by coin cells which was made of CNTs. Maximum-rate ion diffusion was delayed by the improper pore arrangement of electrodes and this occurs because of stagnant frequency response. CNTs display ultra-quick frequency response with a "factor-of-merit" for the working of electrochemical energy depot, because of the development of symmetric coin cells. Without disturbing the rate of charge transfer, a maximum areal capacitance was contributed by doped CNTs—based on the carbon electrodes, the measurement of electrochemical performance was done. Figure 4a represents the pristine carbon electrode's CV plots, at a maximum scan rate of about 0.5 V/S approximately rectangular and symmetric pattern was noticed. Figure 4b–c represents the pristine's galvanostatic CD plots and C doped with 0.044e0.88 mA/cm^2 current densities. These plots have a dependent rectangular CV pattern of a common capacitive behavior and display a moderately definite pattern. Redox reactions in pseudo capacitors and electrostatic adsorption in electrical double-layer capacitors are the most commonly used mechanisms for charge storage [102, 103]. An effective EDLC mechanism in CNTs was represented by galvanostatic CD plots and CV plots. In AC line filters, aluminum electrolytic capacitors are possibly

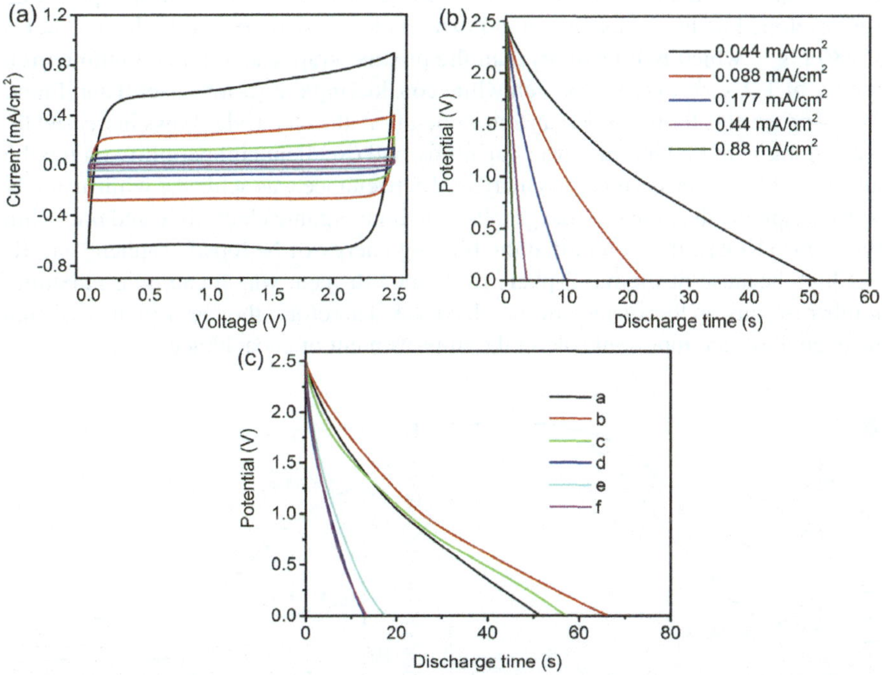

Fig. 4 **a** CV graphs for CNTs at different scan rates, **b** Galvanostatic charge/discharge graphs for CNTs at different current densities, **c** Galvanostatic charge/discharge graphs at 0.88 mA/cm^2 current density: a. Pristine, b. N-doped, c. B-doped, d. P/N-doped, e. B/N-doped, and f. Si-doped (Reproduced with permission from Ref. [78])

replaced by the ultra-quick super capacitors and the advancement of lighter, faster, and smaller electronic equipment are progressed.

Graphene-based materials are predominantly applicable for electrochemical applications, especially for supercapacitors, due to their fascinating properties such as high surface area and electrical conductivity [104]. In the recent years, much advancement is made to attain promising results. It is seen that, graphene is extensively tested and highly used as an electrode material for electrochemical double-layer capacitors (EDLCs). Further, the doping of graphene with heteroatoms such as nitrogen (N), boron (B), phosphorus (P), and sulfur (S) improves the wettability and also induce favorable pseudo-capacitance for electrochemical supercapacitors. The doping of graphene with heteroatoms has additional benefits such as high stability, electrical conductivity, sheet to sheet partition, and chemical reactivity. While doping graphene and reduced graphene oxide (rGO) by exposing them to plasma in ammonia gas or nitrogen atmosphere, there occurs the doping of nitrogen atoms over carbon lattice, eliminating the oxygen groups [105]. The configuration and content of nitrogen can be adjusted by the time of exposure and plasma strength. The capacitance of the nitrogen-doped graphene can be increased by the suitable configuration of nitrogen. Jeong et al. reported that the N-doped rGO as an electrode material of supercapacitor developed using a plasma method exhibits low resistivity and high capacitance of 280 F g^{-1} which is four times than the pristine graphene at a maximum current density of 1 A g^{-1} (Fig. 5). As seen while considering the various useful and fundamental properties for the supercapacitor operation, the electrode shows increased life cycle along with maximum power capability and excellent compatibility on flexible substrates [106]. Furthermore, an increased capacitance was achieved while using N-doped graphene than the pristine graphene in both organic electrolyte and potassium hydroxide (KOH). It is found that the binding energy of N-doped graphene and K+ was high because of the basal-plane pyridinic nitrogen, which can hold maximum number of ions on the surface of the electrode. Therefore, the basal-plane pyridinic nitrogen plays an important role in the improvement of capacitance.

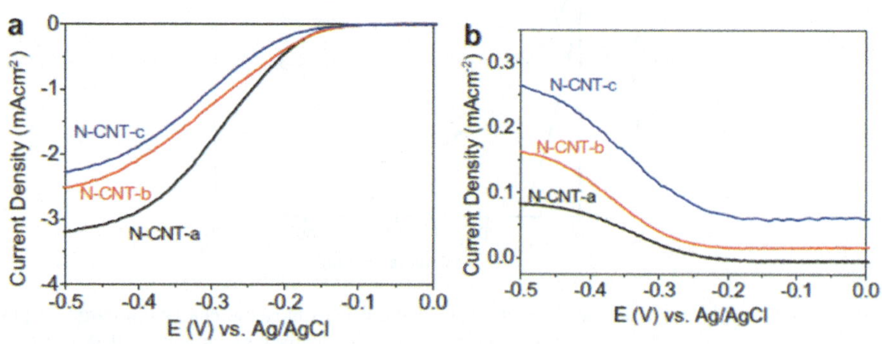

Fig. 5 Polarization graph for N-CNT-a, N-CNT-b, and N-CNT-c. **a** Disk currents, **b** ring currents (Reproduced with permission from Ref. [107])

When Boron was doped on graphene, boron acts a p-type dopant and the boron atoms develops a sp^2 hybridization in carbon lattice. The stability of the in-plane doping is high than the out-of-plane doping. This shows the importance of incorporation methodology in increasing the electrochemical properties of B-doped graphene. Therefore, various strategies have been implemented for doping the boron into carbon lattice. Rao et al. investigated about the various synthesis processes that were recommended [108]. In recent times, Han et al. reported the energy storage of boron doped graphene with large surface area formed from a one-pot solution method for efficient supercapacitors [109]. The graphene was doped using borane (BH3)–tetrahydrofuran (THF) under reflux and drying them in a vacuum while incorporating boron into graphene platelets. The formed compound have a large specific surface area of 466 m2 g^{-1}, which exhibits a exceptional specific capacitance of about 200 F g^{-1} from two-electrode configuration and 193 Fg^{-1} from a three-electrode configuration along with excellent performance due to the use of aqueous electrolytes. It is observed that even after 4500 electrochemical cycles above 95% of the actual capacitance can be retained and there is no distortion of current–voltage curves after the electrochemical cycle test, which demonstrates the stability of electrode material.

Fullerene-based carbon materials were identified as encouraging electrode materials in the field of supercapacitors because of their exclusive carbon structures. By incorporating different functional groups elements on fullerene, the pseudo-capacitance can be increased and double layer capacitances could be produced. Peng et al. [110] explored the ferrocenyl pyrrolidine C60, a fullerene derivative consists of iron and nitrogen as a precursor material. An exclusive microstructure can be formed from a process of liquid–liquid interfacial precipitation. Finally, one-step microstructure annealing at various temperatures were performed. The synthesis of various in situ N- and Fe-codoped 3D hierarchical carbon materials were performed. The prepared nitrogen and Fe-codoped materials were treated under 700 °C shows a maximum specific capacitance of 505.4 F g^{-1} at 0.1 A g^{-1}. For FCL700, there was a maximum current response above −0.8 V which results in high specific capacitance. The enhancement of electrochemical behavior can be achieved by both LLIP and pyrolysis processes. The current–voltage curves were in the shape of quasi-rectangular indicating the double capacitive layer. The faradic reactions of irons and various functional groups elements leads to the redox peaks of FCL700 and FCL600, which in-turn rises the pseudo capacitance [111]. FCL700 possess maximum current density which implies the maximum storage capacity of energy. It is seen that the electrochemical properties can be improved by the process of pyrolysis and LLIP crystallization. Among the fullerene-based electrode materials, FCL700 indicates the maximum specific capacitance. The high pseudo-capacitance is due to addition of iron and nitrogen through the C_{60} functionalization. The utilization of fullerene derivatives in energy storage applications leads to a new pathway for producing electrode materials for super capacitors with maximum performance.

4 Doping of Carbon Nanostructures for Fuel Cells

Fuel cells are considered as one of the most encouraging energy conversion technologies because they possess maximum efficiency and satisfy emission norms. Acceleration of the fuel cell economy method would naturally done by improving low-cost electro catalysts with maximum oxygen reduction reaction activity [112]. To develop long-term balance and decrease in storing capacity of platinum, carbon nanotubes was preferred because of their maximum surface area and chemical stability [113]. By using CVD method, N-doped CNTs are incorporated and they were visible by correlating pristine carbon nanotubes and maximum ORR activities [114, 115]. Graphite design has unified N atoms, improvements in carbon nanotubes catalytic action and platinum removal are lesser when compared with N-doped CNTs and it was represented in numerous groups [9, 10]. Adsorption of platinum in ordinary carbon nanotubes is less stronger then N-doped CNTs which was denoted in first-principles calculations [107].

Chen et al. [116] evaluated the electrocatalytic activities of nitrogen-doped CNTs at various percentages (i.e., 3.28, 2.84, and 2.51%) by the RRDE voltammetry analysis in alkaline conditions. Figure 5 shows the smaller ring current density and maximum disc current density samples with maximum N content. From polarization graphs of N-CNT-a, N-CNT-b and N-CNT-cit is clear that the increase in content of nitrogen CNTs increases the performance of oxygen reduction reaction.

As observed from the Table 2, it is seen that there is only little increase up to ($\sim$ 0.01 V) in half wave potential with the increase in the content of nitrogen. The performance of oxygen reduction reaction noted by RRDE voltammetry analysis is correlated with nitrogen content of N-doped carbon nanotubes. CNTs may contain nitrogen as dopant atoms due to its encapsulation is bamboo-structured cells as nitrogen gas or the creation of C atoms by the chemical bonds. Based on improvements of parameters such as H_2O selectivity and current density limit, the doping of nitrogen has a major effect on electrocatalytic sites and oxygen reduction reaction.

By using RDE process, the electro reduction under oxygen saturated 0.1 M KOH on Fe–nitrogen carbon nanotubes was calculated by Ratso et al. [117]. CNT-based metal catalysts was developed for better oxygen reduction reaction. By utilizing FeCl3 and dicyandiamide (DCDA), low coat precursors, second pyrolysis step and on

Table 2 ORR performance indicators at different percentages of nitrogen [116]

S.No	Percentage of Nitrogen in N-CNTs	Half wave potential (V)	Limiting current density (mA cm^{-2}, at 2500 rpm)	H_2O selectivity (%, at 2500 rpm, 0.5 V)	No. of electrons transferred (at 2500 rpm, 0.5 V)
1	3.28	−0.29	−3.19	81.82	3.63
2	2.84	−0.30	−2.52	59.88	3.20
3	2.51	−0.31	−2.26	38.16	2.76

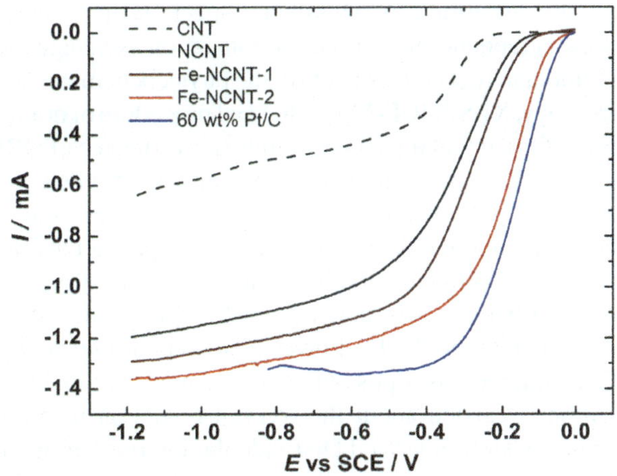

Fig. 6 Oxygen reduction reaction polarization graphs of CNTs, NCNTs, Fe-NCNTs, and Pt/C catalysts (Reproduced with permission from Ref. [116])

subsequent acid analysis, the catalyst incorporation was done by elementary pyrolysis. The material resulted from this process possess high stability and methanol tolerance and better performance of electrocatalytic for oxygen reduction reaction. Figure 6 illustrates the polarization graphs for ORR and the comparison of materials treated with the acids with CNTs, NCNTs, and normal Pt/C catalyst. The sample NCNT-1 possess maximum oxygen reduction reaction, half wave potential of −0.31 V and onset potential of 0.12 V. The values obtained are increased by 0.25 V than the metal-free NCNTs. The Fe-NCNT-2 has the onset potential of -0.04 V and half wave potential of -0.20 V as they are 0.15 V and 0.35 V decreased than that of Pt/C.

Yang et al. reported the performance of oxygen reduction reaction while doping boron at different percentages such as 0.86%, 1.33%, and 2.24% denoted as B_1CNTs, B_2CNTs, and B_3CNTs, respectively. CVD method was used for synthesis with precursors such as ferrocene, benzene, and triphenylborane (TPB) at different concentrations. The maximum peak current for undoped CNT is 2.8 $mAmg^{-1}$ and the doping of boron delivers current density as follows: 3.2 $mAmg^{-1}$ for B_1CNT, 3.8 $mAmg^{-1}$ for B_2CNT, and 8.0 $mAmg^{-1}$ for B_3CNTs. It is also observed that the peak potential shifts positively. The peak potential obtained for undoped CNT is 0.43 V and doping of boron results in 0.41 V, 0.38 V, and 0.35 V for B_1CNT, B_2CNT, and B_3CNT, respectively, with reference to saturated calomel electrode.

Sibul et al. [118] investigated a simple method to produce an high-performance ORR catalyst based on nitrogen- and iron-doped graphenes for fuel cell applications. To synthesis ORR electro catalysts, two distinct graphene-based materials such as graphene and graphene oxide were utilized as substrates. It is revealed that nitrogen- and iron-doped graphenes exhibits a maximum ORR than nitrogen- and iron-doped graphene oxides from the half-cell experiments using the rotating disc electrode (RDE) in alkaline medium. This leads to the larger surface area of nitrogen- and iron-doped graphenes than nitrogen- and iron-doped graphene oxides as determined by physicochemical methods. It is seen that 50% of iron was present in active

Fe–Nx confirmed by Mossbauer spectroscopy. Therefore, the nitrogen- and iron-doped graphenes were used for both anion exchange membrane and proton exchange membrane fuel cell tests. From the physicochemical analyses (57Fe Mössbauer spectroscopy, XPS & ICP-MS), a normal procedure of doping can be very successful and it is confirmed that there is maximum performance of ORR than the undoped graphene. The Fe–N-doped graphene shows maximum ORR in terms of half-wave potential and onset potential than Fe–N-doped graphene oxides in alkaline medium. The half-cell experiments of the Fe–N-doped graphene have low electrocatalytic activity in acid solution than in alkaline solution. It is also seen that the Fe–N-doped graphene exhibits better performance from AEM fuel cell tests than PEMFC results.

Feng Gao et al. [119] investigated about the electrocatalytic activity of n-doped C60 fullerene as a possible cathode material for hydrogen fuel cells. The investigation is carried about the oxygen reduction reaction on fullerene utilizing first-principle spin-polarized DFT calculation. Further the reaction pathways, elementary catalytic reaction steps and activation of energy barriers for water formation reaction on fullerene was studied to understand the electrocatalytic activity on the fuel cells cathode. It is identified that catalytic sites of N–C complex sites on N–C60 and the electrocatalytic activity of the fuel cells can be executed by direct pathways with Ea $= 0$ or indirect pathways with minimum Ea ≈ 0.22–0.34 eV. From the first-principles results, it is revealed that O_2 molecule could be adsorbed and there is a partial reduction on N_C complex sites of N-doped fullerene in the absence of activation barrier. Further, the partially reduced O_2 react with the H_+t hrough a direct pathway and accomplish the water formation reaction in the absence of activation energy barrier. Further, through indirect pathway, partially reduced O_2 react with H_+ and forms H_2O over transition state with minimum activation barrier (0.22–0.37 eV). H_+ could obtain a kinetic energy of ~0.95–3.68 eV during intermediate state to a transition state due to coulomb electric interaction which obviously eliminates the activation energy barrier on water formation reaction. The overall electro catalytic reaction cycles can be finished successfully and N–C60 fullerene retains the actual structure for upcoming electro catalytic reaction cycle. N–C60 fullerene is identified as a possible cathode catalyst for hydrogen fuel cells.

5 Doping of Carbon Nanostructures for Lithium-Ion Batteries (LIBs)

Increased adoption of portable electronics, versatile energy storage systems such as the LIBs attracted strong interests. Because of their longer life and increased energy densities LIBs are mainly preferred as economic batteries, particularly for portable devices [120, 121]. While an enormous attempt have been concerned to develop applicable flexible electrodes [122, 123], electrochemical efficiency also hinders its advancement toward high-scale development. Usually, LIBs depends on well-established compounds like $LiTi_5O_{12}$, $LiCo_2O_4$, and $LiMnO_2$ as active materials.

The theoretical calculation of specific capacities for LIBs are limited to 100–300 mAh g^{-1}[124, 125]. Development of carbon nanotubes become more encouraging candidates among the future Li storage devices. CNTs possess distinct structures and properties like increased tensile strength and electrical conductivity enhances Li storage while using them as anode materials [126]. CNTs used as alternative for the graphite anode due to their enhanced properties like chemical stability, increased electrical conductivity, excellent mechanical properties, and maximum surface area. By the introduction of substitution heteroatoms on the layer of graphite leads to activate the carbon nanotubes. The atomic sizes of the carbons are nearly equal to the Ni and B which are considered as highly desirable hetero elements.

Xifei Li et al. [127] investigated the synthesis of N-doped carbon nanotubes by the method of floating catalyst CVD with nitrogen at 16.4% for enhancing the properties of electrode materials in Li batteries. The electrochemical storage performance for LIBs is compared between pristine CNTs and N-doped CNTs as anode material. As seen the maximum reversible capacity and increased rate capabilities attained than the undoped CNTs in LIBs when there is a higher concentration of nitrogen in N-doped CNTs. The galvanostatic charge/discharge cycle is used for determining the efficiency of the lithium storage while using nitrogen-doped CNTs as anode materials. Certainly, it is studied that during the process of discharge/charge both the CNT and nitrogen-doped CNTs shows better cyclability, but the specific discharge of normal CNTs is very low (~266 mAh g^{-1}), whereas the high concentration nitrogen-doped CNTs possess specific discharge about 397 mAh g^{-1} after 100 cycles. The movement speed of Li+ and its electrons contribute to the speed of charging and discharging of LIBs.

Bulusheva et al. [128] reported the electrochemical performance of nitrogen-doped multi-walled carbon nanotubes. The result obtained from the calculation of quantum-chemical shows the strong binding of lithium ions and the irreversible capacity has been due to pyridinic structure of N_2 atoms. The lower content of nitrogen (~at 1%) at a 0.2 mA cm^{-2} current density shows the high reversible capacity of 270 mAh^{-1}. As seen, there is maximum value of current density shows that the nitrogen atom incorporation enhances the electrochemical performance of the CNTs. The effect of various content of nitrogen on the carbon nanotubes electrochemical performance is shown in Fig. 7. It is observed that there is slight increase in specific capacity up to 20% of nitrogen-doped CNTs. Although the maximum value is resulted for CNTs doped with nitrogen at high concentration, the undoped carbon nanotubes possess maximum irreversible capacity which is clearly understood from the results.

Ren et al. prepared a composite material $Li_4Ti_5O_{12}$ denoted as LTO for electrodes with multiwalled CNTs which is doped with boron and nitrogen. It is represented as N-B-C-LTO and it is resulted from a mixture of LTO and B,N-CNT after the process of calcination. Five different samples such as LTO, CNT-LTO (C-LTO), N-C-LTO, B-C-LTO, and B-N-C-LTO are produced and the results are compared and presented in the Fig. 8. At the initial stage of the cycle, the discharge capacity of the samples are maximum and is being reduced after 150 cycles. There exhibits a high

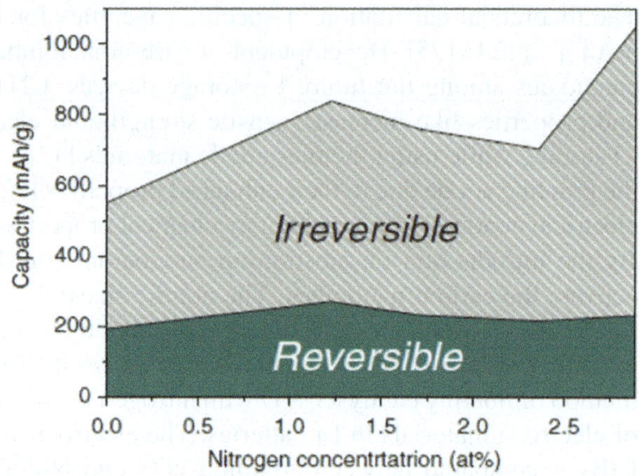

Fig. 7 Specific capacity of CNT electrodes at different nitrogen percentages (Reproduced with permission from Ref. [128])

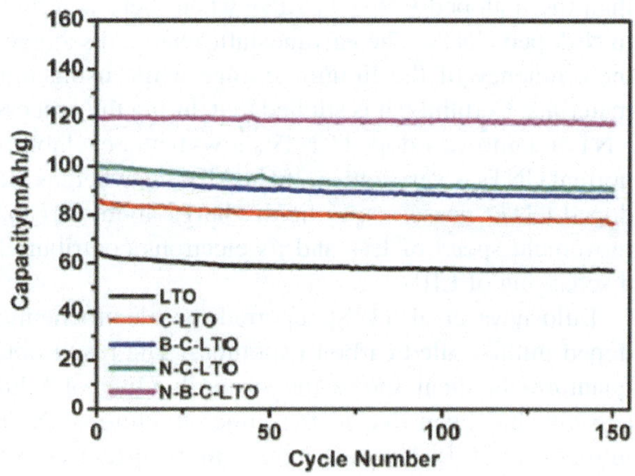

Fig. 8 Cycling performance of LTO, C-LTO, B,C-LTO, N,C-LTO, and N-B-C-LTO (Reproduced with permission from Ref. [129])

electrochemical performance while using N-B-C-LTO. The doping of nitrogen and boron increases the discharge capacity and also enhances the stability of cycling.

Shuai Wu et al. [47], studied the electrochemical properties of nitrogen- and boron-doped graphene at minimum current of about 50 mA g^{-1} in a 1 M LiPF6 electrolyte. It is noted that the cyclic performance and reversible capacity of the doped graphene were highly increased than the pristine graphene [130]. The faradaic efficiency for the pristine graphene is 43.8% and its rises while doping nitrogen to 49.0% and for boron doping, it is increased to 55.6% in the first cycle. This shows that the doping of boron or nitrogen quench electrolyte decomposition and graphene surface side reactions with electrolyte to produce SEI film. The boron-doped and nitrogen-doped graphene has reversible capacities of 1549 and 1043 mAh g^{-1} for the first cycle and 1227 and 872 mAh g^{-1} after 30 cycles, respectively. This is very high than the pristine graphene which results 955 mAh g^{-1} for the first cycle and 638 mAh g^{-1} after 30 cycles [130]. Subsequently, the reversible capacity of N-doped graphene is

83.6% and for the B-doped graphene is for the B-doped graphene which is very high compared to pristine graphene (66.8%) after 30 cycles.

The exceptional charging and discharging performance are achieved in LIBs while using N- and B-doped graphene as electrode materials. The samples were initially charged and discharged at 0.5 A g^{-1} for ten cycles and there is stepwise increase in current rate 25 A g^{-1} for ten cycles at each rate. It is observed that at high current rates, the N- and B-doped graphene electrodes exhibits good stability and capacity without any capacity fluctuations as noticed in pristine graphene. The N- and B-doped graphene could be charged reversibly to 493 and 611 mAh g^{-1} within 1 h at a current rate of 0.5 A g^{-1}. While charging several minutes at current rate of 0.5 A g^{-1} the reversible capacity reaches 380 mAh g^{-1} for the born doped graphene at 273 s and 296 mAh g^{-1} for the nitrogen-doped graphene at 212 s.

Most significantly, the doped graphene electrodes were fastly charged within seconds. As seen, at high current rate of about 25 A g^{-1}, the reversible capacity of N-doped graphene reach 199 mAh g^{-1} at a charging time of ~28 s (126 C) and for B-doped graphene reversible capacity reaches 235 mAh g^{-1} at a charging time of ~33 s (106 C). The results obtained were very impressive than pristine graphene (~100 mAh g^{-1} at current rate 25 A g^{-1}) and other electrode materials, such as porous carbon monoliths [131], graphite [132], graphitized carbon nanobeads [133], carbon nanofibers [134], and carbon nanotubes [135]. This evidently proves that the N- and -doped graphenes were encouraging high capacity and high rate electrodes for LIBs. The cyclic stability of the N- and B-doped graphene electrodes is very excellent at a high current rate. The cells possess better capacity retention while cycling at high current rates from 1 A g^{-1} to 25 A g^{-1}. Further, a capacity of 500 mAh g^{-1} could be retained to initial 0.5 A g^{-1}, demonstrates a good reversibility.

Fullerene is considered as relatively advanced member in carbon allotropes [45] with a closed-cage structure [44]. In precise, the increased electron affinity [46–48] of the fullerene recognized them as a cathode material with high redox potential. It is also found that fullerene-based carbon materials possess excellent structural and electronic properties [49] and also increased chemical stability [44], that could be tamed to meet precise needs of electrochemical devices by doping other elements [53–55] and adding functional groups [50–52]. It is predicted that N-doping could raise the redox potential of fullerene [40] by maximizing the electron affinity. It is anticipated that strongly electron withdrawing functional groups and oxygen-containing functional groups could raise the redox potential of fullerene. In addition, oxygen-containing functional groups could have Faradaic reaction with lithium ions [34]. Anyhow the required increase in redox potential can be achieved by quantitative investigation to design materials with maximum redox potential. Parveen Sood et al. [136] reported the nitrogen doping with fullerene (C60) and also the fullerene's functionalization with strongly electron withdrawing functional groups (SEWFGs) and oxygen-containing functional groups (OCFGs). The electronic properties of fullerene were investigated using first-principles density functional theory (DFT) and procedure [57] for the basic understanding of relationship between molecular structure and electrochemical property. This will be helpful for deriving molecular design guidelines for desirable cathode materials. From the DFT calculations of N-doped fullerene (C60 & C59N), it is clear that there is increase in redox potential up to 12%. The HOMO in C59N is partially filled and the N-dopant generates symmetrical C60 molecule. As seen when the cyano group present at the meta position is much effective in maximizing the redox potential. In comparison with ortho and para positions, the meta position OCFGs, the SEWFG-functionalization of C59N increases the redox potential. It is also reported that there is not much effective increasing of redox potentials, while SEWFG-functionalization of C59N. Precisly, there is a decrease in redox potential during the SEWFG-functionalization of C59N present at ortho and para position. As observed, there is a better correlation between the electron affinity and redox potential and the region nearer to nitrogen is more reactive during functionalization.

Parveen Sood et al. [137] investigated the electronic properties and redox potentials of fullerene (C60 & C59B). It is observed that there is an improvement in redox potential from 2.462 to 3.709 V while doping fullerene with one boron atom, which is associated with the formation of open shell system. The functionalization of C59B with oxygen-containing functional groups (OCFGs) was performed to increase the redox and electronic properties. During this functionalization, it is seen that there is a reduction in redox potential due the change of open shell structure to closed shell, although the redox potentials are high than the pristine fullerene (C60). The fullerene's electron affinity coupled with rich chemical properties of carbon causes fullerene (C60) as most promising cathode materials for LIB applications. The boron present in the fullerene (C60) generates electron deficiency because the boron contains one less electron than carbon and it is necessary to improve the electron affinity of fullerene (C60).

6 Conclusion

The doping methods of carbon nanostructures and the applications of doped carbon nanostructures in energy storage applications such as supercapacitors, fuel cells, and LIBs has been briefly discussed. Carbon nanostructures are doped with nitrogen or boron through different to change its surface property, performance and also the wettability of the carbon materials. These dopings introduce p-type or n-type behavior makes to attain the faradaic response in the systems and improve the energy density and capacitance. If it is possible achieve controlled doping, carbon nanostructures become most promising in energy storage applications. A large amount of power can be generated while using these doped carbon nanostructures than the traditional electrodes. This study clearly indicates that the doping carbon nanostructures will concurrently improve oxygen reduction reaction and electrochemical performance of energy storage devices. The commercialization of carbon nanostructures could be achieved by the advancements in synthesis and processing of carbon nanostructures at reduced cost.

7 Future Outlook

As seen there has been much research taking place for both energy conversion and storage, the application of carbon nanostructures for electrode materials have increased and they are better while tailoring it for specific application. Carbon nanostructures become more viable while it possess high power density and also fast charging and discharging efficiency. It is viewed that energy and power density of carbon nanotubes can be enhanced while incorporating them with composite materials, metal oxides, and polymers. These materials tend to provide best pseudo-capacitance. The future enhancements are not only for the portable electronics and automobiles, they are hindering in the areas of defense, medicine, and also consumer applications. Carbon nanostructures can be implemented in future for the medical applications. Recently, wide range of devices possesses required biocompatibility with human body and many other living things. The superior chemical properties and their excellent mechanical flexibility of doped carbon nanostructures could be a promising one for the interface of artificial and living things.

References

1. Simon, P. and Y. Gogotsi, Materials for electrochemical capacitors, in Nanoscience and technology: a collection of reviews from Nature journals. 2010, World Scientific. p. 320–329.
2. Wang, G., L. Zhang, and J. Zhang, A review of electrode materials for electrochemical supercapacitors. Chemical Society Reviews, 2012. 41(2): p. 797–828.
3. Hall, P.J., et al., Energy storage in electrochemical capacitors: designing functional materials to improve performance. Energy & Environmental Science, 2010. 3(9): p. 1238–1251. https://doi.org/10.1039/c0ee00004c.
4. Hall, P.J. and E.J. Bain, Energy-storage technologies and electricity generation. Energy policy, 2008. 36(12): p. 4352–4355.
5. Ibrahim, H., A. Ilinca, and J. Perron, Energy storage systems-Characteristics and comparisons. Renewable and sustainable energy reviews, 2008. 12(5): p. 1221–1250.
6. Manthiram, A., et al., Nanostructured electrode materials for electrochemical energy storage and conversion. Energy & Environmental Science, 2008. 1(6): p. 621–638. https://doi.org/10.1039/b811802g.
7. Davies, A. and A. Yu, Material advancements in supercapacitors: from activated carbon to carbon nanotube and graphene. The Canadian Journal of Chemical Engineering, 2011. 89(6): p. 1342–1357.
8. Sherman, L.M., Carbon nanotubes: Lots of potential-If the price is right. Plastics technology, 2007. 53(7): p. 68–73.
9. Kierzek, K., et al., Electrochemical capacitors based on highly porous carbons prepared by KOH activation. Electrochimica Acta, 2004. 49(4): p. 515–523.
10. Ding, K., et al., Electrocatalytic activity of multi-walled carbon nanotubes-supported PtxPdy catalysts prepared by a pyrolysis process toward ethanol oxidation reaction. Electrochimica Acta, 2013. 100: p. 147–156. https://doi.org/10.1016/j.electacta.2013.02.130.
11. Rajasekar, R., Nayak, G.C., Malas, A., Das, C.K.M.D., Development of compatibilized SBR and EPR nanocomposites containing dual filler system, 2012. 35, pp.878–885.
12. Nayak, G.C., Rajasekar, R., Bose, S., Das, C.K.J.N.T., Effect of MWNTs and SiC-coated MWNTs on properties of PEEK/LCP blend, 2009.

13. Nayak, G.C., Sahoo, S., Rajasekar, R., Das, C.K.C.P.A.A.S.M.: Novel approach for the selective dispersion of MWCNTs in the Nylon/SAN blend system. 43(8), 1242–1251 (2012).
14. Gordillo-Vazquez, F.J., V.J. Herrero, and I. Tanarro, From Carbon Nanostructures to New Photoluminescence Sources: An Overview of New Perspectives and Emerging Applications of Low-Pressure PECVD. Chemical Vapor Deposition, 2007. 13(6–7): p. 267–279.
15. Ghasempour, R. and H. Narei, CNT basics and characteristics, in Carbon nanotube-reinforced polymers. 2018, Elsevier. p. 1–24.
16. Rafique, I., et al., Exploration of epoxy resins, hardening systems, and epoxy/carbon nanotube composite designed for high performance materials: A review. Polymer-Plastics Technology and Engineering, 2016. 55(3): p. 312–333.
17. Meyyappan, M., et al., Carbon nanotube growth by PECVD: a review. Plasma Sources Science and Technology, 2003. 12(2): p. 205.
18. Terranova, M.L., V. Sessa, and M. Rossi, The world of carbon nanotubes: an overview of CVD growth methodologies. Chemical Vapor Deposition, 2006. 12(6): p. 315–325. https://doi.org/10.1002/cvde.200600030
19. Eklund, P., J. Holden, and R. Jishi, Vibrational modes of carbon nanotubes; spectroscopy and theory. Carbon, 1995. 33(7): p. 959–972. https://doi.org/10.1016/0008-6223(95)00035-C.
20. Thess, A., et al., Crystalline ropes of metallic carbon nanotubes. Science, 1996. 273(5274): p. 483–487.
21. Georgakilas, V., et al., Broad family of carbon nanoallotropes: classification, chemistry, and applications of fullerenes, carbon dots, nanotubes, graphene, nanodiamonds, and combined superstructures. Chemical reviews, 2015. 115(11): p. 4744–4822.
22. Wang, Y., et al., Multifunctional carbon nanostructures for advanced energy storage applications. Nanomaterials, 2015. 5(2): p. 755–777.
23. Guldi, D.M. and N. Martín, Carbon nanotubes and related structures: synthesis, characterization, functionalization, and applications. 2010: John Wiley & Sons.
24. Lota, G., K. Fic, and E. Frackowiak, Carbon nanotubes and their composites in electrochemical applications. Energy & Environmental Science, 2011. 4(5): p. 1592–1605.
25. Kitano, T., Y. Maeda, and T. Akasaka, Preparation of transparent and conductive thin films of carbon nanotubes using a spreading/coating technique. Carbon, 2009. 47(15): p. 3559–3565.
26. Cho, J., et al., Characterisation of carbon nanotube films deposited by electrophoretic deposition. Carbon, 2009. 47(1): p. 58–67.
27. Niu, C., et al., High power electrochemical capacitors based on carbon nanotube electrodes. Applied physics letters, 1997. 70(11): p. 1480–1482.
28. Khomenko, V., E. Frackowiak, and F. Beguin, Determination of the specific capacitance of conducting polymer/nanotubes composite electrodes using different cell configurations. Electrochimica Acta, 2005. 50(12): p. 2499–2506.
29. Shen, J., et al., How carboxylic groups improve the performance of single-walled carbon nanotube electrochemical capacitors? Energy & Environmental Science, 2011. 4(10): p. 4220–4229.
30. Duclaux, L., Review of the doping of carbon nanotubes (multiwalled and single-walled). Carbon, 2002. 40(10): p. 1751–1764.
31. Nair, R., et al., Graphene as a transparent conductive support for studying biological molecules by transmission electron microscopy. Applied Physics Letters, 2010. 97(15): p. 153102.
32. Liu, X.-L., et al., Desulfurization performance of iron supported on activated carbon. Fuel, 2014. 123: p. 93–100.
33. Valota, A.T., et al., Electrochemical behavior of monolayer and bilayer graphene. ACS nano, 2011. 5(11): p. 8809–8815.
34. Ekiz, O.O., et al., Reversible electrical reduction and oxidation of graphene oxide. Acs Nano, 2011. 5(4): p. 2475–2482.
35. Brabec, C.J., et al., Organic photovoltaics: concepts and realization. Vol. 60. 2003: Springer Science & Business Media.
36. Zheng, S., H. Ju, and X. Lu, A High-Performance Supercapacitor Based on KOH Activated 1D C70 Microstructures. Advanced Energy Materials, 2015. 5(22): p. 1500871.

37. Shrestha, L.K., et al., Nanoporous carbon tubes from fullerene crystals as the π-electron carbon source. Angewandte Chemie, 2015. 127(3): p. 965–969.
38. Bairi, P., et al., Mesoporous graphitic carbon microtubes derived from fullerene C 70 tubes as a high performance electrode material for advanced supercapacitors. Journal of Materials Chemistry A, 2016. 4(36): p. 13899–13906.
39. Bairi, P., et al., Mesoporous carbon cubes derived from fullerene crystals as a high rate performance electrode material for supercapacitors. Journal of Materials Chemistry A, 2019. 7(20): p. 12654–12660.
40. Tang, Q., et al., Quasi 2D mesoporous carbon microbelts derived from fullerene crystals as an electrode material for electrochemical supercapacitors. ACS applied materials & interfaces, 2017. 9(51): p. 44458–44465.
41. Benzigar, M.R., et al., Highly crystalline mesoporous C60 with ordered pores: a class of nanomaterials for energy applications. Angewandte Chemie International Edition, 2018. 57(2): p. 569–573.
42. Yang, W., et al., Polymerization-dissolution strategy to prepare Fe, N, S tri-doped carbon nanostructures for Zn-Air batteries. Carbon, 2019. 147: p. 83–89.
43. Shen, M., et al., Transition metal-nitrogen-carbon nanostructured catalysts for the oxygen reduction reaction: from mechanistic insights to structural optimization. Nano Research, 2017. 10(5): p. 1449–1470.
44. Fu, S., et al., Two-dimensional N, S-codoped carbon/Co9S8 catalysts derived from Co (OH) 2 nanosheets for oxygen reduction reaction. ACS applied materials & interfaces, 2017. 9(42): p. 36755–36761.
45. Yang, L., et al., Boron-doped carbon nanotubes as metal-free electrocatalysts for the oxygen reduction reaction. Angewandte Chemie International Edition, 2011. 50(31): p. 7132–7135.
46. Guo, M.-q., et al., Hydrothermal synthesis of porous phosphorus-doped carbon nanotubes and their use in the oxygen reduction reaction and lithium-sulfur batteries. New Carbon Materials, 2016. 31(3): p. 352–362.
47. Wu, Z.-S., et al., Doped graphene sheets as anode materials with superhigh rate and large capacity for lithium ion batteries. ACS nano, 2011. 5(7): p. 5463–5471.
48. Babeł, K. and K. Jurewicz, Electrical capacitance of fibrous carbon composites in supercapacitors. Fuel processing technology, 2002. 77: p. 181–189.
49. Inagaki, N., et al., Implantation of amino functionality into amorphous carbon sheet surfaces by NH3 plasma. Carbon, 2007. 45(4): p. 797–804.
50. Li, W., et al., Nitrogen enriched mesoporous carbon spheres obtained by a facile method and its application for electrochemical capacitor. Electrochemistry Communications, 2007. 9(4): p. 569–573.
51. Stein, A., Z. Wang, and M.A. Fierke, Functionalization of porous carbon materials with designed pore architecture. Advanced Materials, 2009. 21(3): p. 265-293. https://doi.org/10.1002/adma.200801492.
52. Inagaki, M., H. Konno, and O. Tanaike, Carbon materials for electrochemical capacitors. Journal of power sources, 2010. 195(24): p. 7880–7903.
53. Peigney, A., et al., Specific surface area of carbon nanotubes and bundles of carbon nanotubes. Carbon, 2001. 39(4): p. 507–514.
54. Niu, J.J., et al., An approach to carbon nanotubes with high surface area and large pore volume. Microporous and Mesoporous Materials, 2007. 100(1–3): p. 1–5.
55. Zhang, Y., et al., Preparation and electrochemical properties of nitrogen-doped multi-walled carbon nanotubes. Materials letters, 2011. 65(1): p. 49–52.
56. Lee, K.-Y., et al., Influence of the nitrogen content on the electrochemical capacitor characteristics of vertically aligned carbon nanotubes. Physica E: Low-dimensional Systems and Nanostructures, 2010. 42(10): p. 2799–2803.
57. Koos, A.A., et al., Comparison of structural changes in nitrogen and boron-doped multi-walled carbon nanotubes. Carbon, 2010. 48(11): p. 3033–3041.
58. Wang, D.-W., et al., Synthesis and electrochemical property of boron-doped mesoporous carbon in supercapacitor. Chemistry of materials, 2008. 20(22): p. 7195–7200.

59. Gu, W., et al., Hybrid of g-C3N4 Assisted Metal-Organic Frameworks and Their Derived High-Efficiency Oxygen Reduction Electrocatalyst in the Whole pH Range. ACS Applied Materials & Interfaces, 2016. 8(51): p. 35281–35288.
60. Stoyanov, S.R., A.V. Titov, and P. Král, Transition metal and nitrogen doped carbon nanostructures. Coordination Chemistry Reviews, 2009. 253(23-24): p. 2852–2871.
61. Palaniselvam, T., et al., Zeolitic Imidazolate Framework (ZIF)-Derived, Hollow-Core, Nitrogen-Doped Carbon Nanostructures for Oxygen-Reduction Reactions in PEFCs. Chemistry-A European Journal, 2013. 19(28): p. 9335–9342.
62. Yang, W., T.-P. Fellinger, and M. Antonietti, Efficient metal-free oxygen reduction in alkaline medium on high-surface-area mesoporous nitrogen-doped carbons made from ionic liquids and nucleobases. Journal of the American Chemical Society, 2011. 133(2): p. 206–209.
63. Yu, D., Q. Zhang, and L. Dai, Highly efficient metal-free growth of nitrogen-doped single-walled carbon nanotubes on plasma-etched substrates for oxygen reduction. Journal of the American Chemical Society, 2010. 132(43): p. 15127–15129.
64. Qu, L., et al., Nitrogen-doped graphene as efficient metal-free electrocatalyst for oxygen reduction in fuel cells. ACS nano, 2010. 4(3): p. 1321–1326.
65. Yang, S., et al., Graphene-based carbon nitride nanosheets as efficient metal-free electrocatalysts for oxygen reduction reactions. Angewandte Chemie, 2011. 123(23): p. 5451–5455.
66. Liu, R., et al., Nitrogen-doped ordered mesoporous graphitic arrays with high electrocatalytic activity for oxygen reduction. Angewandte Chemie, 2010. 122(14): p. 2619–2623.
67. Dubal, D.P., et al., Synthetic approach from polypyrrole nanotubes to nitrogen doped pyrolyzed carbon nanotubes for asymmetric supercapacitors. Journal of Power Sources, 2016. 308: p. 158–165.
68. Muhulet, A., et al., Fundamentals and scopes of doped carbon nanotubes towards energy and biosensing applications. Materials today energy, 2018. 9: p. 154–186.
69. Xu, H., L. Ma, and Z. Jin, Nitrogen-doped graphene: Synthesis, characterizations and energy applications. Journal of energy chemistry, 2018. 27(1): p. 146–160.
70. Ayala, P., et al., The doping of carbon nanotubes with nitrogen and their potential applications. Carbon, 2010. 48(3): p. 575–586.
71. O'Brien, P. and H. Kroto, RSC Nanoscience & Nanotechnology. 2005: Royal Society of Chemistry.
72. Sharma, R., et al., Field emission properties of boron and nitrogen doped carbon nanotubes. Chemical physics letters, 2006. 428(1-3): p. 102–108.
73. Czerw, R., et al., Identification of electron donor states in N-doped carbon nanotubes. Nano Letters, 2001. 1(9): p. 457–460.
74. Zhou, O., et al., Defects in carbon nanostructures. Science, 1994. 263(5154): p. 1744–1747. https://doi.org/10.1126/science.263.5154.1744.
75. Maurin, G., et al., Electrochemical intercalation of lithium into multiwall carbon nanotubes. Chemical physics letters, 1999. 312(1): p. 14–18.
76. Koós, A.A., et al., Effect of the experimental parameters on the structure of nitrogen-doped carbon nanotubes produced by aerosol chemical vapour deposition. Carbon, 2009. 47(1): p. 30–37.
77. Meysami, S.S., et al., Aerosol-assisted chemical vapour deposition synthesis of multi-wall carbon nanotubes: II. An analytical study. Carbon, 2013. 58: p. 159–169.
78. Han, Z.J., et al., High-frequency supercapacitors based on doped carbon nanostructures. Carbon, 2018. 126: p. 305–312.
79. Ewels, C. and M. Glerup, Nitrogen doping in carbon nanotubes. Journal of nanoscience and nanotechnology, 2005. 5(9): p. 1345–1363.
80. Panchakarla, L., A. Govindaraj, and C. Rao, Nitrogen-and boron-doped double-walled carbon nanotubes. ACS nano, 2007. 1(5): p. 494–500.
81. Golberg, D., et al., Large-scale synthesis and HRTEM analysis of single-walled B-and N-doped carbon nanotube bundles. Carbon, 2000. 38(14): p. 2017–2027.
82. Wang, B., et al., Direct and large scale electric arc discharge synthesis of boron and nitrogen doped single-walled carbon nanotubes and their electronic properties. Carbon, 2009. 47(8): p. 2112–2115.

83. Misra, A., et al., FTIR studies of nitrogen doped carbon nanotubes. Diamond and related materials, 2006. 15(2-3): p. 385–388.
84. Droppa Jr, R., et al., Incorporation of nitrogen in carbon nanotubes. Journal of Non-Crystalline Solids, 2002. 299: p. 874–879.
85. Meyyappan, M., Carbon nanotubes: science and applications. 2004: CRC press.
86. Journet, C., et al., Large-scale production of single-walled carbon nanotubes by the electric-arc technique. nature, 1997. 388(6644): p. 756–758.
87. Zhang, Y., et al., Heterogeneous growth of B–C–N nanotubes by laser ablation. Chemical Physics Letters, 1997. 279(5-6): p. 264–269.
88. Glerup, M., et al., Synthesis of N-doped SWNT using the arc-discharge procedure. Chemical Physics Letters, 2004. 387(1–3): p. 193–197. https://doi.org/10.1016/j.cplett.2004.02.005.
89. Ambrosi, A., et al., Capacitance of p-and n-doped graphenes is dominated by structural defects regardless of the dopant type. ChemSusChem, 2014. 7(4): p. 1102–1106.
90. Zhang, L.L., et al., Nitrogen doping of graphene and its effect on quantum capacitance, and a new insight on the enhanced capacitance of N-doped carbon. Energy & Environmental Science, 2012. 5(11): p. 9618–9625.
91. Izadi-Najafabadi, A., et al., Ion diffusion and electrochemical capacitance in aligned and packed single-walled carbon nanotubes. Journal of the American Chemical Society, 2010. 132(51): p. 18017–18019.
92. Honda, Y., et al., Aligned MWCNT sheet electrodes prepared by transfer methodology providing high-power capacitor performance. Electrochemical and Solid State Letters, 2007. 10(4): p. A106. https://doi.org/10.1149/1.2437665.
93. Lin, J., et al., 3-dimensional graphene carbon nanotube carpet-based microsupercapacitors with high electrochemical performance. Nano letters, 2013. 13(1): p. 72–78.
94. Miller, J.R., R. Outlaw, and B. Holloway, Graphene double-layer capacitor with ac line-filtering performance. Science, 2010. 329(5999): p. 1637–1639.
95. Bo, Z., et al., Emerging energy and environmental applications of vertically-oriented graphenes. Chemical Society Reviews, 2015. 44(8): p. 2108–2121.
96. Yoon, Y., et al., Vertical alignments of graphene sheets spatially and densely piled for fast ion diffusion in compact supercapacitors. Acs Nano, 2014. 8(5): p. 4580–4590.
97. Wan, L., et al., ZIF-8 derived nitrogen-doped porous carbon/carbon nanotube composite for high-performance supercapacitor. Carbon, 2017. 121: p. 330–336.
98. Lota, G., B. Grzyb B, H. Machnikowska, J. Machnikowski, E. Frackowiak. Chem. Phys. Lett, 2005. 404: p. 53.
99. Lota, G., K. Lota, and E. Frackowiak, Nanotubes based composites rich in nitrogen for supercapacitor application. Electrochemistry Communications, 2007. 9(7): p. 1828–1832.
100. Shiraishi, S., et al., Electric double layer capacitance of multi-walled carbon nanotubes and B-doping effect. Applied Physics A, 2006. 82(4): p. 585–591.
101. Deng, Y., et al., Review on recent advances in nitrogen-doped carbons: preparations and applications in supercapacitors. Journal of Materials Chemistry A, 2016. 4(4): p. 1144–1173.
102. Seo, D.H., et al., Structure-controlled, vertical graphene-based, binder-free electrodes from plasma-reformed butter enhance supercapacitor performance. Advanced Energy Materials, 2013. 3(10): p. 1316–1323.
103. Simon, P. and Y. Gogotsi, Capacitive energy storage in nanostructured carbon-electrolyte systems. Accounts of chemical research, 2013. 46(5): p. 1094–1103.
104. Kumar, N.A. and J.-B. Baek, Doped graphene supercapacitors. Nanotechnology, 2015. 26(49): p. 492001.
105. Nolan, H., et al., Nitrogen-doped reduced graphene oxide electrodes for electrochemical supercapacitors. Physical Chemistry Chemical Physics, 2014. 16(6): p. 2280–2284.
106. Jeong, H.M., et al., Nitrogen-doped graphene for high-performance ultracapacitors and the importance of nitrogen-doped sites at basal planes. Nano letters, 2011. 11(6): p. 2472–2477.
107. Li, Y.-H., T.-H. Hung, and C.-W. Chen, A first-principles study of nitrogen-and boron-assisted platinum adsorption on carbon nanotubes. Carbon, 2009. 47(3): p. 850–855.

108. Rao, C., K. Gopalakrishnan, and A. Govindaraj, Synthesis, properties and applications of graphene doped with boron, nitrogen and other elements. Nano today, 2014. 9(3): p. 324–343.
109. Han, J., et al., Generation of B-doped graphene nanoplatelets using a solution process and their supercapacitor applications. ACS nano, 2013. 7(1): p. 19–26.
110. Peng, Z., et al., Fullerene-based in situ doping of N and Fe into a 3D cross-like hierarchical carbon composite for high-performance supercapacitors. Advanced Energy Materials, 2019. 9(11): p. 1802928.
111. Cheng, E., et al., Porous ZnO/Co3O4/N-doped carbon nanocages synthesized via pyrolysis of complex metal-organic framework (MOF) hybrids as an advanced lithium-ion battery anode. Acta Crystallographica Section C: Structural Chemistry, 2019. 75(7): p. 969–978.
112. Zhang, L., et al., Progress in preparation of non-noble electrocatalysts for PEM fuel cell reactions. Journal of Power Sources, 2006. 156(2): p. 171–182.
113. Tang, H., et al., High dispersion and electrocatalytic properties of platinum on well-aligned carbon nanotube arrays. Carbon, 2004. 42(1): p. 191–197.
114. Matter, P.H., et al., Oxygen reduction reaction activity and surface properties of nanostructured nitrogen-containing carbon. Journal of Molecular Catalysis A: Chemical, 2007. 264(1-2): p. 73–81.
115. Shao, Y., et al., Nitrogen-doped carbon nanostructures and their composites as catalytic materials for proton exchange membrane fuel cell. Applied Catalysis B: Environmental, 2008. 79(1): p. 89–99.
116. Chen, Z., D. Higgins, and Z. Chen, Nitrogen doped carbon nanotubes and their impact on the oxygen reduction reaction in fuel cells. Carbon, 2010. 48(11): p. 3057–3065.
117. Ratso, S., et al., Enhanced oxygen reduction reaction activity of iron-containing nitrogen-doped carbon nanotubes for alkaline direct methanol fuel cell application. Journal of Power Sources, 2016. 332: p. 129–138.
118. Sibul, R., et al., Iron-and nitrogen-doped graphene-based catalysts for fuel cell applications. ChemElectroChem, 2020. 7(7): p. 1739–1747.
119. Gao, F., et al., Nitrogen-doped fullerene as a potential catalyst for hydrogen fuel cells. Journal of the American Chemical Society, 2013. 135(9): p. 3315–3318.
120. Armand, M. and J.-M. Tarascon, Building better batteries. nature, 2008. 451(7179): p. 652–657.
121. Goodenough, J.B. and K.-S. Park, The Li-ion rechargeable battery: a perspective. Journal of the American Chemical Society, 2013. 135(4): p. 1167–1176.
122. Jiang, Y., et al., Flexible energy storage devices based on carbon nanotube forests with built-in metal electrodes. Sensors and Actuators A: Physical, 2013. 195: p. 224–230.
123. Wang, X., et al., Flexible energy-storage devices: design consideration and recent progress. Advanced materials, 2014. 26(28): p. 4763–4782.
124. Ren, J., et al., Elastic and wearable wire-shaped lithium-ion battery with high electrochemical performance. Angewandte Chemie, 2014. 126(30): p. 7998–8003.
125. Shen, L., et al., In situ growth of Li 4 Ti 5 O 12 on multi-walled carbon nanotubes: novel coaxial nanocables for high rate lithium ion batteries. Journal of Materials Chemistry, 2011. 21(3): p. 761–767.
126. Landi, B.J., et al., Carbon nanotubes for lithium ion batteries. Energy & Environmental Science, 2009. 2(6): p. 638–654.
127. Li, X., et al., High concentration nitrogen doped carbon nanotube anodes with superior Li+ storage performance for lithium rechargeable battery application. Journal of Power Sources, 2012. 197: p. 238–245.
128. Bulusheva, L., et al., Electrochemical properties of nitrogen-doped carbon nanotube anode in Li-ion batteries. Carbon, 2011. 49(12): p. 4013–4023.
129. Ren, B., et al., Boron and nitrogen co-doped CNT/Li4Ti5O12 composite for the improved high-rate electrochemical performance of lithium-ion batteries. Journal of Alloys and Compounds, 2018. 740: p. 784–789.
130. Wu, Z.-S., et al., Graphene anchored with Co3O4 nanoparticles as anode of lithium ion batteries with enhanced reversible capacity and cyclic performance. ACS nano, 2010. 4(6): p. 3187–3194.

131. Hu, Y.S., et al., Synthesis of hierarchically porous carbon monoliths with highly ordered microstructure and their application in rechargeable lithium batteries with high-rate capability. Advanced functional materials, 2007. 17(12): p. 1873–1878.
132. Buqa, H., et al., High rate capability of graphite negative electrodes for lithium-ion batteries. Journal of the Electrochemical Society, 2005. 152(2): p. A474.
133. Wang, H., et al., Graphitized carbon nanobeads with an onion texture as a lithium-ion battery negative electrode for high-rate use. Advanced materials, 2005. 17(23): p. 2857–2860.
134. Subramanian, V., H. Zhu, and B. Wei, High rate reversibility anode materials of lithium batteries from vapor-grown carbon nanofibers. The Journal of Physical Chemistry B, 2006. 110(14): p. 7178–7183.
135. Xu, Y.J., et al., A Comparative Study on the Lithium-Ion Storage Performances of Carbon Nanotubes and Tube-in-Tube Carbon Nanotubes. ChemSusChem: Chemistry & Sustainability Energy & Materials, 2010. 3(3): p. 343–349.
136. Sood, P., K.C. Kim, and S.S. Jang, Electrochemical and electronic properties of nitrogen doped fullerene and its derivatives for lithium-ion battery applications. Journal of energy chemistry, 2018. 27(2): p. 528–534.
137. Sood, P., K.C. Kim, and S.S. Jang, Electrochemical Properties of Boron-Doped Fullerene Derivatives for Lithium-Ion Battery Applications. ChemPhysChem, 2018. 19(6): p. 753–758.

Defected Carbon Nanotubes and Their Application

Sathish Kumar Palaniappan, Moganapriya Chinnasamy, Rajasekar Rathanasamy, Veerakumar Chinnasamy, and Santhosh Sivaraj

Abstract Some of the challenges and specific outcomes of the doped carbon nanotubes have been reviewed in detail. In particular, the doping techniques and the implementations of doped CNTs depending on the specific applications such as sensors, electronic devices, nanorobots, composites, and so on have been reported. Various synthetic approaches have been utilized to dope the novel CNT materials. Recently, the transport behavior of doped CNTs was explored by several researchers. In this chapter, the current advancement on carbon nanotubes doping has been highlighted. The recent progress of the functionalization, fluorination, and electrochemical modifications of doped CNTs have been briefly discussed.

Keywords Defected carbon nanotubes · Functionalization of carbon nanotube and applications

1 Introduction

Fascinating and empowered physicists, chemicalists, biologists, and fresh medicinal professionals have investigated since the appearance of carbon nanotubes (CNTs) in 1991. Such involvement in CNTs is due to the attractive characteristics, which comprise the simplified shape of a single, smoothly folded graphene layer. Both are relied on various fundamental elements that distinguish CNTs: (i) a high aspect and a delocalized electron structure and (ii) some all-carbon chemical composition on surfaces. Firstly, the excellent optical and magnetic characteristics of CNTs are

Present Address:
S. K. Palaniappan · S. Sivaraj
Department of Mining Engineering, Indian Institute of Technology Kharagpur, Kharagpur, West Bengal 721302, India

M. Chinnasamy · R. Rathanasamy (✉)
Department of Mechanical Engineering, Kongu Engineering College, Erode, Tamil Nadu 638060, India

V. Chinnasamy
Department of Mechanical Engineering, Chosun University, Gwangju, South Korea

S. Sahoo et al. (eds.), *Defect Engineering of Carbon Nanostructures*, Advances in Material Research and Technology, https://doi.org/10.1007/978-3-030-94375-2_5

accountable. This nano-sized surface morphology does have high surface to volume ratio which causes major containment consequences. This offers a distinct 1D character of the particle mechanism and the CNTs phonon spectrum. The interrelationship among the particles and phonons is quite small because of the 1D structure. As a result, load transportation is accomplished at room-temperature in CNTs with quite moderate thermal conductivity [1, 2] in close proximity to diamond (the world's highest thermo-physical properties). Notably, CNTs are very resistant to grid faults owing to the conical geometry and the delocalized electron mechanism. CNTs are strongly engaged in basic study on the basis of such characteristics as a paradigm for 1D structures as transistors with field effects [3–5], electrodes, gigahertz actuators [4], interconnecting in electronic components, logic gates [3–5], and modern solid-state storage modules [5]. Secondly, the greater level of attention in CNTs among engineers, geneticists, and public health researchers is due to its carbon-based, all-surface molecular structure [5–8]. Possibly, one of the most widely investigated areas of study is bio-chemistry of carbon primarily due to the extremely massive proportion of carbon-based substances in the physical surroundings. CNTs provide great information of divergence and structural components for a new type of up-bottom substance, since they are an all-carbon molecular structure in conjunction with a broad accessibility for surface modification. Such characteristics of the CNTs provided insight on usage in gasoline [9–11] and various nano-sensors [12], drug transceivers, electron ions across the membrane [13, 14], and genetic transport trucks [15].

Although the electrical conductivity of CNTs is closely related to the molecule electron systems, those characteristics can easily be influenced mostly by alteration of CNTs. The electrical conductivity of a CNT may therefore be intentionally described by the right approach of the form of alteration. The establishment of electronic characteristics is also known as CNT doping. For clarification, three major types of doping will be defined: 1. exohedral doping, 2. endohedral doping, and 3. substitutional doping. However, such doped structures could expose unforeseen quantum phenomena and the charge transport of CNTs should be tailored to lower doping concentrations.

2 Doping Techniques for CNTs

The distinctive geometry of CNTs means that many possibilities of doping are open and thus shift in certain physical and electronic characteristics has occurred, this is shown in Fig. 1. Most of these are benefits of the molecularity of CNTs, but it is still necessary to adjust carbon atoms with contaminants like N_2 in the more conventional context of the manufacturing sector. The separation performance and the molecular working conduct of N_2 doped CNTs are discussed in detail. Organic compounds can also be added to the substrate of a CNT such that the characteristics of the CNT and the attachment can be integrated, as shown in Fig. 1f–h. The subsequent section explores and explains the different strategies to surface functionalization of

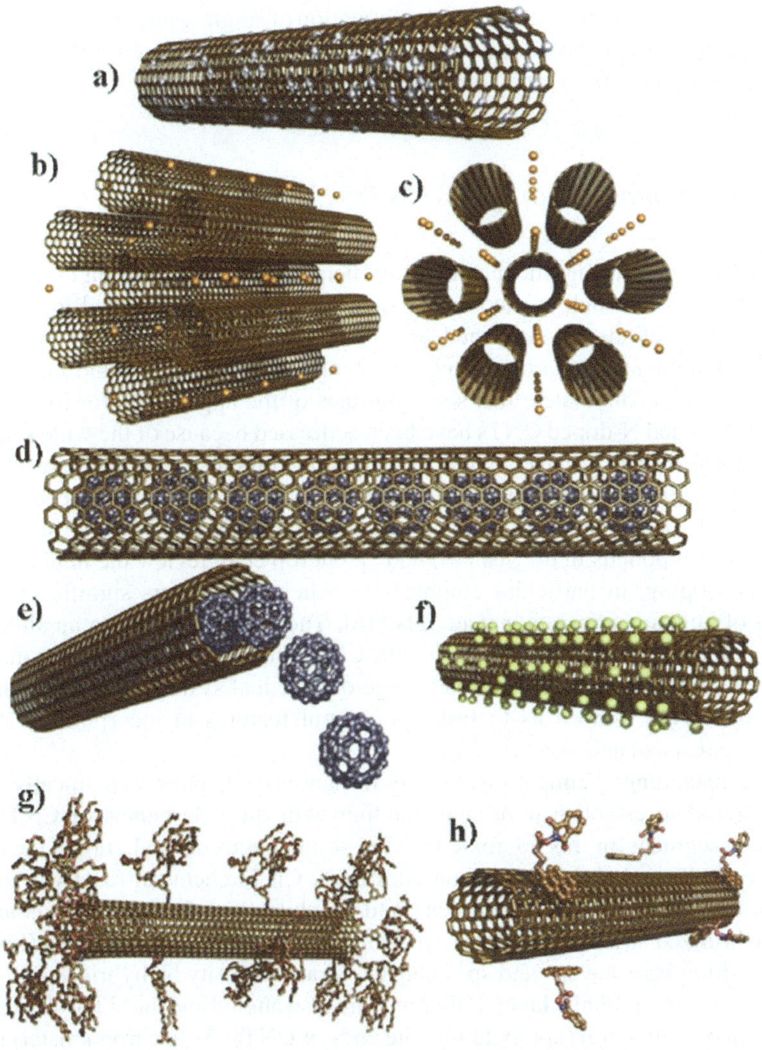

Fig. 1 Different approaches to chemical modification of carbon nanotubes. **a** substitutional doped single-walled nanotubes (either during synthesis or by post-growth ion-implantation), **b**, **c** nanotube bundles intercalated with atoms or ions, **d**, **e** peapods: SWNTs filled with fullerenes (other endohedral fillings are possible), **f** fluorinated tubes, **g** covalently functionalized tubes and **h** functionalized nanotubes via π-stacking of the functionality and the tubes (Reproduced with permission from Ref. [16])

CNTs. Researchers concentrate on polymerization of single-walled carbon nanotubes (SWCNTs) in specific because it is a highly advanced sector which covers synthesis, characterization methods, and simulation aid.

2.1 *Functionalization of SWCNTs*

SWCNTs have been a significant tool in the chemical utilization of their distinct material characteristics, such as electricity conductivity, shear stress, and high-specific substrates. Functions for hardened and biocompatible fabrics, solar energy and camera components, research strategies for microscopy samples, and many others are the potential candidates. Just some outlines of the approaches for ionic derivation of CNTs and N-doped CNTs have been addressed because of the wide variety of practical approaches. Chen et al. described first-ever molecular bonding of SWCNT substrate to organic compounds by the formation of amide structure between carboxyl groups of oxidized nanostructures and surfactants to have CNT derivatives dissolve in organic compounds in the year 1998 [17]. Glerup et al. review the field of carbon nanotube doping, in particular emphasizing where this differs significantly from doping of bulk semiconductor materials [16]. The capacity to create an alternative for nanotubes has made it possible to handle CNTs on a single wall as a mass material. With this scientific research, a broad range of practical systems have been applied to soluble CNTs as well as to fasten structural features to incorporate SWCNT characteristics and add-ons [18–23].

The outstanding chemical insolubility of nanoparticle allows chemically explosive material to establish hydrogen bonding with the side panels of CNTs. The important segment for this nontoxicity is the homogenous delocalizing of the CNTs. In view of all sp^2 hybridized carbon allotropes, CNTs chemical reactivity may be located between reactives of fullerene and graphite (Fig. 2a). Graphite is an ultimate hybridized structure of planar sp^2 which delocalizes all π-electrons (Fig. 2c). The transition between sp^2 and sp^3 enhances strain intensity by hybridization of the carbon atoms in a graphite layer. Fullerenes are also angled and the sidewall of a CNT twisted in one direction (not including the ends of CNT). As in carbon materials, the SWCNTS sp^2-carbon atoms become stretched just along the tube region. The pyramidalization impact becomes less marked than in fullerenes of the two C=C links, which are parallel and contrary to the loading direction (Fig. 2b). Non-pyramidal but still misaligned π-orbitals of CNTs-carbon bonds perpendicular to its axis [17, 24–32] has been occurred. On one side, in the exohedral region a curved π-system might have an enhanced density of the atomic nucleus in comparison with planar graphite and hence obtained a more reactive effect. The primary driver for additional reactions, but at the other side, is significantly weaker than those of fullerenes because of its lack of strain intensity contained in the C-atoms [33]. The numerical simulations verified the conceptual implication which shows a decrease in excitation of CNTs with growing circumference [34].

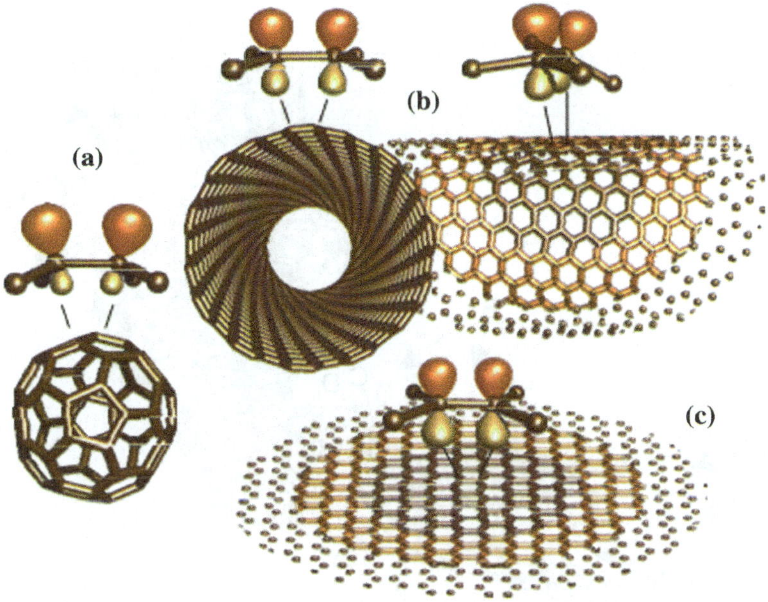

Fig. 2 Schematic presentation of the π-orbitals of pyramidalized C=C double bonds **a** of a C60 molecule, **b** a (15,10) SWCNT, and its π-orbital misalignment in comparison with the π-orbitals of a plane C=C double bond of graphene. **c** The degree of pyramidalization and misalignment indicates the reactivity due to addition reactions [16]

A notable chemical inertness of complete SWCNT needs extensive reactive compounds that can change the CNTs sidewall chemically. CNTs can be known as highly reactive zones because of their rising constraints (Fig. 1). The so-called defect categories can include heptagon-pentagon deficiencies, CNTs lattice vacancies, and sp^3 carbon hybrid molecules, saturated hydrogen or oxygen-treated [19]. CNT oxidation leads to additional defect categories like NO_2, hydrogen atoms, carboxylates, and carboxyls [35–37]. The following is of primary concern as the development of amides or esters will further change such carboxyl functional group (Fig. 3). The effectiveness of such a defect team approach was inspired by chemical oxidations that are primarily formed by COOH clusters from the carbon molecules on the CNTs sidewall [36, 38–42]. Such important accomplishments in favor of lateral flexibility are additional interactions with reacting organic substances such as radicals [43, 44], carbon dioxides [43, 45], phosphorus [43, 46], azomethine ylides [47], lithium alkyls [48–51], fluorine [52], and diazonium salts [53, 54] (Fig. 4a–d and f) as well as molecular improvement of deficiency groups. Less extreme processing parameters for the working of the CNTs sidewall can be used for nitrogen-doped SWCNTs [55] (Fig. 4g).

Interestingly, it's hard to enforce regio- and chemoselectivity. The exfoliation and deterioration of modules during fastening of addends is essential for the homogenous functionality of CNTs. The surfactant can be used to manufacture independent

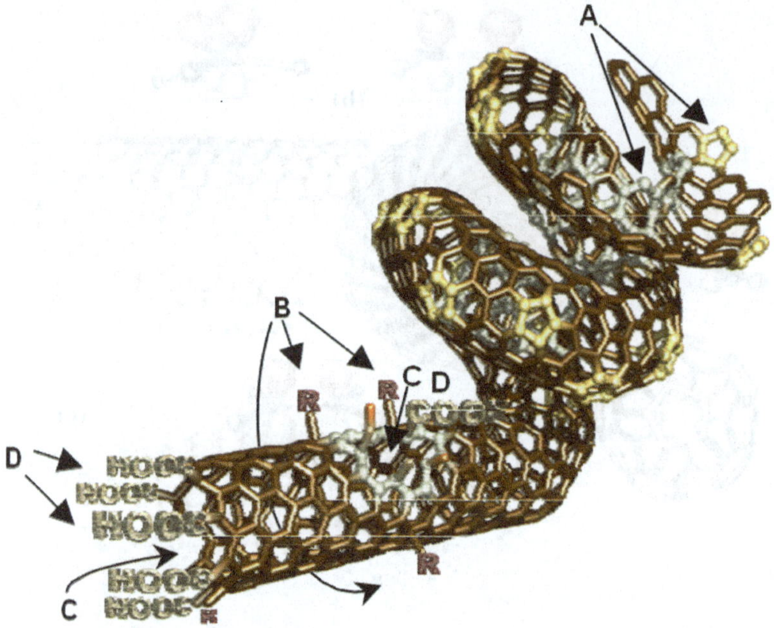

Fig. 3 Possible defect groups in SWCNTs: **a** Pentagon and heptagon defects (disclinations). These defect groups have not yet been chemically exploited but can lead to nice deformation of the tubes such as tube coiling or Y-shaped tubes (not shown). **b** sp^3-defects are formed during the growth process and during oxidation. They are of less interest for functionalization but they are suspected of leading to a partial localization of the π-system leading to a higher accessibility for addition reactions. **c** Vacancies, and their larger cousins holes, lined with functional groups. **d** Carbon oxides, whereas the carboxylic acid groups are of highest interest for SWCNT functionalization (Reproduced with permission from Ref. [16])

nanoparticles in dispersion [56] and subsequently can be functionalized employing diazonium salts [57]. Billups et al. [58, 59] have also given an illustration for relatively homogenous operational CNTs. Due to magnetic repulsive force after particles were charged under Birch reduction parameters, the CNTs may be isolated elsewhere. Through electrophilic additional process, the samples of functional nanotube could have been acquired through alkyl halides.

2.2 *Amide/Ester Formation with Carboxylic Acid Groups of Oxidized SWCNTs*

The initial study of complex binding of the ionic compound to the oxidized CNT sidewall is amide development [17]. SWCNTs with degradation of alkylamines are efficient, allowing the use of possibility that nanoparticles have a high carboxylic acid content preceding the oxidation process [38, 60, 61]. In the process of thionyl

Fig. 4 Different methods for the covalent attachment of functionalities to the SWCNT sidewall carbon atoms. **a** Formation of amides or esters with the carboxylic acid groups of oxidized SWCNTs. **b** addition of nitrenes (B1), carbenes (B2), and radicals (B3). **c** [2 + 3] cycloaddition of azomethine ylides. **d** Electrochemical functionalization with diazonium salts **e** nucleophilic attack of Lithium alkyls. **f** Mannich-type reaction with N-doped single-walled carbon nanotubes. **g** Fluorination (Reproduced with permission from Ref. [16])

chloride, several carboxyl groups were transformed into acyl chloride categories. SWCNTs of acid dissolution are then prone to amide-giving reactions with alcohols (Fig. 5a). The related physically usable CNTs have considerably increased solubility in polar compounds than the reaction product. The modules are grouped into smaller individuals and different channels in the context of their surface modifications. Some recent findings have showed that a process through amine at elevated temperatures (120–130 °C) to create non-covalently bound zwitterionic substances in increased levels will prevent the intermediate stage of acidic content development [61, 62] (Fig. 5c).

The 'steglich' approach offers an elegant substitute to the development of amides or esters with the oxidized carboxylic acid chains of SWCNT. Amides arise by diimide compounds as a binding solvent from carboxylic acid chains and amines. By forming an intermediary O-acyl urea that has toxicity equivalent to anhydrides, the diimide triggers the carboxylic chains. The amine nuclear attack contributes to the amide and urea derivatives as a by-product, which could be extracted after the interaction via multiple bleaching processes. Another benefit of this aldehyde relationship is that the response is positive and moderate [63]. Some other value

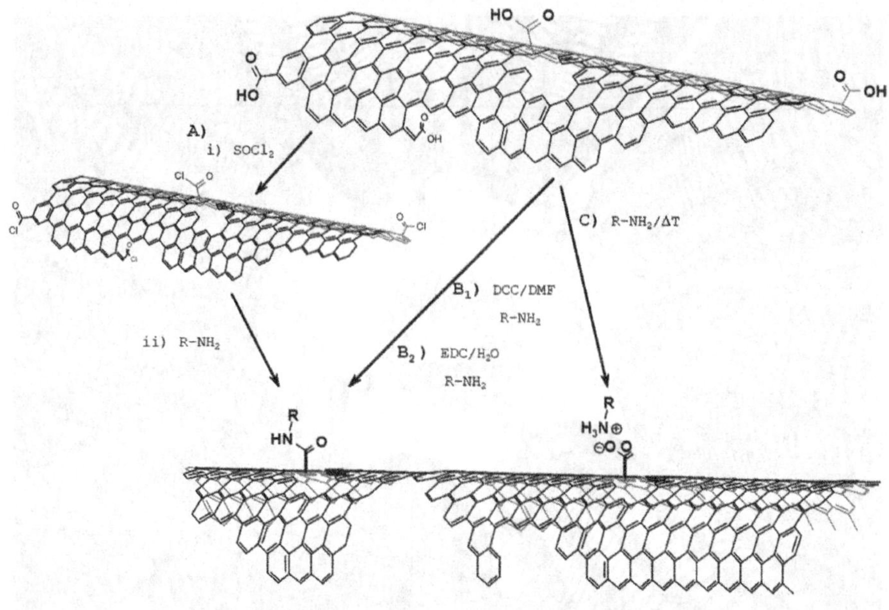

Fig. 5 Reaction scheme of the formation of amides with the carboxylic acid groups of oxidized carbon nanotubes and amines via **a** Activation of the carboxylic acid groups by forming acid chlorides with $SOCl_2$ in an additional reaction step. **b** Activation of the carboxylic acid groups using carbodiimides as coupling reagents and **c** Formation of ammonium carboxylates with amines at elevated temperature (Reproduced with permission from Ref. [16])

is that oxygenated CNTs can be distributed for this form of process at appropriate contents in suitable solvents. A reversible compound of carbodiimide is required for each product. Dicyclohexylcarbodiimide (DCC) is now a popular binding agent for amidation processes by oxidized CNTs in the DMF (Fig. 5b) [64]. SWCNTs with extremely-oxidized carboxylic bands have been formed underneath oxidative conditions because of its capacity to hold highly oxidized SWCNTs in deionized water or sustainable utilizations [65]. Such target could have been accomplished by means of EDC as water-soluble primary ingredient like (1-ethyl-3-(3-dimethylamino-propyl) carbodiimide [66, 67]. Also, chemically responsive biomolecules may be installed onto the sidewall of the CNT following these pathways [68]. The wide range of these working methods, usable reactants, and the possibility to achieve stable CNTs under mild and simple to manage circumstances render amide formulation among nanoparticles and surfactants as the most popular process category.

2.3 Reaction of SWCNTs with Nitrenes

SWCNTs have been distributed in ideal detergents like 1,1,2,2-tetrachloroethane (TCE), or 1,2 dichlorobenzene (ODCB) for the inclusion of nitrenes [43]. The nitrenes were produced by thermally induced extruder of N2 by the addition of a substantial quantity of an alkyl azodiformate and response to the alkoxycarbonyl aziridine-SWCNT sidewall. For the nitrene incorporation, use of azido carbonates is expected. During nucleophilic reactions, formation of amines or isocyanates generally occurs with azidoalkyls or acids. Azido carbonate emulsion of N_2 produces nitrene in individual and multiple conditions. Individual nitrene is a cycloaddition that targets the sidewall of the CNT, and it can be turned into triple nitrene through the inter-system crossing (ISC). A triple phase nitrene interacts as bi-radical with the clearly indicated system of the sides of the nanoparticle. The introduction of nitrene to CNT sides contributes to the forming of an aziridine ring, irrespective of a transesterification process (Figs. 3 and 6).

A significant amount of synthetic components can be added with this form of interaction [46]. Dendrons, which specifically improved the viscosity of SWCNTs in organic compounds, are still the most prominent add-ons that could be applied using this functionalization process. Chelating substances such as transition ethers and alkyl chains [46], which have been terminated at either end with azido carbonates

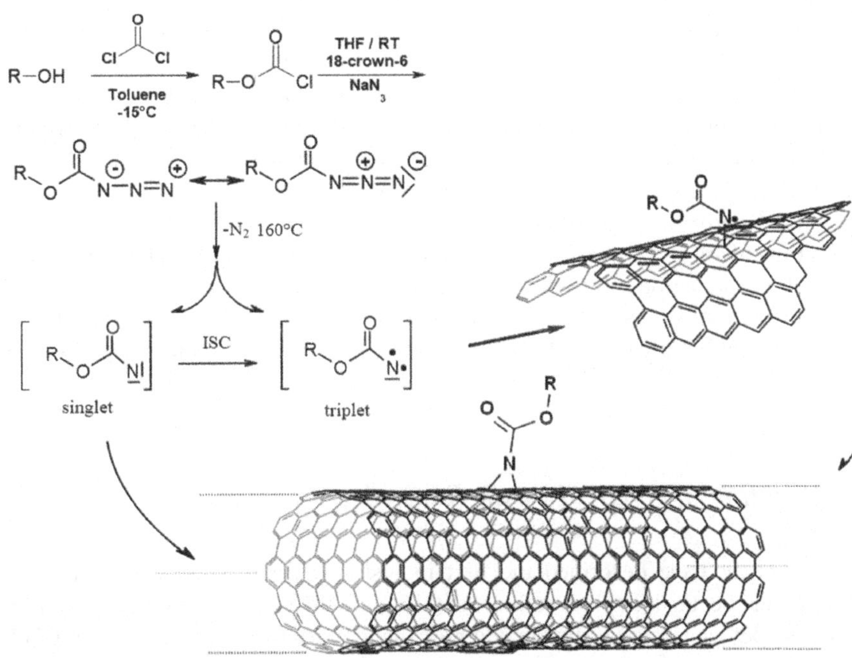

Fig. 6 Reaction mechanism of the [2 + 1] cycloaddition to the SWCNT sidewall (Reproduced with permission from Ref. [16])

in order to inter-link the SWCNTs in response [69], were effectively bound to the sidewall of SWCNT using this sort of interactions. Nitrene flexibility is theoretically used to produce nanoparticle devices or nanotube-based hardened components. This nitrene mechanism is a safe process for nanotubes working and has been thoroughly characterized. Although only SWCNTs with relatively inactive addenders are able to survive under intense experimental parameters and the strong reactivity of nitrene.

2.4 *Reaction of SWCNTs with Carbenes*

Chen et al. have described incorporating carbenes to the sidewall of CNT [17, 38]. In the existence of a catalyst support, a SWCNT specimen was immersed into the chloroform mixture solution. Potassium hydroxide was then developed for the carbenes. The reaction was performed with soluble SWCNTs which have already been conducted by amide process with octadecyl amine (ODA) [38]. Toluene and phenyl (bromodichloromethyl) mercury ($PhCCl_2HgBr$), the carbene component, were distributed into customized CNTs. The dichlorocarbenes are derived precisely by the removal of chlorine from the tissue and PhHgBr. The carbenic complexes produced are highly electrophilic and react alongside the CNT. In specific, XPS examined the corresponding SWCNT variant, which calculated an additional degree of around 1% (2% chlorine).

A recent phenomenon for the incorporation of carbene is the reaction of dipyridyl imidazolidine. This formulation is the blueprint for a nucleophilic carbene that is capable of self-stumping [70–72]. The dipyridyl imidazolium system with potassium tert-butanolat is formed by deprotonation of carbene (Fig. 6) [73]. Except from the incorporation of dichlorocarbenes, this carbene interacts with the CNTs as a nucleophile which offer 1:1 adducts instead of the cyclopropanes forming from conventional carbenic cycloadditions to the zwitterionic model. The exceptional stabilization of the resulting aromatic 14 bis-perimeter of dipyridyl imidazolium cation is the explanation for this univalent binding. The protective layer for this zwitterionic formation was passed to the sidewall of the N-type CNT. This provides an additional variable for a controlled change in the electronic characteristics of CNTs (Fig. 6).

2.5 *Reaction of SWCNTs with Radicals*

Holzinger et al. have described the first inclusion of radicals to CNTs sidewall [43]. Radicals occur in heptadecafluorooctyl iodide through photo-induced homolysis where they generate perfluorinated radicals to react with the sidewalls of CNT. The photo-stimulated radical process was transmitted from fullerene composition like most of the other reliability assessment of SWCNTs [74]. Radical compounds could therefore shape as initiators for thermo extended processes utilizing peroxides such as benzoyl peroxide or persulfate potassium. This theory was used to polymerize vinyls such as styrene [75] that contributed in increased solubility of toluene nanotubes and to have a significant potential for the production of high-performance composite nanoparticle components. Benzoyl peroxides themselves acted as precursor for phenyl radicals that are now being incorporated to the CNTs sidewall [76]. An ultrasonic-induced emulsification polymerization is a further development of reactive species with CNTs for composites with large nanotubes filtration [77].

2.6 Reduction of SWCNTs with Alkali Metals

A substantial negative charge on SWCNT sidewalls contributes to a gross linkage between the nanotube packets. Under 'birch conditions', a sidewall of nanotube can be equipped adversely [78, 79]. The reaction mechanism begins with ammonia condensation and alkaline metal dissipation. Due to solved charged particles, uncolored condensed ammonia transforms into dark blue. The gradual breakdown of the SWCNTs, under these conditions, suggests that the nanoparticles are reduced rapidly [80]. With the addition of alkyl halides, the strongly loaded independent nanoparticles will act reversibly [58, 59]. Petit et al. have documented an innovative approach for transferring electrons to the sidewall of nanotube [81]. The SWCNT specimens were mounted in a quartz optical cell that is attached to a U-shaped glass container, with an ampoule of complex chemicals as an electron transporter on the second branch. These compounds are anthraquinone, fluorenone, and naphthalene (Fig. 7).

The addition of a lithium component contributes to the radical anion state of a molecule with Li^+ as the counter ion as the electron transport from the lithium to the complex chemicals. Such a research procedure built for the n-doping of polyacetylene [82] enables the SWCNT specimen to be minimized by interaction between the substance and the film. The resultant n-doped CNTs have random tendency to dissolve in polar aprotic solutions such as sulfur, 1-methyl-2-pyrrolidone (NMP), and dimethyl-formamides (DMF) [83].

2.7 Reduction of SWCNTs with Lithium Alkyls

Processing of organolithic nanomaterials result in the development of intermediate RnSWNTn—negative-charged SWCNTs. This form of reaction was used initially to initiate polymerization processes or react using structural composites for the functional groups of the CNTs [48–50]. Graupner et al. actually reported detailed inquiries into this n-type doping of SWCNTs [51]. The SWCNTs have been processed in anhydrous benzene by tert-butyllithium (t-BuLi). The unsteady distribution becomes dark and homogenous during the response as can be seen when n-type nanoparticles

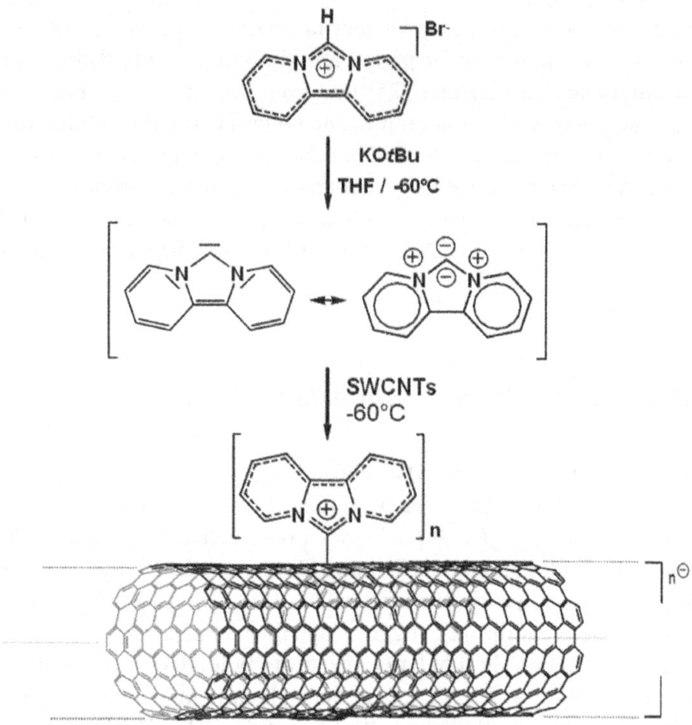

Fig. 7 Reaction scheme of the addition of dipyridyl imidazolidine to the nanotube sidewall (Reproduced with permission from Ref. [16])

have been doped with alkaline metals [83]. This finding is in line with the transference of the electrons into the SWCNTs, along with a nuclear bombardment of tert-butyllithium to the sidewalls.

The process compounds of t-BunSWCNTsn were reoxidized with oxygen to provide t-BunSWCNT, which precipitates after degradation, the undoped and soluble intermediate nanoparticles. Fortunately, related studies are famous for fullerene chemistry [83]. The degradation of the lower carbon sidewall of the CNTs is predicted to be much more preferable in accordance with fullerenes. The degree in the C-atom in SWCNTs pyramidization is smaller than the amount of fullerenes and carbanions favor pyramidized C-atomic geometry. Consequently, for CNTs, it is further pronounced to modify the preliminary straining of pyramidized sp^2-carbon molecules than those for fullerenes [33]. The level of t-butyls connected covalently enhances after each step up to the oxide-to-addend maximum rate of 31 by executing this series of interaction. Photoelectron spectroscopies have analyzed this recent interaction series in particular. The doping of SWCNTs with t-BuLi in n-type suggests that the energy gap of C1 valence electrons is specifically transferred to excitation frequencies. After material degradation, the stability characteristics distribution of the homogeneous catalyst for carbon 1s is almost similar and shows the reproducibility

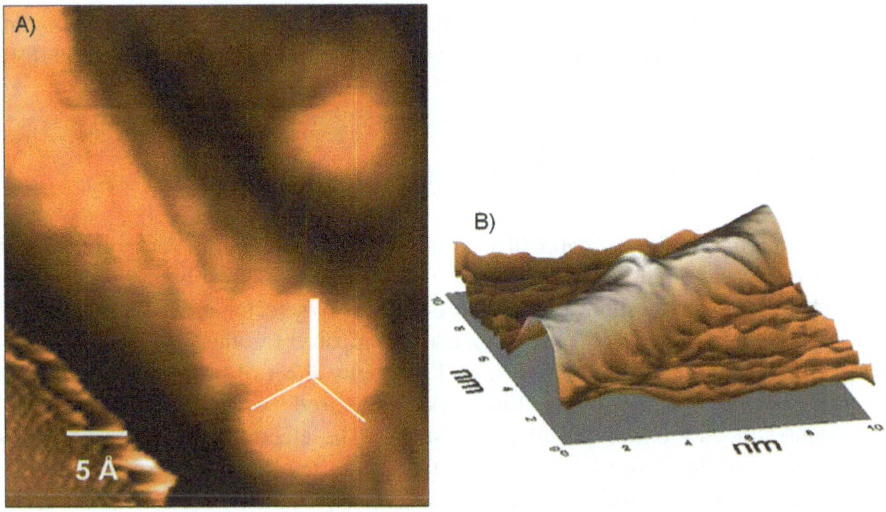

Fig. 8 **a** Topographical STM image of t-BunSWNT with a linear background removed. **b** Perspective view of an STM image of an individual t-BunSWNT. In both images, the three-fold symmetry of the functional group is evident (Reproduced with permission from Ref. [50])

of this oxidation phase. The t-butyl addition can be visualized with the subsequent three-fold symmetry (Fig. 8) with the scanning tunnel microscopy (STM).

2.8 Dipolar (1,3) Cycloaddition of Azomethine Ylides to SWCNTs

The 1,3-dipolar cycloaddition of azomethine ylides (Fig. 8) produced by the polymerizing of a glycine compound and an aqueous phase has also been a characteristic of a strong fullerene reaction to the surface synthesis of CNTs [83, 85]. The glycine derivatives and aldehyde are the intermediary iminium carbohydrates, form the instable azomethine ylide, and interact with the CNT mechanism after deformation of carbon dioxide. By extracting DMF reaction companion for many days at 130 °C, pyrrolidine-nanotubes can indeed be summed up with a full number of diverse functionalities.

Biomolecules could have been connected to the sidewall of CNT which served as biosensor receptors using this derivatization approach [86]. The adequacy of active CNTs of photovoltaic cells is demonstrated by light inducted transfer of electrons from associated electron acceptors [87]. This cross-linking process is the result of a water-soluble SWCNT product that includes an amino-terminated movement of pyrrolidine nitrogen oligo ethylene glycol. This mixture forms supramolecular compounds with electrostatic interactions of the plasmid DNA. Compounds may infiltrate the molecules of the cells [88]. The CNTs may be used as a key component for nano- and biotechnology purposes, powered by a 1,3-dipolar cycloaddition [89].

2.9 Electrochemical Modification of CNTs

The tightening of electrochemically modified aryl diazonium compounds, first mentioned by Bahr et al. is a very effective method for the direct binding of organic addends [53]. Such reaction is considered for highly selectable pyrolitic graphites (HOPG) and glassy carbon (GC) electrodes, compared to other practical methods derived from fullerene chemistry [90–94]. A bucky sheet of paper was used as an interface in the 3-electrode cell, the working electrode was the $Ag/AgNO_3$, and the counter electrode was the platinum string. The electromagnetic process was conducted in a tetrafluoroborate mixture that had a tendency of -1.0 V potential containing diazonium salts and electrodes. The aryl radical reacts with the CNT sidewalk by reducing diazonium salts. The CNT itself then becomes a fundamental mechanism such as interfacial polymerization processes, and the substance may react further to quench them. The trend of the original aryl radical in solvent is reduced by a revolutionary surface morphology at which response is desired. The radical is reduced to the minimal. In contrast with a solution-phase process wherein the reactive substance produced can often be quenched by the substance or certain other compounds. This is the key benefit of electrochemical processing (Fig. 9).

McDermott et al. [95] suggested transesterification growth for derivatives of the HOPG with 4-nitrophenyl diazonium salt, with several aryl radicals attacking the carbon substrate and the subsequent radicals binding to the polyphenylene chains (Fig. 11). Kooi et al. have explored the electrochemical polymerization of phenylene dramatically on the sidewalls of SWCNTs [96]. Specific nanoparticles and small packages were mounted on a crystal membrane and electric field lithography was used to target membranes. Reduction functioning was carried out as an aryl radical precursor in DMF at -1.3 V versus Pt, while NBu4 + BF4− was used as an electrode material. The process was accompanied by a growth of the stiffness leading to the incorporation of the synthetic laminate only in nanomaterials interacted with the electrode surface (Fig. 10). The synthesized nanoparticles in oxidative environments have shown similar findings. The reactants were 4-aminobenzylamine and 4-aminobenzoic acid. Oxidative relation with LiClO4 as reference electrode was conducted in ethanol at a capacity of $+0.85$ V versus Pt. The density of nanotubes coated with polymers increases linearly to the life of the actual utilization. The first method of electrochemical modification of nanotubes carbon in nanometer-sized electronic equipment was the observations (Fig. 11).

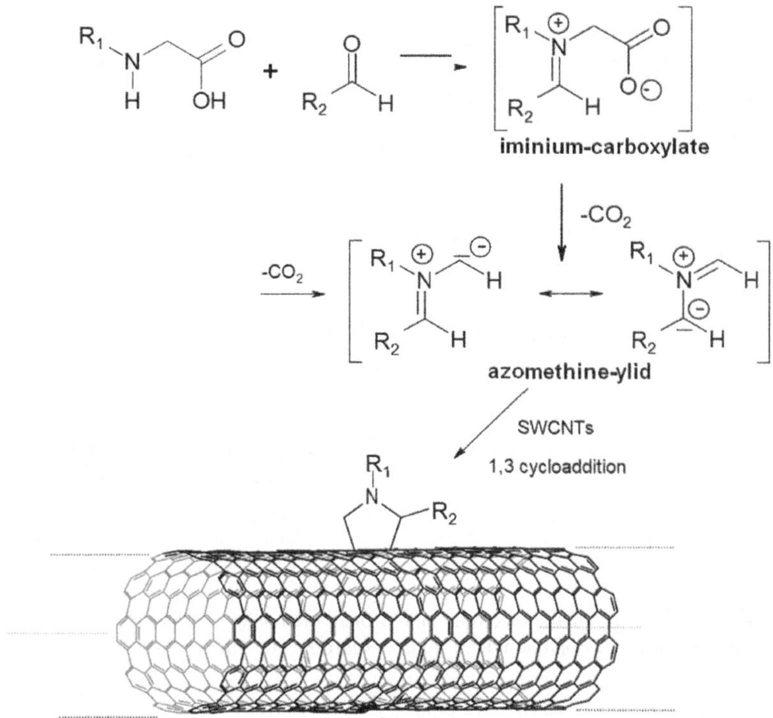

Fig. 9 Reaction scheme of the 1,3-dipolar cycloaddition of azomethine ylides to the nanotube sidewall (Reproduced with permission from Ref. [16])

3 Functionalization of N-Doped SWCNTs

Although SWCNT doped with nitrogen can now be acquired in large quantities [97], extensive study into this substance has now been made necessary by different materials. The nitrogen is being used in nanotubes as normal defects that affect the tubes chemical behavior. This can be determined that the responses of these N-doped CNTs are more sensitive than undoped CNTs of the same size. A location of unpaired electrons all around nitrogen deficiency in semiconducting hetero CNTs is determined with mathematical models [98]. The performance of doped SWCNT materials in this scenario can be reduced toward less rigid experimental parameters than flat SWCNTs [55]. The covalent trapping of the C–H acidic substance into a Mannich-like process, implemented for the production of heterofullerene substances is an effective process for the functionality of N-doped CNTs (Fig. 12) [99]. The tetrachloroethane (TCE) and p-toluene sulphonic acid and acetylacetone have been sprinkled into the pristine N-doped CNT produced by arc discharge process. The scattering was warmed to the air circulation point. The electron orbitals in the nanotube crystal structure, adjacent to the nitrogen atoms, would be oxidized under certain situations. The subsequent iminium carbo-cations are added by acetylacetone.

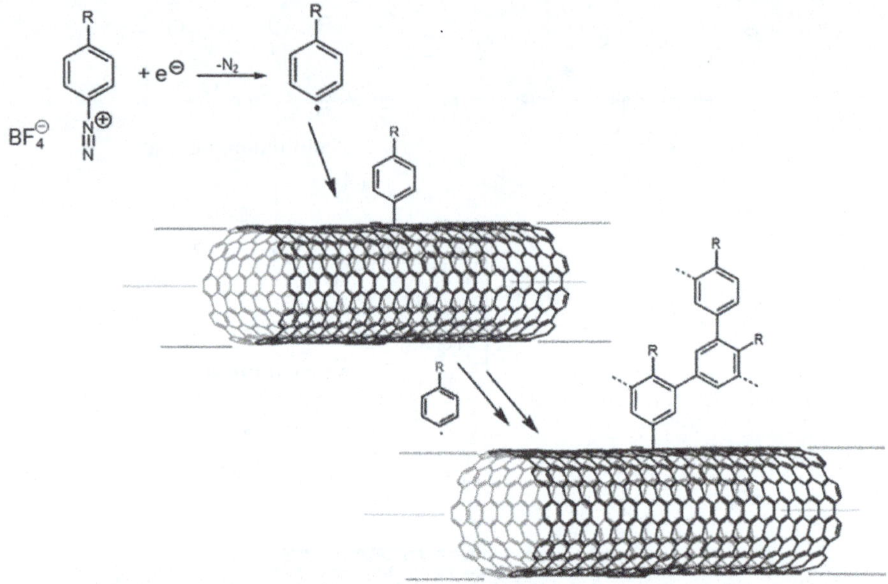

Fig. 10 Reaction scheme of the electrochemical functionalization of SWCNTs with diazonium salts and the formations of polyphenylene chains (Reproduced with permission from Ref. [16])

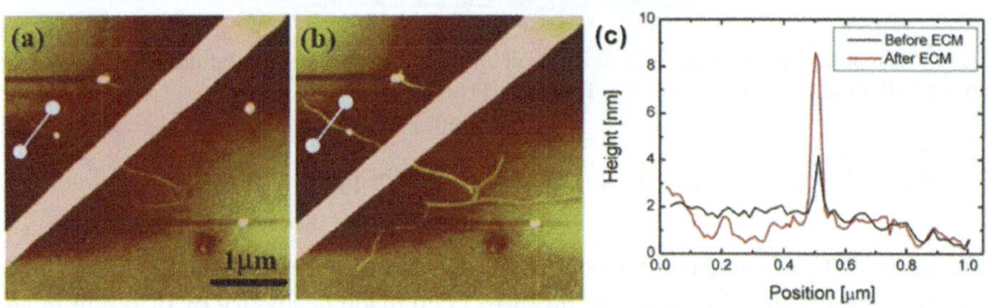

Fig. 11 Electrochemical modification (ECM) of an individual electrically contacted SWCNT using a diazonium salt: Atomic force microscope (AFM) images before (**a**) and after (**b**) modification; **c** line profiles along the lines marked in the AFM images before and after ECM (Reproduced with permission from Ref. [22])

3.1 Fluorination of CNTs

The very first functionalization of multiconducting CNTs did take place a decade ago, accompanied by single-walled polymerization of CNTs [100, 101]. After 2 years, fluorination is an essential method to work and to chemically activate CNTs [52, 102]. The functional approaches have an algorithm to optimize, since they guarantee a high interface density of functional groupings up to C2F before damaging the nanotubes structure. This is an simple and fast exothermic process and repulsive

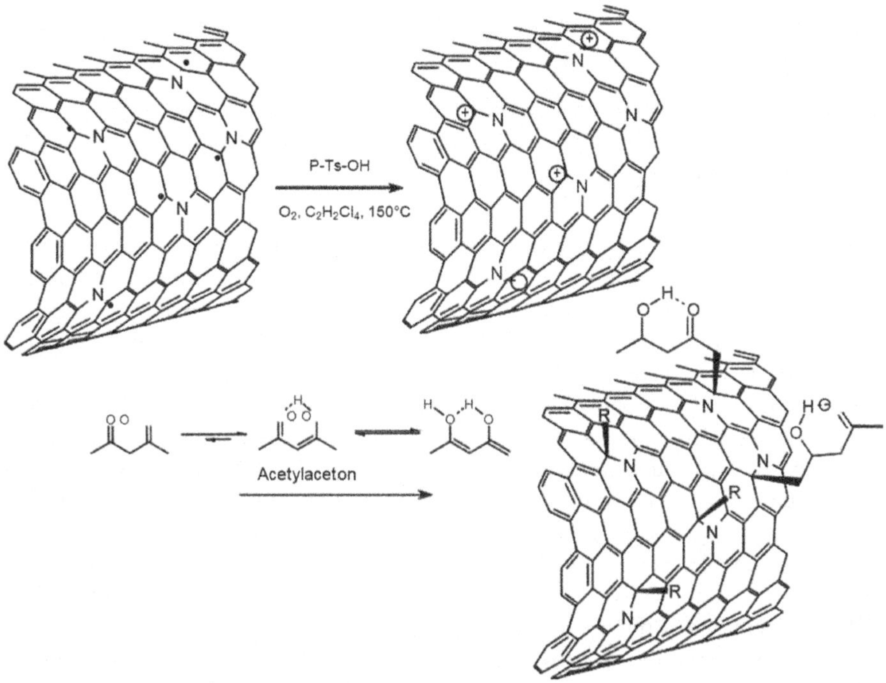

Fig. 12 Reaction scheme of the covalent functionalization of acetylacetone, a C–H acidic compound, to the carbon atoms close-by the nitrogen atoms of the N-doped carbon nanotube sidewall (Reproduced with permission from Ref. [16])

collisions of fluorine atoms on the substrate monitor the nanotubes which normally helps to disperse. Hydrophobic interactions in alcoholic and other organic solvents between both the hydroxyl group and electrons lead to enhanced solubility, providing solutions carried with chemicals, and allowing CNTs to be purified and processed [52, 103]. Fluorinated CNT can be used as catalysts for more functionalization by nuclear suspension since they are stronger electron receivers as pristine CNTs and thus effective interactions with nucleophilic solvents [103–109]. Post-defluoration of anhydrous hydrazine as a chemical compound can be actually achieved by sonication [52, 103].

There was a considerable focus upon its features of the distribution of usable CNTs. The absorption coefficient for metallic nanoparticles with a lateral fluorine inclusion sequence [110, 111], pending observations of reduced conductivity for fluorinated nanomaterials [110], has indeed been projected to disappear. Consequently, the examination of potential trends of inclusion continues to be a significant concern and will be addressed herein. Promising usage in lithium-ion batteries [112–115], supercapacitors [116, 117], and lubricants [118–120] are developments of fluorinated CNTs. The applications of fluorinated CNTs as a new path for polymer enhancement have recently been explored in which they greatly increase structural properties of the materials in contrast to pure CNTs [121, 122].

3.2 Fluorination Techniques

The application of metallic fluorine [102], F2 gasoline [123], CF4 electromagnetic wave [124], and RF fluid [108, 125], resulting to a coating of up to C2F in fluoride, while CNTs are damaged in a tough environment, is a complementary process for polymerization of CNTs [123]. A slightly reduced protection of fluorine with BrF3 [126–128] or XeF2 [109, 129] is ascertained. Crosslinking is one of the very few methodologies of functioning not generally solvent based. The distribution increases at higher temperatures in the event of fluorinating process of F2 gas. The spectral efficiency of C2F is now between 250 and 300 °C for SWCNTs [130, 131]. As the temperature increases slightly, CNTs divide into a graphical substance at a fluorination frequency of around CF [132]. The F1s peak based at 688 eV for temperatures up to 200 °C, which represents the sub-ionic bonding, was seen to be the frequency variant X-ray photoelectron (XPS), while at 250 °C, the covalent bond with such an XPS peak centered at 691 eV is shifted [130, 131].

In electro-energy loss spectroscopy (EELS), all kinds of bonds are often found in energy loss near-end structures (ELNES) [117–119]. The F/C ratio has been improved by roughly 0.25 at 200–250 °C, only with material resistivity improvement by such a factor of 3 is expressed. In FTIR, peaks of 1100 and 1210–1250 cm^{-1} are seen, which correspond to XPS F1s for sub-ionic and intermolecular C–F, accordingly [130, 131]. Tight-binding fluorine diffusion boundary measurements on large parts of the nanotube revealed that this change from 200 to 250 °C leads to the restriction of fluorine groups to propagate through the surrounding interfacial area of the CNTs [132]. Plasma functionalization is the quickest process of CNTs; CF4-gas plasma radiation of CNTs between seconds to several minutes and its both feasible ways to polymerize the membrane on a CNT surface or to conduct covalent functionality with exposure up to C2F depends on plasma circumstances [125].

3.3 Fluorine Addition Patterns

Till now, theoretical experiments have mainly focused on C2F polymerization [104, 106, 132, 134]. Apparently, in complete contrast with theoretical aspects reported STM circumferential banding with heavy loading, lateral fluorine substitution trend [18, 101, 104, 106, 111]. The "contiguous fluorine additive," in which the sequence for fluorine axial applied to form sharp circumference arches of unfluorinated nanotube sections was projected which mainly describes this difference [135]. Thus, the more robust axial incorporation contributes to the circulatory banding of fluorine (Fig. 13a–b).

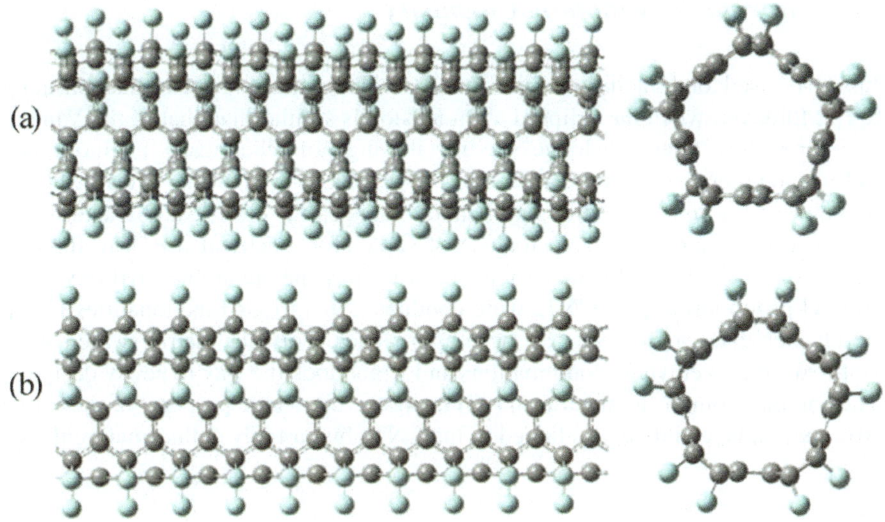

Fig. 13 Axial **a** versus circumferential **b** fluorine addition pattern for a (5,5) carbon nanotube (Reproduced with permission from Ref. [16])

3.4 Conductance of Fluorinated CNTs

Fluorinated CNT shield is forecasted to stay metallic [101, 104, 111], although fluorination is already being proposed to isolate absorption paths on the substrate of CNT, which will in turn contribute to enhanced conductivity [111]. Even then, tests of two-probe sensitivity of the "buckypaper" fluorinated SWCNT have demonstrated a 6-order rise in intensity response. Instead of modified stiffness in tubes, this has already been proposed because of decreased electrical interaction among CNTs [52]. Lately, four-stage intensity tests showed a reduction in conductivity of CNT by up to two orders of magnitude following XeF2 fluorination just at ambient temperature, offering 0.8 fluorine at a percentage basis, while the fluorinated CNTs are still metallic in temperature dependency for the capacitance [110]. Even then, this isn't the issue in fluorinated peapods with a C60, which had the ability to serve as a superconductor [110]. Zhao et al. demonstrated a decline in permittivity due to possible defluorination of CNTs with a rising temperature [136]. In order to address the problem of fluorinated tube capacitance, the mobility tests will be requested for specific fluorinated tubes. The electrical characteristics of a system could also be altered by fluorinated CNTs, such as nanoparticles in SWCN/PAN; composite materials with pristine SWCNTs were improving quality, whereas materials with fluorinated SWCNTs had been highly conducting [122].

3.5 *Fluorinated Nanotube Chemistry*

This can be used for the addition of fluorinated CNTs for their elevated nucleophilicity [104]. Moreover, when the eclipsed atom tension is similar than that of the Van der Waals range, it is easier to dislodge or extract it using moist chemicals, and fluorinated CNT would otherwise be impossible for use as a guide to even further versatile [121]. Even as vast fluorine flexibility significantly changes the electrical characteristics, innovative ideas are opened up for use of CNTs for electrical use. The fluorine penetration of fluorinated nanoparticles in order to generate amino, hydroxyl, and carboxyl group-terminated CNTs were conducted in various functionalities [121]. The chapter gives a summary of these approaches. Polymer matrix, which were identified for polyethylene combinations and which lock the way to newly designed CNT-polymer composite materials [137] using the electronic property of these two active materials, could also be linked to the CNTs by partially defluorinating them.

3.6 *Implementations of Doped CNTs*

In the succeeding sections, we will discuss briefly a few potential applications in different fields, such as electronics, biotechnology, composite materials, and catalytics, for various forms of doped CNTs. As powdered nanoparticles cannot be chosen on request for the formulation process, the production of effective purification and isolation technologies is currently a quite critical aspect in nanoparticles technologies [138–141].

3.6.1 Field Emission Sources

The analytical and experimental confirmation of the improved transmission spectra of B-doped MWNTs compared with the pure carbon MWNTs as explained by Charlier et al. [142]. The existence of B molecules at the surface of nanoparticles contributes to an increment in the amount of conditions closer at the Fermi level. This is assumed to be the cause of significant and guided conceptual measurements of ab-initio, suggest that the role of the B-doped SWNT is slightly less than that seen in pure carbon MWNTs. Likewise, it was shown to release electrons with N-doped MWNTs with small turn-on voltages and with a strong current [143]. More recently for each N-doped MWNT, outstanding field emissions at 800 K, testing work functions of 5 eV and emission currents of roughly 100 nA was achieved at ± 10 V [144].

3.7 Li-Ion Batteries

Currently, many technological vendors produce these batteries on personal devices, smartphones, portable scanners, etc. Surprisingly, Endo et al. [145] showed that the B-doped vapor-grown carbon fibers (VGCFs) and carbon nanofibers are much better than almost any type of carbon mostly in graphitic anode within the Li-ion batteries. This may be attributed to the rising affinity of the community of Li-ions at the B-doped sites, which could contribute to greater storage capability. The effective sustainability of flexible Li (480 mAh/g) has also been demonstrated by N-doped CNTs and nanofibers; considerably better than standard carbon products for Li+ (330 mAh/g) [146].

3.8 Gas Sensors

Since 1998, different community [9, 147, 148] studies showed the identification of poisonous gases and other contaminants which can be carried out by pure carbon SWCNTs and MWCNTs [9]. In this case, however, N-doped MWCNTs tend to be more effective because of the incorporation of toxic gases and bio-solvents which will show a swift reaction (order of milliseconds) [149]. The concentration of substances on E_f, suggesting reduced deformation and chemisorbed, is reported from its theory.

3.9 Polymer Composites with Doped Nanotubes

The development of robust nanotubes/polymer surfaces is a key to produce nanotube composite materials with enhanced reliability. For this instance, the substrate of strong nanostructures, graphite-like MWCNTs appears to have been chemically "non-reactive". Substrate shift therapies are indeed important to create successful relationships between nanotube and matrices [150]. The development of nanotubes in the hexagon-shaped conventional network contains a limited quantity of suitable atoms, such as N or B, may thus bypass these concerns. In certain situations, the mechanical characteristics of such "doped" systems will never be modified substantially, when the degree of doping is smaller (<1–2%). In addition, a 20 °C rise was seen in glass transitional temperature with just an addition of 2.5 wt% value by the tightly mechano-analyzed DMTA of CNx MWCNT in preliminary experiments upon this production of composite materials employing N-doped MWCNT [151]. Most importantly, the use of atomic transfer radical polymerization (ATRP) [152] and radical polymerization medicating nitroxides [153] without any acid therapy has been shown to promote polystyrene (PS) to expand on the substrate of the N-doped MWCNT. Latest engineering and manufacturing observations have shown that the exhibited properties of CNx nanotubes with PS grafts relative to blends of PS and

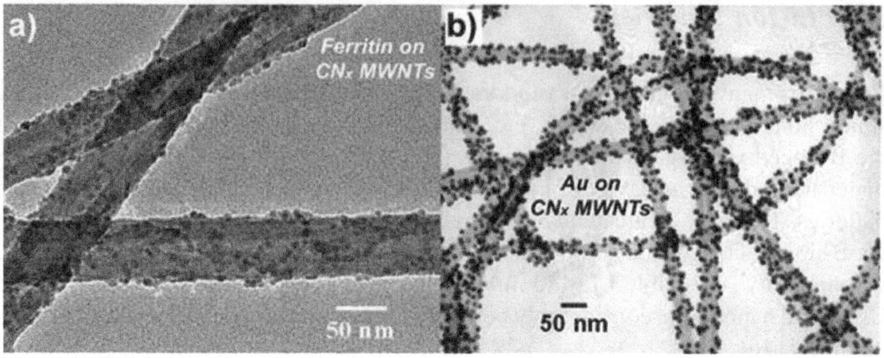

Fig. 14 **a** TEM image of ferritin-CNx MWNT conjugates [156] and **b** TEM photograph of gold nanoparticle-CNx nanotube hybrid structures (Reproduced with permission from Ref. [157])

pure CNx nanotubes have improved [154]. CNTs are moved to the pendant ferrocene groups by means of photoexcitation [155], which transforms renewable power into an electromagnetic field.

3.9.1 Metal Surfaces for Anchoring Molecules

In past years, oxidation states for nucleophilic effective link of proteins have been identified [156], Au [157], Ag [158], Fe, and Pt [159] on the N-doped MWCNT surfaces (Figs. 3 and 14). The areas are rich in nutrients and have been identified in the past. In contrast to plain CNTs, it was shown that the doped CNTs are far more effective for reinforcing organic compounds.

4 Toxicity of Doped CNTs

Latest discoveries found that CNx [160] tubes showed up much less dangerous relative to earlier toxicology tests using SWCNTs. In comparison to previous studies of undoped MWCNTs or SWCNTs, no mortal effects were found in CNx nanotubes at very large concentrations (e.g., 5 mg/kg) [161]. In both cases, the dose and time-varying pathological modifications produced by pure MWCNTs are perhaps more extreme than those caused by CNx MWCNTs [160]. CNx tubes could perhaps also be included as pharmaceutical goods, improved enzyme concentration, biofilters, inhibitors of viruses, gene exchanges, etc. For the development and development of such nanoparticles into advanced materials, appropriate, and rigorous infection control mechanisms must be created. In contrast, certain forms of doped CNTs should be examined on toxicological results in the coming years. In recent times, a comprehensive survey of amoeba and various kinds of nanomaterials (pure carbon MWCNTs and N-doped MWCNTs) was performed by Elias et al. [162]. The findings showed that they persisted and no improvements in cellular behavior or morphology existed at any dosage whenever cells were stimulated with CNx MWCNT. Most amoebas, however, died with undoped MWCNTs after 8 h of fermentation (100 μg) (Figs. 14 and 15). These findings indicate that the combination of nitrogen improves (as stated before) biocompatibility and that it could be investigated for use in many applications, including cell fusion and pharmaceutical transport methods.

5 Conclusions

In this chapter, the review on the field of carbon nanotube doping, in particular emphasizing where this differs significantly from doping of bulk semiconductor materials have been discussed in detail. Also, some of the challenges and solutions specific to carbon nanotube use in electronics have been discussed. The variety of possible approaches to modify the electronic properties of carbon nanotubes has been described in detail. In particular nanotube fictionalization is also covered, taking nanotube fluorination as a specific in-depth example. Finally, more detail on the transport behavior of carbon nanotubes and doped nanotube systems has been analyzed and explained, before drawing some general conclusions. Upon facing several disputes, the doping of CNTs finds its application in developing nanotube composites, sensors, composites, electronic devices, nanorobots, etc.

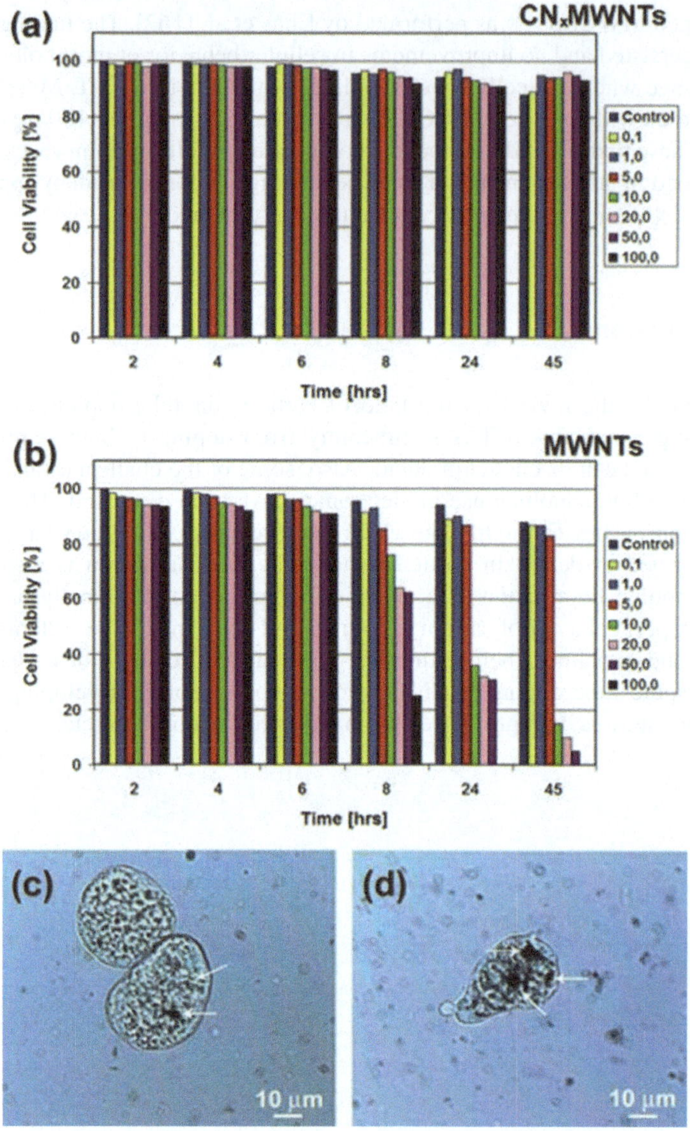

Fig. 15 Cell viability of Entamoeba histolytica cells incubated with **a** CNx MWNTs; **b** undoped MWNTs. Optical microscope images of living amoebas incubated with CNx MWNTs for **c** 45 **min** and **d** 22 h (Reproduced with permission from Ref. [162])

References

1. Ruoff, R. S. & Lorents, D. C. Mechanical and thermal properties of carbon nanotubes. Carbon **33**, 925–930 (1995).
2. Kim, P., Shi, L., Majumdar, A. & McEuen, P. L. Thermal transport measurements of individual multiwalled nanotubes. Phys. Rev. Lett. **87**, 215502 (2001).
3. Javey, A. et al. High-κ dielectrics for advanced carbon-nanotube transistors and logic gates. Nat. Mater. **1**, 241–246 (2002).
4. Javey, A., Guo, J., Wang, Q., Lundstrom, M. & Dai, H. Ballistic carbon nanotube field-effect transistors. Nature **424**, 654–657 (2003).
5. Tans, S. J., Verschueren, A.R.M. & Dekker, C. Room-temperature transistor based on a single carbon nanotube. Nature **393**, 49–52 (1998).
6. Rajasekar, R., Nayak, G.C., Malas, A., Das, C.K.M.D. Development of compatibilized SBR and EPR nanocomposites containing dual filler system. **35**, 878–885 (2012).
7. Nayak, G.C., Rajasekar, R., Bose, S., Das, C.K.J.N.T. Effect of MWNTs and SiC-coated MWNTs on properties of PEEK/LCP blend. (2009).
8. Nayak, G.C., Sahoo, S., Rajasekar, R., Das, C.K.C.P.A.A.S.M. Novel approach for the selective dispersion of MWCNTs in the Nylon/SAN blend system. **43**(8), 1242–1251 (2012).
9. Kong, J. et al. Nanotube molecular wires as chemical sensors. Science **287**, 622–625 (2000).
10. Robinson, J. A. et al. Role of defects in single-walled carbon nanotube chemical sensors. Nano Letters **6**, 8, 1747–1751 (2006).
11. Ong, K.G., Zeng, K. & Grimes, C.A. A wireless, passive carbon nanotube-based gas sensor. IEEE Sens. J.**2**, 82–88 (2002).
12. Lin, Y., Lu, F. & Wang, J. Disposable carbon nanotube modified screen-printed biosensor for amperometric detection of organophosphorus pesticides and nerve agents. Electroanalysis**16**, 145–149 (2004).
13. Cai, D. et al. Highly efficient molecular delivery into mammalian cells using carbon nanotube spearing. Nat. Methods **2**, 449–454 (2005).
14. Singh, R. et al. Binding and condensation of plasmid DNA onto functionalized carbon nanotubes: toward the construction of nanotube-based gene delivery vectors J. Am. Chem. Soc. **127**, 4388–4396 (2005).
15. Park, K.H., Chhowalla, M., Iqbal, Z. & Sesti, F. Single-walled carbon nanotubes are a new class of ion channel blockers. J. Biol. Chem. **278**, 50212–50216 (2003).
16. Glerup, M., Krstic, V. Ewels, C. Holzinger, M. &Van Lier. G. Doping of Carbon Nanotubes. In Eds., Doped Nanomaterials and Nanodevices – Volume 3 by Wei Chen (2007).
17. Chen. Y. et al. Chemical attachment of organic functional groups to single-walled carbon nanotube material. J. Mater. Res. **13**, 2423–2431 (1998).
18. Bahr, J. L. & Tour, J. M. Covalent chemistry of single-wall carbon nanotubes. J. Mater. Chem. **12**, 1952–1958 (2002).
19. Hirsch. A. Functionalization of single-walled carbon nanotubes. Angew. Chemie - Int. Ed. **41**, 1853–1859 (2002).
20. Banerjee, S., Kahn, M. G. C. & Wong, S. S. Rational chemical strategies for carbon nanotube functionalization. Chem. Eur. J. **9**, 1898–1908 (2003).
21. Dyke, C. A. & Tour. J. M. Covalent functionalization of single-walled carbon nanotubes for materials applications. J. Phys. Chem. A **108**, 11151–11159 (2004).
22. Banerjee, S., Hemraj-Benny, T. & Wong. S. S. Covalent surface chemistry of single-walled carbon nanotubes. Adv. Mater. **17**, 17–29 (2005).
23. Balasubramanian, K. & Burghard, M. Chemicallyfunctionalized carbon nanotubes1. Small**2**, 180–192 (2005).
24. Hamon, M. A. et al. Effect of rehybridization on the electronic structure of single-walled carbon nanotubes. J. Am. Chem.Soc., **123**, 11292–11293 (2001).
25. Haddon, R. C. Chemistry of the fullerenes: The manifestation of strain in a class of continuous aromatic molecules. Science **261**, 1545–1550 (1993).

26. Haddon, R. C. Measure of nonplanarity in conjugated organic molecules: which structurally characterized molecule displays the highest degree of pyramidalization? J. Am. Chem. Soc. **112**, 3385–3389 (1990).
27. Rabideau, P. W. & Sygula, A. Buckybowls: Polynuclear aromatic hydrocarbons related to the buckminsterfullerene surface. Acc. Chem. Res. **29**, 235–242 (1996).
28. Scott, L. T., Bratcher, M. S. & Hagen. S. Synthesis and characterization of a $c_{36}h_{12}$ fullerene subunit. J. Am. Chem. Soc. **118**, 8743–8744 (1996).
29. Haddon, R. C. & Raghavachari, K. Electronic structure of the fulleroids: Homoconjugation in bridged C_{60} derivatives. Tetrahedron **52**, 5207–5220 (1996).
30. Weedon, B. R.,Haddon, R. C., Spielmann, H. P. & Meier, M. S. Fulleroid addition regiochemistry is driven by π-orbital misalignment. J. Am. Chem. Soc. **121**, 335–340 (1999).
31. Bodwell, G.J. et al. 1,7-Dioxa[7](2,7)pyrenophane: The pyrene moiety is more bent than that of C_{70}. Chem. Eur. J. **5**, 1823–1827 (1999).
32. Srivastava, D. et al. Predictions of enhanced chemical reactivity at regions of local conformational strain on carbon nanotubes: Kinky chemistry. J. Phys. Chem. B **103**, 4330–4337 (1999).
33. Chen, Z. Thiel, W. & Hirsch, A. Reactivity of the convex and concave surfaces of single-walled carbon nanotubes (SWCNTs) towards addition reactions: dependence on the carbon-atom pyramidalization. Chem. Phys. Chem. **4**, 93–97 (2003).
34. Chen, Z. et al. Side-wall opening of single-walled carbon nanotubes (swcnts) by chemical modification: A critical theoretical study†.Angew. Chemie - Int. Ed. **116**, 1578–1580 (2004).
35. Holzinger, M. et al. Characterization of oxidized SWCNTs by XPS. AIP Conf. Proc. **633**, 96–99 (2002).
36. Hiura, H., Ebbesen, T.W. & Tanigaki, K. Opening and purification of carbon nanotubes in high yields. Adv. Mater.**7**, 275–276 (1995).
37. Ebbesen, T.W. et al. Decoration of carbon nanotubes. Adv.Mater. **8**, 155–157 (1996).
38. Chen, J. et al. Haddon.Solution properties of single-walled carbon nanotubes. Science **282**, 95–98 (1998).
39. Liu, J. et al. Fullerene pipes. Science **280**, 1253–1256 (1998).
40. Mawhinney, D.B. et al. Infrared spectral evidence for the etching of carbon nanotubes: ozone oxidation at 298 kJ. Am.Chem. Soc. **122**, 2383–2384 (2000).
41. Kuznetsova, A. et al. Enhancement of adsorption inside of single-walled nanotubes: opening the entry ports. Chem. Phys. Lett. **321**, 3–4, 292–296 (2000).
42. Ajayan, P.M. et al. Opening carbon nanotubes with oxygen and implications for filling. Nature **361**, 333 (1993).
43. Holzinger, M. et al. Sidewall functionalization of carbon nanotubes. Angew. Chemie - Int. Ed. **40**, 21, 4002–4005 (2001).
44. Peng, H., Reverdy, P., Khabashesku, V. N. & Margravea, J. L. Sidewall functionalization of single-walled carbon nanotubes with organic peroxides†. Chem. Commun. **3**, 362–363 (2003).
45. Chen, H.S., Kortan, A. R., Haddon, R. C. & Kopylov, N. Formation energy of alkali-metal-doped fullerite compounds A_6C_{60}. J. Phys. Chem. B **97**, 3088–3090 (1993).
46. Holzinger, M. et al. Functionalization of single-walled carbon nanotubes with (R-) oxycarbonyl nitrenes. J. Am. Chem. Soc. **28**, 125, 8566–8580 (2003).
47. Georgakilas, V. et al.Organic functionalization of carbon nanotubes. J. Am. Chem. Soc. **124**, 760–761 (2002).
48. Viswanathan, G. et al. Single-step in situ synthesis of polymer-grafted single-wall nanotube composites. J. Am. Chem. Soc. **125**, 9258–9259 (2003).
49. Blake, R. et al. A generic organometallic approach toward ultra-strong carbon nanotube polymer composites. J. Am. Chem.Soc. **126**, 10226–10227 (2004).
50. Chen, S., Shen, W., Wu, G., Chen, D. & Jiang, M. A new approach to the functionalization of single-walled carbon nanotubes with both alkyl and carboxyl groups. Chem. Phys. Lett. **402**, 312–317 (2005).
51. Graupner, R. et al. Nucleophilic−alkylation−reoxidation: A functionalization sequence for single-wall carbon nanotubes. J. Am. Chem. Soc. **128**, 6683–6689 (2006).

52. Mickelson, E.T. et al. Fluorination of single-wall carbon nanotubes. Chem.Phys. Lett., **296**, 188–194 (1998).
53. Bahr, J.L. et al. Functionalization of carbon nanotubes by electrochemical reduction of aryl diazonium salts: A bucky paper electrode. J. Am. Chem. Soc. **123**, 6536–6542 (2001).
54. Bahr, J.L. & Tour, J.M. Highly functionalized carbon nanotubes using in situ generated diazonium compounds. Chem. Mater. **13**, 3823–3824 (2001).
55. Holzinger, M. et al. Purification and functionalisation of nitrogen-doped single-walled carbon nanotubes. AIP Conf. Proc. **786**, 211–214 (2005).
56. O'Connell, M. J. et al. Band gap fluorescence from individual single-walled carbon nanotubes. Science **297**, 593–596 (2002).
57. Dyke, C. A. & Tour, J. M. Unbundled and highly functionalized carbon nanotubes from aqueous reactions. Nano Lett. **3**, 1215–1218 (2003).
58. Liang, F. et al. A convenient route to functionalized carbon nanotubes. Nano Lett. **4**, 1257–1260 (2004).
59. Chattopadhyay, J. et al. Carbon nanotube salts. Arylation of single-wall carbon nanotubes. Org. Lett. **7**, 4067–4069 (2005).
60. Hamon, M. A. et al. Dissolution of single-walled carbon nanotubes. Adv.Mater. **11**, 834–840 (1999).
61. Chen, J. et al. Dissolution of full-length single-walled carbon nanotubes. J. Phys. Chem. B **105**, 2525–2528, (2001).
62. Haddon, R.C. & Chen. J. Solubilizing single-walled carbon nanotubes by direct reaction with amines and alkylaryl amines.US Patent 6187823 (2001)
63. Neises, B. & W. Steglich. Simple method for the esterification of carboxylic acids†. Chem. Int. Ed. **17**, 522–524, (1978).
64. Wong, S.S.,Woolley, A. T., Joselevich, E., Cheung, C. L. & Lieber, C. M. Covalently-functionalized single-walled carbon nanotube probe tips for chemical force microscopy. J. Am. Chem. Soc. **120**, 8557–8558, (1998).
65. Holzinger, M. et al. Novel purification procedure and derivatization method of single-walled carbon nanotubes (SWNTs). AIP Conf. Proc. **544**, 246–249 (2000).
66. Desai, M. C. & Stramiello, L. M. S. Polymer bound EDC (P-EDC): A convenient reagent for formation of an amide bond. Tetrahedron Lett. **34**, 7685–7688 (1993).
67. Kohasaka, H. & Carson. D. A. Solid-Phase polymerase chain reaction. J. Clin. Lab. Anal. **8**, 452–455 (1994).
68. Williams, K.A., Veenhuizen, P.T., Beatriz, G., Eritja, R. & Dekker, C. Carbon nanotubes with DNA recognition. Nature **420**, 761, (2002).
69. Holzinger, M. et al. [2+1] cycloaddition for cross-linking SWCNTs. Carbon **42**, 941–947, (2004).
70. Weiss, R., Reichel, S., Handke, M. & Hampel, F. Generation and trapping reactions of a formal 1:1 complex between singlet carbon and 2,2'-bipyridine. Angew. Chemie - Int. Ed. **37**, 344–347 (1998).
71. Stapfner, S., Ost, L., Hunger, D., Reichel, J., Favero, I. and Weig, E.M., 2013. Cavity-enhanced optical detection of carbon nanotube Brownian motion. Applied Physics Letters, 102(15), p.151910.
72. Weiss, R. & Reichel, S. Novel urea derivatives as two-step redox systems.Eur. J. Inorg. Chem. **9**, 1935–1939 (2000).
73. Calder, I. C., Spotswood, T. M. & Sasse, W.H.F. The dipyrido [1,2-e:2',1'-e] imidazolium cation, a new aromatic ring system. Tetrahedron Lett. **65**, 95–100 (1963).
74. Fagan, P. J. et al. Production of perfluoroalkylated nanospheres from buckminsterfullerene. Science **262**, 404–407 (1993).
75. Shaffer, M. S. P. & Koziol, K. Polystyrene grafted multi-walled carbon nanotubes. Chem. Commun. **18**, 2074–2075 (2002).
76. Ying, Y., Saini, R. K., Liang, F., Sadana, A. K., & Billups, W. E. Functionalization of carbon nanotubes by free radicals. Org. Lett. **5**, 1471–147, (2003).

77. Xia, H.,Wang, Q. & Qiu, G. Polymer-encapsulated carbon nanotubes prepared through ultrasonically initiated in situ emulsion polymerization. Chem. Mater. **15**, 3879–3886 (2003).
78. Wooster, C.B. & Godfrey, K. L. Mechanism of the reduction of unsaturated compounds with alkali metals and water. J. Am. Chem. Soc. **59**, 596–597 (1937).
79. Birch, A. J.The reduction of organic compounds by metal-ammonia solutions. Quart. Rev. **4**, 69–93 (1950).
80. Hirsch, A. and Vostrowsky, O., 2005. Functionalization of carbon nanotubes. Functional molecular nanostructures, pp.193–237.
81. Petit, P., Mathis, C., Journet, C. & Bernier, P.Tuning and monitoring the electronic structure of carbon nanotubes. Chem. Phys. Lett. **305**, 370–374 (1999).
82. Mathis, C. & François, B. Experimental studies of n-doped $(CH)_x$: D.c. electrical conductivity. Synth. Met. **9**, 347–354 (1984).
83. Pénicaud, A. Poulin, P., Derré, A., Anglaret, E. & Petit, P. Spontaneous dissolution of a single-wall carbon nanotube salt. J. Am. Chem. Soc. **127**, 8–9 (2004).
84. Maggini, M. Scorrano, G. & Prato, M. Addition of azomethine ylides to C60: synthesis, characterization, and functionalization of fullerene pyrrolidines. J. Am. Chem. Soc. **115**, 9798–9799 (1993).
85. Prato, M. & Maggini, M. Fulleropyrrolidines: A family of full-fledged fullerene derivatives. Acc. Chem. Res. **31**, 519–526 (1998).
86. Pantorotto, D. et al. Synthesis, structural characterization, and immunological properties of carbon nanotubes functionalized with peptides. J.Am.Chem. Soc. **125**, 6160–6164 (2003).
87. D. M. Guldi et al. Single-Wall Carbon Nanotube–Ferrocene Nanohybrids: Observing intramolecular electron transfer in functionalized swnts. Angew. Chemie - Int. Ed. **115**, 4338–4341, (2003).
88. D. Pantarotto et al. Functionalized carbon nanotubes for plasmid DNA gene delivery. Angew. Chemie - Int. Ed. **43**, 5242–5246 (2004).
89. Tagmatarchis, N. & Prato. M. Functionalization of carbon nanotubesvia 1,3-dipolar cycloadditions. J. Mater. Chem. **14**, 437–439 (2004).
90. Delamar, M. et al. Modification of carbon fiber surfaces by electrochemical reduction of aryl diazonium salts: Application to carbon epoxy composites. Carbon **35**, 801–807 (1997).
91. Allongue, P. et al. Covalent Modification of carbon surfaces by aryl radicals generated from the electrochemical reduction of diazonium salts. J. Am Chem. Soc. **119**, 201–207 (1997).
92. Ortiz, B., Saby, C., Champagne, G. Y. & Bélanger, D. J. Electrochemical modification of a carbon electrode using aromatic diazonium salts. 2. Electrochemistry of 4-nitrophenyl modified glassy carbon electrodes in aqueous media. Electroanal. Chem. **455**, 75–81 (1998).
93. Saby, C., Ortiz, B., Champagne, G. Y. & Bélanger, D.Electrochemical Modification of Glassy Carbon Electrode Using Aromatic Diazonium Salts. 1. Blocking Effect of 4-Nitrophenyl and 4-Carboxyphenyl Groups. Langmuir **13**, 6805–6813 (1997).
94. Delamar, M. Hitmi, R. Pinson, J. & Savéant, J. M. Covalent modification of carbon surfaces by grafting of functionalized aryl radicals produced from electrochemical reduction of diazonium salts. J. Am. Chem. Soc. **114**, 5883–5884 (1992).
95. Kariuki, J. K. &McDermott. M. T. Nucleation and growth of functionalized aryl films on graphite electrodes. Langmuir **15**, 6534–6540 (1999).
96. Kooi, S.E., Schlecht, U., Burghard, M. & Kern, K. Electrochemical modification of single carbon nanotubes.Angew. Chemie - Int. Ed.**41**, 1353–1355 (2002).
97. Glerup, M. et al. Synthesis of N-doped SWNT using the arc-discharge procedure. Chem. Phys. Lett. **387**, 193–197 (2004).
98. Nevidomskyy, A. H., Csányi, G. & Payne, M. C. Chemically active substitutional nitrogen impurity in carbon nanotubes.Phys. Rev. Lett. **91**, 105502 (2003).
99. Hauke, F. & Hirsch, A. Mannich functionalization of $C_{59}N$. J. Chem. Soc. Chem. Commun. **21**, 2199–2200 (1999).
100. Khabashesku, V. N., Billups, W. E. & Margrave, J. L. Fluorination of single-wall carbon nanotubes and subsequent derivatization reactions. Acc. Chem. Res. **35**, 1087–1095 (2002).

101. Bettinger, H. F. Experimental and computational investigations of the properties of fluorinated single-walled carbon nanotubes. Chem. Phys. Chem. **4**, 1283–1289 (2003).
102. Nakajima, T., Kasamatsu, S. & Matsuo, Y. Synthesis and characterization of fluorinated carbon nanotube. Eur. J. Inorg. Chem. **33**, 831–840 (1996).
103. Mickelson, E. T. et al. Solvation of fluorinated single-wall carbon nanotubes in alcohol solvents. J. Phys. Chem. B **103**, 4318–4322 (1999).
104. Kudin, K. N., Bettinger, H. F. & Scuseria, G. E. Fluorinated single-wall carbon nanotubes.Phys. Rev. B Condens. Matter Mater. Phys. **63**, 045413 (2001).
105. Boul, P. J. Reversible sidewall functionalization of buckytubes. Chem. Phys. Lett. **310**, 367–372 (1999).
106. Kelly, K. F. et al. Insight into the mechanism of sidewall functionalization of single-walled nanotubes: An STM study.Chem. Phys. Lett. **313**, 445–450 (1999).
107. Saini, R. K. et al. Covalent sidewall functionalization of single wall carbon nanotubes.J. Am. Chem. Soc. **125,** 3617–3621 (2003).
108. Stevens, J. L. Sidewall amino-functionalization of single-walled carbon nanotubes through fluorination and subsequent reactions with terminal diamines. Nano Lett. **3**, 331–336 (2003).
109. Valentini, L., Puglia, D., Armentano, I. & Kenny, J. M. Sidewall functionalization of single-walled carbon nanotubes through CF_4 plasma treatment and subsequent reaction with aliphatic amines.Chem. Phys. Lett. **403**, 385–389 (2005).
110. Dettlaff-Weglikowska, U. et al. Effect of fluorination on electrical properties of single walled carbon nanotubes and C_{60} peapods in networks. Curr. Appl. Phys.**7**, 42–46 (2006).
111. Seifert, G., Kohler, T. & Frauenheim, T. Molecular wires, solenoids, and capacitors by sidewall functionalization of carbon nanotubes. Appl. Phys. Lett. **77**, 1313–1315 (2000).
112. Hamwi, A., Gendraud, P., Gaucher, H., Bonnamy, S. & Beguin, F. Electrochemical properties of carbon nanotube fluorides in a lithium cell system.Mol. Cryst. Liq. Cryst. Sci. Technol. Sec. A: Mol. Cryst. Liq. Cryst. **310**, 185–190 (1998).
113. Mukhopadhyay, I. et al. Effect of chemical modification on electrochemical Li insertion in highly ordered multi-wall carbon nanotubes. Proceedings - Electrochemical Society **2014**, 37–40 (2001).
114. Peng, H. et al. Fluorotubes as cathodes in lithium electrochemical cells. Nano Lett. **1**, 625–629 (2001).
115. Root, M. J. Comparison of fluorofullerenes with carbon monofluorides and fluorinated carbon single wall nanotubes: thermodynamics and electrochemistry. Nano Lett. **2**, 541–543 (2002).
116. Lee, J. Y., An, K.H., Heo, J. K. & Lee, Y. H. Supercapacitors using fluorinated singlewalled carbon nanotube. Proceedings – Electrochem. Soc. **2003–15**, 366–70 (2003).
117. Lee, J. Y., An, K. H., Heo, J. K. & Lee, Y. H. Fabrication of supercapacitor electrodes using fluorinated single-walled carbon nanotubes. J. Phy. Chem. B **107**, 8812–8815 (2003).
118. Hayashi, T. et al. Nano Teflons: Structure and eels characterization of fluorinated carbon nanotubes and nanofibers. Nano Lett. **2**, 491–496 (2002).
119. Gupta, V. Comments on "NanoTeflon: Structure and eels characterization of fluorinated carbon nanotubes and nanofibers". Nano Lett. **4**, 999 (2004).
120. Hayashi, T. Reply to "A comment on 'nanoteflons: structure and eels characterization of fluorinated carbon nanotubes and nanofibers". Nano Lett. **4**, 1001–1002 (2004).
121. Khabashesku, V. N., Margrave, J. L. & Barrera, E. V. Functionalized carbon nanotubes and nanodiamonds for engineering and biomedical applications.Diam. Relat. Mater. **14**, 859–866 (2005).
122. Owens, F. J. Raman and mechanical properties measurements of single walled carbon nanotube composites of polyisobutylene. J. Mater. Chem. **16**, 505–508 (2006).
123. Marcoux, P. R. et al. A spectroscopic study of the fluorination and defluorination reactions on single-walled carbon nanotubes. Phys. Chem. Chem. Phys. **4**, 2278–2285 (2002).
124. Khare, B. N., Wilhite, P. & Meyyappan, M. The fluorination of single wall carbon nanotubes using microwave plasma. Nanotechnology **15**, 1650–1654 (2004).
125. Felten, A., Bittencourt, C., Pireaux, J. J. Van Lier, G. & Charlier, J. C. Radio-frequency plasma functionalization of carbon nanotubes surface O_2, NH_3, and CF_4 treatments. J. Appl. Phys. **98**, 074308 (2005).

126. Yudanov, N. F. et al. Fluorination of arc-produced carbon material containing multiwall nanotubes.Chem. Mater. **14**, 1472–1476 (2002).
127. Okotrub, A. V. et al. Fluorination of CN_x nanotubes, fullerenes, nanotubes, and carbon nanostructures. **12**, 99–104 (2004).
128. Gevko, P. N. et al. Optical absorption and Raman spectroscopy study of the fluorinated double-wall carbon nanotubes.Fuller. Nanotub. Carbon Nanostructures **14**, 233–238 (2006).
129. Unger, E. et al. Fluorination of carbon nanotubes with xenon difluoride. Chem. Phys. Lett. **399**, 280–283 (2004).
130. An, K. H. et al. X-ray photoemission spectroscopy study of fluorinated single-walled carbon nanotubes. Appl. Phys. Lett. **80**, 4235–4237 (2002).
131. Lee, Y. S. et al. Surface properties of fluorinated single-walled carbon nanotubes, J. Fluor. Chem. **120**, 99–104 (2003).
132. Hamwi, A., Alvergnat, H., Bonnamy, S. & Beguin, F. Fluorination of carbon nanotubes. Carbon **35**, 723–728 (1997).
133. Bauschlicher, C. W. Hydrogen and fluorine binding to the sidewalls of a (10,0) carbon nanotube. Chem. Phys. Lett. **322**, 237–241 (2000).
134. Jaffe, R. L. Quantum chemistry study of fullerene and carbon nanotube fluorination. J. Phys. Chem. B **107**, 10378–10388 (2003).
135. Van Lier, G., Ewels, C. P., Zuliani, F., De Vita, A. & Charlier, J.-C. Theoretical analysis of fluorine addition to single-walled carbon nanotubes: Functionalization routes and addition patterns.J. Phys. Chem. B **109**, 6153–6158 (2005).
136. Zhao, W., Song, C., Zheng, B., Liu, J. & Viswanathan, T. Thermal recovery behavior of fluorinated single-walled carbon nanotubes. J. Phys. Chem. B **106**, 293–296 (2002).
137. Shofner, M. L., Khabashesku, V. N. & Barrera, E. V. Processing and mechanical properties of fluorinated single-wall carbon nanotube-polyethylene composites. Chem. Mater. **18**, 906–913 (2006).
138. Georgakilas, V. et al. Purification of HiPCO carbon nanotubes via organic functionalization. J. Am. Chem. Soc. **124**, 14318–14319 (2002).
139. Zheng, M. et al. DNA-assisted dispersion and separation of carbon nanotubes. Nature Mater. **2**, 338–342 (2003).
140. Zheng, M. et al. Structure-based carbon nanotube sorting by sequence-dependent DNA assembly. Science **302**, 1545–1548 (2003).
141. Heller, D. A. et al. Optical detection of DNA conformational poly-morphism on single-walled carbon nanotubes. Science **311**, 508–511 (2006).
142. Charlier, J. C. et al. Enhanced electron field emission in B-doped carbon nanotubes. Nano Lett. **2**, 1191–1195 (2002).
143. Golberg, D. Structure, transport and field-emission properties of compound nanotubes: CN_x vs. BNC_x ($x < 0.1$). Appl. Phys. A-Mater. **76**, 499–507 (2003).
144. Doytcheva, M., Kaiser, M., Reyes-Reyes, M., Terrones, M. & de Jonge, N. Electron emission from individual nitrogen-doped multi-walled carbon nanotubes. Chem. Phys. Lett. **396**, 126–130 (2004).
145. Endo, M. et al. Vapor-grown carbon fibers (VGCFs) - Basic properties and their battery applications. Carbon **39**, 1287–1297 (2001).
146. Zhang, D. Y. et al. Lithium storage in polymerized carbon nitride nanobells. Appl. Phys. Lett. **79**, 3500–3502 (2001).
147. Wong, S. S., Joselevich, E., Woolley, A. T., Cheung, C. L. & Lieber, C. M. Covalently functionalized nanotubes as nanometre-sized probes in chemistry and biology. Nature **394**, 52–55 (1998).
148. Collins, P., Bradley, K., Ishigami, M. & Zettl, A. Extreme oxygen sensitivity of electronic properties of carbon nanotubes. Science **287**, 1801–1804 (2000).
149. Villalpando-Paez, F. et al. Fabrication of vapor and gas sensors using films of aligned CN_x nanotubes. Chem. Phys. Lett. **386**, 137–143 (2004).
150. Calvert, P. Nanotube composites - A recipe for strength. Nature 399, 210–211 (1999).

151. Eitan, A. et al. Processing and thermal characterization of nitrogen doped MWNT/epoxy composites, in Proc. Tenth US-Japan Conf. Compos. Mater., 634–640 (2002).
152. Fragneaud, B., Masenelli-Varlot, K., Gonz'alez-Montiel, A., Terrones, M. & Cavaill'e, J. Y. Efficient coating of N-doped carbon nanotubes with polystyrene using atomic transfer radical polymerization. Chem. Phys. Lett. **419**, 567–573 (2005).
153. Dehonor, M. et al. Nanotube brushes: Polystyrene grafted covalently on CN_x nanotubes by nitroxide-mediated radical polymerization. Chem. Commun. **42**, 5349–5351 (2005).
154. Fragneaud, B., Masenelli-Varlot, K., Gonzalez-Montiel, A., Terrones, M. & Cavaill'e, J.-Y. Electrical behavior of polymer grafted nanotubes/polymer nanocomposites using N-doped carbon nanotubes. Chem. Phys. Lett. **444**, 1–8 (2007).
155. Guldi, D. M. et al. Single-wall carbon nanotube–ferrocene nanohybrids: observing intramolecular electron transfer in functionalized swnts. Angew. Chemie-Int. Ed. **42**, 4338–4341 (2003).
156. Jiang, K. et al. Protein immobilization on carbon nanotubes via a two-step process of diimide-activated amidation. J. Mater. Chem. **14**, 37–39 (2004).
157. Jiang, K. Selective attachment of gold nanoparticies to nitrogen-doped carbon nanotubes. Nano Lett. **3**, 275–277 (2003).
158. Zamudio, A. et al. Efficient anchoring of silver nanoparticles on N-doped carbon nanotubes. Small **2**, 346–350 (2005).
159. Lepr'o, X. et al. Production and characterization of co-axial nanotube junctions and networks of CN_x/CNT. Nano Lett.**7**, 2220–2226 (2007).
160. Carrero-S'anchez, J. L. et al. Biocompatibility and toxicological studies of carbon nanotubes doped with nitrogen. Nano Lett. **6**, 1609–1616 (2006).
161. Warheit, D. B. et al. Comparative pulmonary toxicity assessment of single-wall carbon nanotubes in rats. Toxicol. Sci. **77**, 117–125 (2004).
162. Elias, A. L. et al. Comparative viability studies of pure carbon and nitrogen-doped multi walled carbon nanotube with amoeba cells: From amoebicidal to biocompatible structures. Small **3**, 1723 (2007).
163. Radosavljevic, M., Freitag, M.,Thadani, K.V. & Johnson, A.T. Nonvolatile molecular memory elements based on ambipolar nanotube field effect transistors. Nano Lett. **2**, 761–764 (2002).
164. Hirsch, A. & Brettreich, M. Fullerenes: chemistry and reactions. John Wiley & Sons (2006).
165. Ewels, C. P., Van Lier, G., Charlier, J. C., Heggie, M. I. & Briddon, P. R. Pattern formation on carbon nanotube surfaces. Phys. Rev. Lett. **96**, 216103 (2006).

Recent Progress in N-Doped Graphene: Properties and Applications

Harikrishna Kumar Mohan Kumar, Rajasekar Rathanasamy, Moganapriya Chinnasamy, and GobinathVelu Kaliyannan

Abstract Graphene is a significant progress in modern research and one of the most hopeful materials used in succeeding electronic devices due to its unique characteristics. Since, graphene possess high surface area, thermal conductivity and Young's modulus, it is mostly preferred in various engineering applications. Chemical doping of graphene using various elements like nitrogen, boron, hydrogen, etc., can lead to increase in electronic properties effectively. Nitrogen and boron doped graphene leads to fresh electronic band structure. Those doped graphene materials results in usage of materials in various application such as nanoelectronics, nanophotonics, sensor devices, green energy technology and alternative eco-friendly material for electrocatalytic applications. Nitrogen doped graphene is good donor of electron and it is used in energy related applications such as supercapacitors, fuel cells, batteries, medical domain, etc. Technical properties of N-doped graphene were determined using various techniques such as STM, TEM, SEM, AFM, XPS, Raman spectroscopy and XRD. This chapter deals with properties and various applications of nitrogen doped graphene in engineering applications.

Keywords Graphene · Nitrogen doped graphene · Supercapacitors · Lithium-ion batteries · Field effect transistor · Electro-catalyst

---

H. K. M. Kumar · R. Rathanasamy (✉)
Department of Mechanical Engineering, Kongu Engineering College, Erode, Tamil Nadu 638060, India

Department of Mechanical Engineering, Sri Krishna Polytechnic College, Coimbatore, Tamil Nadu 641042, India

Present Address:
M. Chinnasamy
Department of Mining Engineering, Indian Institute of Technology, Kharagpur, West Bengal 721302, India

G. Kaliyannan
Department of Mechatronics Engineering, Kongu Engineering College, Erode, Tamil Nadu 638060, India

S. Sahoo et al. (eds.), *Defect Engineering of Carbon Nanostructures*, Advances in Material Research and Technology, https://doi.org/10.1007/978-3-030-94375-2_6

1 Introduction

Graphene corresponds to an important development in research works and it is preferred as most hopeful materials that is used in new model electronic devices [1]. Graphene is recently emerged as a one of nanomaterial due to its unique characteristics such as quantum Hall Effect and electron ballistic conduction at ambient temperature [2–4]. High surface area, thermal conductivity and Young's modulus of single layer graphene are 2630 m^2/g, $\sim$5000 W/mK and $\sim$1 Tpa respectively was seen as an attractive property to use it in various engineering applications [1, 2]. Unique characteristics was achieved for the single layer graphene since it is a gapless material [5]. However, two-dimension and three-dimension graphene also possess novel characteristics but they are not gapless [2]. It is required to produce a band gap for single layer graphene to use it for engineering applications. Increasing the band gap of single layer graphene reduces the electron movement. As a result, such band gap can be produced through surface modification or chemical doping. Chemical doping is proved approach which is observed in doping of carbon nanotubes (CNTs) resulted wider range of applications [6, 50–52]. Chemically doping of graphene was carried out by two methods such as (i) adsorption of gas or metal on to the graphene surface and (ii) introducing heteroatoms into the carbon lattice of graphene by substitutional doping technique. Two methods mentioned above will enhance the electronic properties of graphene [6]. Substitutional doping method can increase the band gap of graphene which leads to metal semiconductor transition and finds potential application to use in engineering field [7]. Doping of graphene with heteroatoms (other than carbon and hydrogen) can change its electronic structure and other intrinsic properties effectively [8]. In-order to trigger and enhance the electronic properties of graphene doping with other elements were seen as a promising method [7]. 2D graphene crystal consist of carbon atoms formatted in a honeycomb structure possess fascinating chemical and physico-mechanical properties arising from its new electronic structure. Doping of graphene leads to fresh electronic band structure which possess possible applications in nanoelectronics, nanophotonics, sensor devices, and green energy technology [9]. Chemically doped graphene was also used as alternative eco-friendly material for electrocatalytic applications [10]. Doping of graphene by nitrogen (N) or boron (B) atoms can obtain n-type or p-type semiconducting graphene. Properties of N-doped graphene would get differed from pristine graphene. Nitrogen dopants would cause the active region on the surface of graphene which influences spin density and charge distribution of carbon atoms. Catalytic reaction such as oxygen reduction reaction would take place in active region on the graphene surface [6]. Doping of graphene with N moves the Dirac point below the EF and energy gap was observed at high symmetric K-point in its band structure [11]. Thus, N or B doped graphene materials has its unique characteristics. It is also proved that N-doped graphene possess higher electro-catalytic activity when compared with other electro-catalyst [12]. N doped graphene is good donor of electron, it is used in supercapacitors, fuel cells, batteries, medical domain, etc. [2, 12, 13]. Doping of

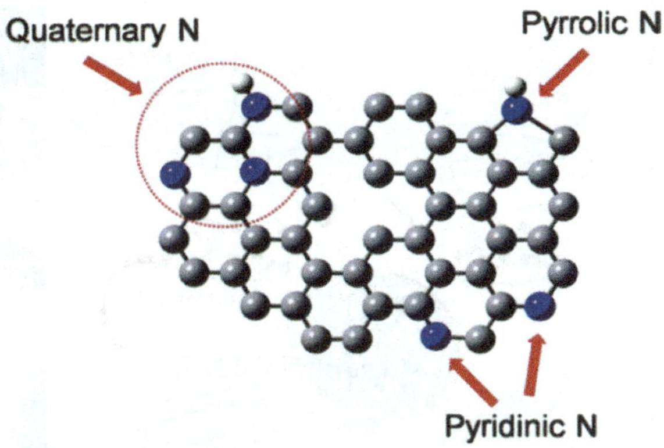

Fig. 1 Bonding configuration of N-doped graphene (Reproduced with permission from Ref. [1])

graphene with nitrogen atom forms three bonding configurations in the carbon lattice such as quaternary N, pyridinic N and pyrrolic N as shown in Fig. 1.

Doping of nitrogen into graphene has been carried out using following methods such as CVD, thermal annealing, pyrolysis, arc-discharge, plasma treatment, N2 H4 treatment, hydrothermal method, solvothermal method, microwave assisted hydrothermal, wet chemical synthesis, microwave treatment, flame treatment, supercritical reaction method and lyophilization-assisted heat treatment [6]. Despite the number of methods are available to produce N-doped graphene, CVD was most preferred method to grow graphite or the doped graphite. Synthesis of graphene was widely carried out by the CVD technique [7]. Figure 2 represents the formation nitrogen doped foam.

2 Determination of Properties

2.1 Microscopic Techniques

STM, TEM, SEM and AFM are the significant tools used for capturing the images of nitrogen doped graphene under microscopic tools with varying magnification factors [2, 6]. STM is a powerful technique for studying N-doped graphene material surfaces at the atomic level. The STM images directly provide the atomic information of the N-dopants in N-doped graphene [12, 15, 16]. Electronic properties of N-doped graphene can be investigated using STM. Charge density of N-doped graphene were examined at Fermi level using STM. Lowest unoccupied and highest occupied of electrons can be determined when bias voltage is employed in among tip and the sample. When bias voltage is positive lowest unoccupied states of electrons can be examined. Similarly, if negative bias voltage is applied highest occupied states of electrons are examined. Atomic resolution images can be obtained due to the sharp tip in the STM [6]. Topography of doped nitrogen was found using STM images.

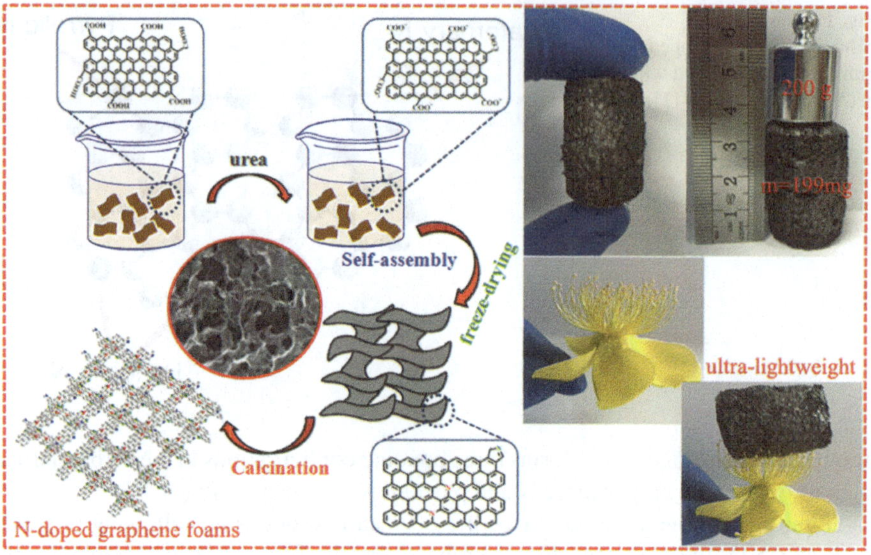

Fig. 2. Synthesis of nitrogen doped foam and its applications (Reproduced with permission from Ref. [14])

From the Fig. 3, it was observed that nitrogen doping was obtained from honeycomb lattice with brighter bumps in the images. It was evidently proved that electronic properties were triggered from the brighter bumps seen in images [2].

Transformation of graphene into N-doped graphene can be examined utilizing TEM images [1]. Figure 4 depicts the TEM images of synthesized graphene and N-doped graphene materials. Figure 4a, showing high surface/volume ratio when compared to Fig. 4b. Nitrogen doped graphene TEM image clearly shows that

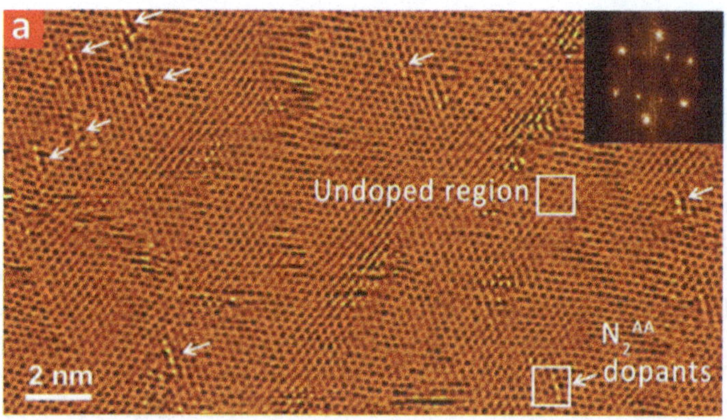

Fig. 3 STM images of nitrogen doped graphene (Reproduced with permission from Ref. [2])

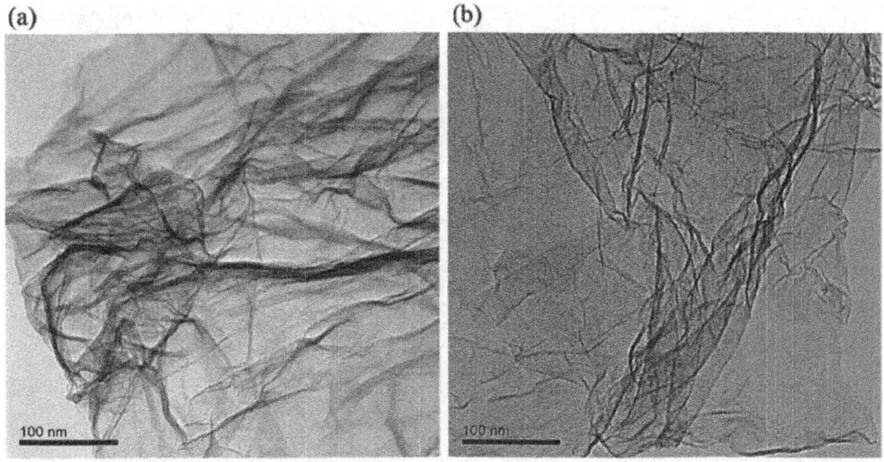

Fig. 4 **a** TEM images of synthesized graphene **b** TEM images of N-doped graphene (Reproduced with permission from Ref. [17])

graphene planar sheets clearly and 2D structure of graphene are asserted [17]. Morphological structure can be examined accurately using TEM images. High resolution TEM images are used to calculate the number of graphene layers based on the cross section of the sample [6].

The morphology of nitrogen doped graphene is widely studied using SEM tool. Interfacial morphology has a major influence on interfacial morphology, polarization, diffuse scattering and dielectric losses. Hence investigation of nitrogen doped graphene is needed and the same was analyzed through SEM images [14]. Figure 5a and b indicates the SEM images of graphene foams and nitrogen doped graphene

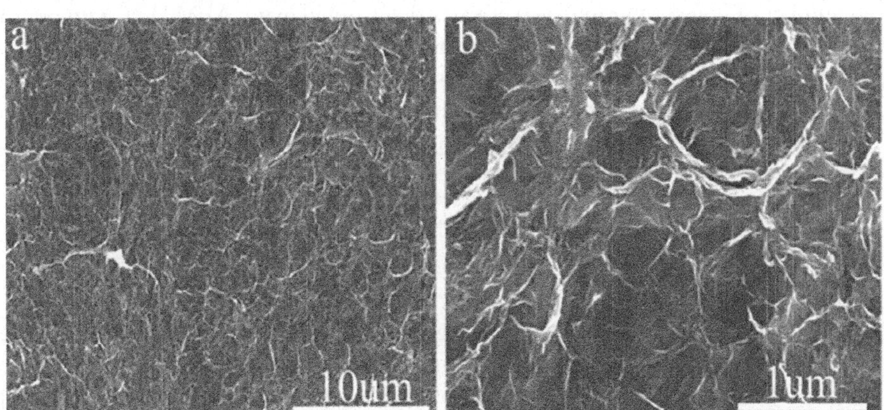

Fig. 5 **a** SEM image of graphene foam and **b** SEM images of N-doped graphene (Reproduced with permission from Ref. [14])

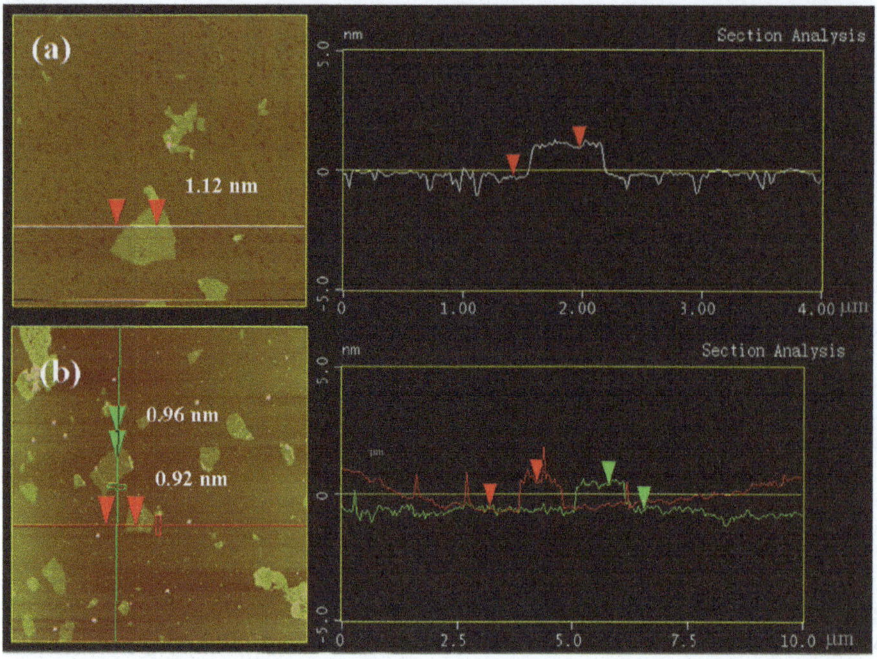

Fig. 6 **a** AFM image of graphene sheets **b** With height profile (Reproduced with permission from Ref. [25])

foams which were calcinated at 800 °C. It was observed that randomly enclosed walls with distributed reticulum-like porous structure. From SEM morphological structure of materials can be determined [14].

AFM was utilized to find out the structure of graphene samples. The thickness and number of layers of graphene were also determined using AFM [12]. Number of graphene layers was calculated based on the interlayer distance [6]. From the Fig. 6a, depicts the AFM images of graphene sheets, Fig. 6b depicts the AFM images nitrogen doped graphene with the height determination. It was estimated that mean thickness of graphene as 5 nm while nitrogen doped graphene mean thickness as 12 nm [18].

2.2 XPS Technique

XPS is a technique used for determining the concentration of dopants in a base material [1, 2]. Using this technique doping effect of nitrogen on graphene can be analyzed. Elemental compositions, nature of dopant with chemical and elemental states in a base material can be analyzed. Also, with the help of binding energy, bonding between C and N can be confirmed [12, 19, 20].

Figure 7a and b shows the deconvoluted peaks of supercritical fluids assisted dimethyl glyoxime doped graphene oxide for nitrogen (N 1s) and carbon (C 1s) energy levels. Similarly, Fig. 7c and d shows the deconvoluted peaks of hydrothermal assisted dimethyl glyoxime doped graphene oxide for nitrogen and carbon energy levels. Figure 7a shows four binding energy peak values at 398.9, 399.6, 400.4 and 401.2 eV corresponding to pyridinic-N, nitrile-N, pyrrolic-N, and quaternary-N atoms at lattice. Similarly, Fig. 7c shows three different binding energy peaks at 399.5, 400.6 and 401.3 eV nitrile-N, pyrrolic-N, and quaternary-N atoms at lattice. From these peaks, it is clear that nitrogen doped graphene oxide can be distinguished by two methods based on binding energies [21–23]. Among the obtained peaks, the pyrrolic-N peak was prominent, indicating that the effect of N-doping is in the form of pyrrolic nitrogen. The bonding of a nitrogen atom with two carbon atoms in pyrrolic nitrogen results in the addition of two p electrons to the π system. However, in the case of pyridinic nitrogen, this results in the addition of one p electron to the π system. In nitrile nitrogen, three atoms are bonded to carbon with nitrogen [21, 24].

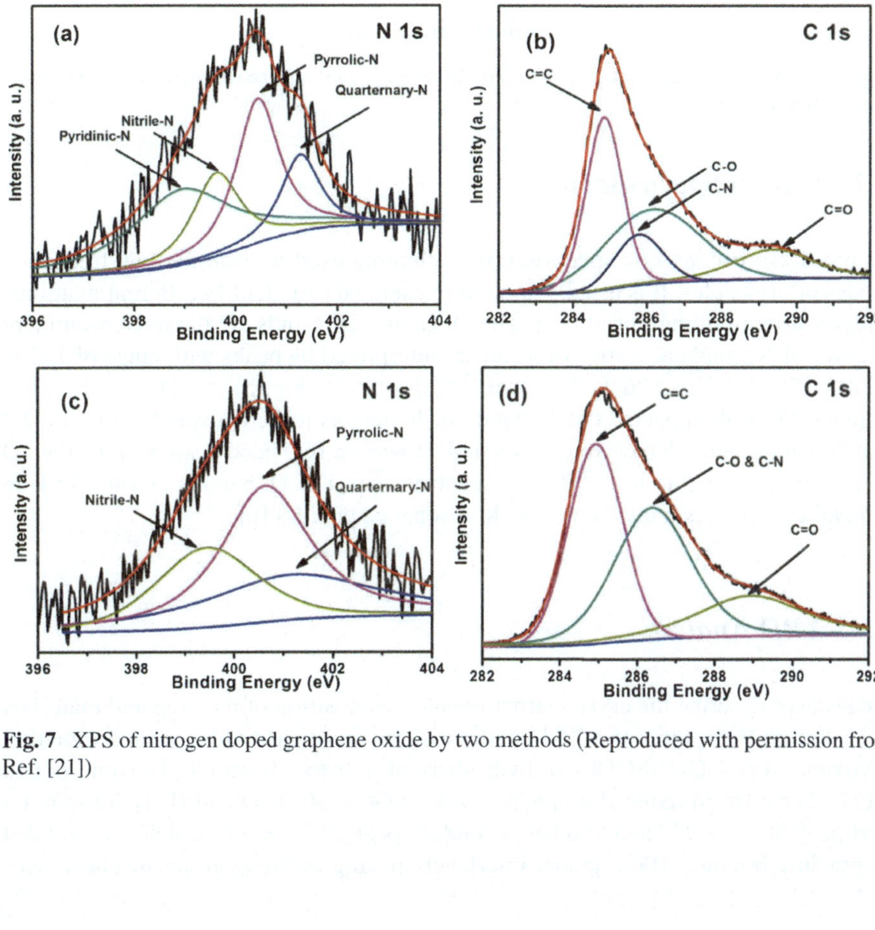

Fig. 7 XPS of nitrogen doped graphene oxide by two methods (Reproduced with permission from Ref. [21])

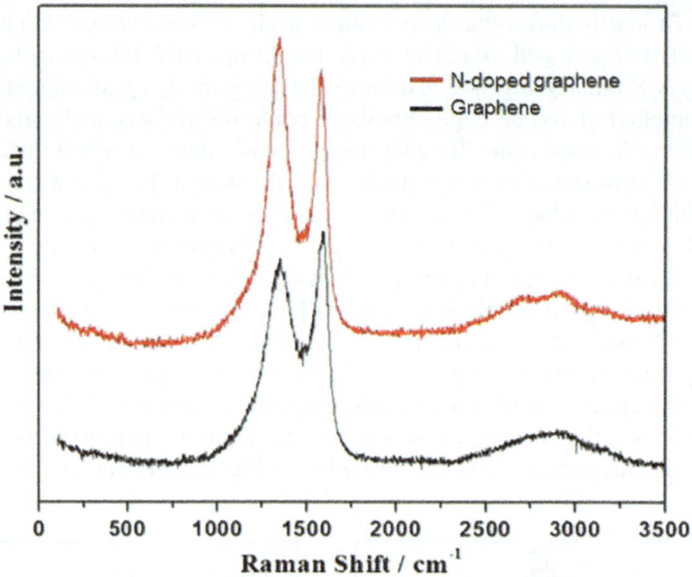

Fig. 8 Raman spectroscopy of graphene and nitrogen doped graphene (Reproduced with permission from Ref. [12])

2.3 Raman Spectroscopy

Raman spectroscopy is a spectroscopic technique used to examine low-frequency modes of molecules. It is usually applied in chemistry to identify different materials with its structural fingerprint [12]. The D, G, and 2D bands are the most prominent features of N-graphene's spectrum and are interpreted by peaks with range of 1320–1350, 1570–1585 and 2640–2680 cm^{-1} respectively (Fig. 8) [6]. Due to defect in D mode D′ peak appears. The D band needs defects to be activated while the 2D band do not require defects to be activated. However, if D band is not visible, the 2D band is always visible in the Raman spectroscopy. The D′ band is produced by an intravalley which is induced by double-resonance process [6].

2.4 XRD Analysis

XRD curve describe the crystal structure and composition of the prepared materials [12]. Figure 9 explores the XRD pattern of supercritical fluids assisted dimethyl glyoxime doped GO (SCDGO), hydrothermal assisted dimethyl glyoxime doped GO (HTDGO) and reduced graphene oxide (RGO). SCDGO and HTDGO shows a first peak at $2\theta = 25°$ and another secondary peak at $2\theta = 43°$ which is correlated to graphite 002 and 100th plane. Interlayer spacing between graphene sheets was

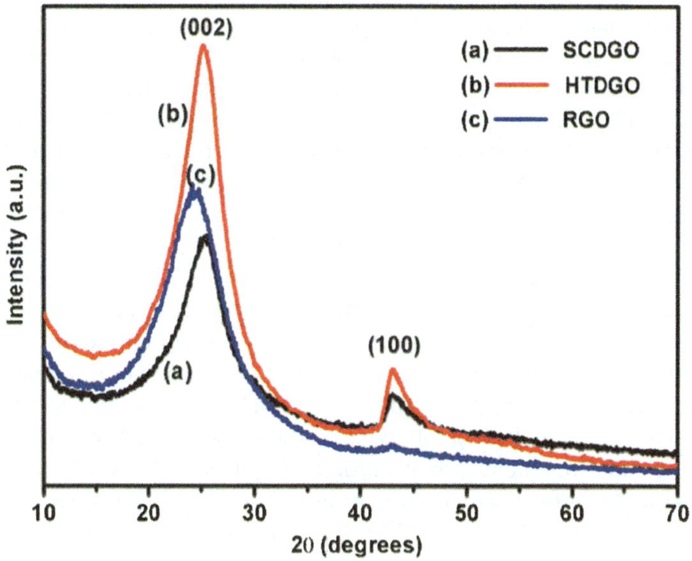

Fig. 9 XRD pattern of SCDGO, HTDGO and RGO (Reproduced with permission from Ref. [21])

calculated as 0.36 nm for 002 plane and is larger when compared to interlayer spacing of graphite (0.33 nm). Whereas interlayer spacing of SCDGO and HTDGO was less when compared to RGO (0.37 nm). The absence of peak at $2\theta = 12°$ in both SCDGO and HTDGO materials is similar to of RGO. This clearly describes that GO was reduced in both the materials [21].

3 Applications

3.1 Supercapacitors

Supercapacitors store energy in the terms of electrochemical form. Storage and discharge of energy occurs in between electrode and electrolyte via reversible adsorption and desorption of ions. Supercapacitors are also known as ultracapacitor or electrochemical capacitors which has high capacitance value with lesser voltage limits. Supercapacitor are used bridge the space among electrolytic capacitors and rechargeable batteries which will store 10 to 100 times than electrolytic capacitors. Supercapacitors are classified based on the mechanism of storage of energy such as pseudo and electrical double-layer capacitors [12, 26]. It is especially used in rapid charging and discharging applications like regenerative braking in automobiles, cranes, mobile electronics, hybrid vehicles, etc. [27]. Electrochemical capacitors are preferred in various applications due to its high power, high effectivity in repeated charging cycles, energy density and low cost [1, 27]. Carbon grounded supercapacitors demonstrate excellent capacitive capacity, electrical conductivity and mechanical behavior due to larger surface area [1, 28, 29]. Since graphene has also larger surface area and good electron movement in room temperature, it is also widely used in supercapacitors. Carbon nanomaterials are also used as electrode materials to enhance the performance of supercapacitors [1, 30]. First supercapacitor developed with graphene has shown that specific capacitance can reach 117 F/g and 75 F/g in aqueous and ionic electrolytic solutions [31].

N-doped graphene typically has a wide surface region, and it has been observed that the functionalization of N may alter the characteristics of the graphene donor/acceptor [12, 32, 33]. High structural defects were produced due to N-doping and these defects help in energy storage. These qualities lead to enhancement in capacitive efficiency of N-doped graphene [12]. 3D N-doped graphene with carbon nanotubes were developed using following methods such as hydrothermal treatment, freeze drying and subsequent carbonization of graphene oxide. Using the developed material as electrode in supercapacitor it resulted in good capacitive capacity, 180 F/g specific capacitance and retained 96% of capacitance after 3000 cycles. Hence author has concluded that developed material was an assuring material for energy storage applications [34].

Nitrogen doped graphene was prepared using polyethylene terephthalate (PET) material with urea at various temperatures. Specific capacitance, energy density and maximum power density was observed that 405 F/g, 68.1 W h/kg and 558.5 W/kg respectively. After 5000 cycles of charging cycles 87.7% of capacitance was retained and thus author has concluded that n-doped graphene prepared with PET could be used as an efficient energy storage material [35]. Nitrogen doped graphene was also developed using supercritical fluid assisted processing and hydrothermal heat treatment techniques. Specific capacitance and specific capacity retention were estimated as 286 F/g and 100% for 1000 cycles for the sample prepared using fluid assisted processing technique [21]. Nitrogen-doped graphene with a versatile coating tantalum foil is made as an aluminum-ion hybrid supercapacitor cathode. The capacitance and coulombic performance of nitrogen-doped graphene cathode was observed as 130 F/g and 100% for 2000 life cycle at a higher current density. Aluminium ion hybrid supercapacitors developed with nitrogen doped graphene shows superior energy storage mechanism when compared with double layered capacitance [18].

3.2 Lithium Ion Batteries

Because of its low reduction potential (3.04 V) and extremely high theoretical power (3400 mAh/g), lithium metal is regarded as the primary anode materials. [29, 30]. However, the use of lithium metal is dangerous owing to electrochemically instability with organic electrolytes. Since Sony's first commercial application in 1991, graphene has been a viable and alternative option for anode material in lithium-ion batteries [36]. Lithium-ion batteries was preferred among all other batteries due to high electrical properties with good surface area, reversible capacity enhanced mechanical behavior and long-life cycle. Evolution of alternative cathode material for lithium-ion batteries becomes essential in-order to develop sustainable and low-cost lithium-ion batteries [37]. However, nitrogen doped graphene has reversible capacity of 1013–1054 mAh/g at lower charging rate but still it has high discharge rate as 500 mA/g [1, 38, 39]. Reversible discharge capacity was doubled owing to the elevated Lithium-ion interaction when nitrogen doped graphene is used with lithium batteries [1, 40]. N-doped graphene nanosheets were employed as anode material in lithium batteries which displayed superior charging cycles. Specific capacity of such batteries was evaluated at 100th and 501st cycle as 684 and 452 mAh/g respectively. Higher specific capacity was correlated to higher charging cycles which was attributed to larger structural defects due to doping of nitrogen in graphene nanosheets [23].

N-doped porous graphene material was developed using freeze drying method which possess higher surface area with 1170 m^2/g. Prepared N-doped porous graphene was used as a cathode in lithium-ion batteries and shown higher charging cycle stability, discharge capacity as 672 mAh/g with a current density of 100 mA/g. Discharge capacity of N-doped lithium-ion batteries was much higher than N-doped porous graphene material, this was due to the superior interaction between N-doped graphene and lithium-ion material and 3D porous graphene structure [41].

N-doped graphene with functionalization groups was used as a cathode material for lithium-ion batteries. It was observed that reversible capacity was retained as 344 and 146 mAh/g after 200 and 1000 charging cycles respectively with a current density of 50 mA/g and 1 A/g. From the results, author concluded that nitrogen doping increases the cathode performance of lithium-ion batteries [37]. Nitrogen doped graphene hybrid nanosheets was developed using one step pyrolysis technique demonstrates a static reversible specific capacity of 741.8 mAh/g when used as anode in lithium-ion batteries. It is also observed that nitrogen doped graphene hybrid nanosheets has excellent electrochemical stability of 90.38% after 1000 cycles with 5 A/g current capacity [42].

3.3 Field Effect Transistor

In-order to have electricity flow in semiconductor material, heat or light absorption activation is required to fill the gap between valance bond and conduction band [1, 43]. If such kind of semiconductor is activated by external energy then it is referred as field emission transistor (FET). Field emission properties of heteroatoms doped graphene are similar to those of carbon nanotubes. N-doped graphene possess low current density at which field emission begins [40, 44]. The efficient nitrogen doping effect shifts the dirac point to about −140 V. Appropriate annealing allows an increased electron mobility of 1150 $cm^2/V\,s$ at a low temperature of 80 °C [45]. Negative differential resistance of double gate graphene field effect transistor achieved a peak to valley current ratio as 105 with −0.6 μA at room temperature. Author has also simulated of field effect of negative differential resistance to use it electronic applications [46].

Nitrogen doped graphene prepared by in-situ technique with pristine was analyzed with transmission electron microscopy to determine the field emission characteristics. The voltage transfer required to draw a 1 nA emission current from pristine and N-doped graphene was found to be 230 and 110 V, respectively. Field emission current was increasing to approximately 6.9 μA when nitrogen doped graphene sheets undergo structural deformation or contraction or buckling effect. The effect is correlated to joule heating effect [47]. A chemical vapor deposition process was used to synthesize N-doped graphene. The recombination of carbon atoms through chemical vapor deposition process into graphene atoms was achieved by substitutional doping into the graphene lattice. Electrical measurements of N-doped graphene through substitutional doping can increase graphene's electrical properties [7].

3.4 Electrochemical Application

The chemical reaction between cathode and anode is converted into an electrical signal (current, voltage or impedance) by electrochemical instruments. An electrochemical system consists of a fuel cell that produces electricity through the reduction of oxygen content in cathode. Oxygen reduction reaction (ORR) should break the O=O bond to achieve notable current density and thereby lowering the activation energy [1]. Thus, ORR and oxygen evolution reaction (ORR) plays a critical part in several energy conversion and storage applications for example fuel cells, photocatalytic water splitting, hydrogen production from water, rechargeable batteries, etc. [48]. Platinum material was used as the catalyst material in ORR. Due to low availability of platinum material and high-cost researchers start looking for alternative material. It was found that N-doping in carbon structure increases the ORR activity. Thus N-doped graphene was found as a promising electrocatalytic material owing to increase in electron density beyond the fermi level [1]. Due to their unique electronic properties gained from coupling effect from nitrogen electrons and graphene 2D structure, it exhibited brilliant characteristics to use it in electrochemical reactions [48].

Nitrogen doped graphene was used as electrocatalyst and lowest over-potentials in OER and ORR were evaluated as 0.405 and 0.445 V respectively. Those values are also comparable to results obtained while platinum was used as a catalyst. Thus, author suggested that N-doped graphene can be applied as a catalyst for batteries, water splitting and fuel cells [48]. Thermal chemical vapour synthesis of single layer graphene doped with pure pyridinic nitrogen via deposition of hydrogen and ethylene on Cu foils in front of ammonia. Through UV photoemission spectroscopy analysis, it was found that valance gap of graphene structure was altered and current density increases nearer to the fermi level. So, author concludes that nitrogen doping in carbon materials increases the ORR activity [9]. Nitrogen doping of graphene is also synthesized using nitrogen plasma treatment. Nitrogen doped graphene has showed superior electrocatalytic activity to reduce hydrogen peroxide and to transfer energy to glucose oxidation. Thus, author concluded that N-doped graphene can be employed for glucose biosensing applications at low concentration 0.01 mM [17].

N-doped with graphene has been formulated for its use in acidic media in ORR. In addition to that boron and phosphorus atom is also doped along with nitrogen into graphene, to improve the operation of ORR. N-doped graphene, boron-nitrogen doped graphene, phosphorus-N-doped graphene displays onset potential as 0.84, 0.86 and 0.87 V respectively. Similarly, it was also observed that it generates mass activity as 0.45, 0.53 and 0.80 mA/mg respectively [49].

4 Summary

This chapter discusses about various properties and applications of nitrogen doped graphene.

Graphene was considered as most predominant materials to use in new model electronic devices owing to its large surface area, thermal conductivity and Young's modulus. Chemical doping of graphene using various elements like nitrogen, boron, hydrogen, etc., can leads to fresh electronic band structure. Doping of graphene by nitrogen (N) or boron (B) atoms can obtain n-type or p-type semiconducting graphene. N or B doped graphene materials has its unique characteristics. It is also proved that N-doped graphene possess higher electro-catalytic activity when compared with other electro-catalyst. Nitrogen doped graphene was used in various application such as nanoelectronics, nanophotonics, supercapacitors, fuel cells, batteries, medical domain, electro-catalytic applications. Despite the number of methods are available to produce N-doped graphene, CVD was most commonly preferred method to grow graphite or the doped graphite.

The STM tool is used to image the surfaces of materials at the atomic level. Transformation of graphene in to nitrogen doped graphene can be examined using TEM images. SEM tool was utilized to analyze the morphological properties of N-doped graphene. AFM was used to determine the thickness as well as number of layers of graphene and structure. Dopant concentration on base material was studied using XPS. Raman spectroscopy technique used to examine the low-frequency modes of molecules. XRD patterns describe the crystal structure and composition of the N-doped materials. Those characterization techniques were used to examine the technical properties of N-doped graphene.

Numerous research works have been executed using N-doped graphene material. Supercapacitors, lithium-ion batteries, field effect transistor and in electrocatalyst applications N-doped graphene material was used and enhancement in corresponding technical properties is described in this chapter.

References

1. Yadav, R. & Dixit, C. Synthesis, characterization and prospective applications of nitrogen-doped graphene: A short review. J.Sci-Adv. Mater. Dev. **2**, 141–149 (2017).
2. Rao, C., Gopalakrishnan, K. & Govindaraj, A. Synthesis, properties and applications of graphene doped with boron, nitrogen and other elements. Nano Today. **9**, 324–343 (2014).
3. Rao, C. N. R., Sood, A. K., Subrahmanyam. K. S. & Govindaraj, A. Graphene: The new two-dimensional nanomaterial. Angew. Chemie-Int. Ed. **48**, 7752–7777 (2009).
4. Geim, A. K. & Novoselov, K. S. The rise of graphene. Nanoscience and technology: A review. Nanosci. Technol. 11–19 (2010).
5. Balandin, A. A. et al. Superior thermal conductivity of single-layer graphene. Nano Lett.**8**, 902–907 (2008).
6. Wang, H., Maiyalagan, T. & Wang, X. Review on recent progress in nitrogen-doped graphene: Synthesis, characterization, and its potential applications. ACS Catal. **2**, 781–794 (2012).
7. Wei, D. et al. Synthesis of N-doped graphene by chemical vapor deposition and its electrical properties. Nano Lett. **9**, 1752–1758 (2009).
8. Panchakarla, L. et al. Synthesis, structure, and properties of boron- and nitrogen-doped graphene. Adv. Mater. **21**, 4726–4730 (2009).
9. Luo, Z. et al. Pyridinic N doped graphene: synthesis, electronic structure, and electrocatalytic property. J. Mater. Chem. **21**, 8038–8044 (2011).
10. Ito, Y. et al. Bicontinuousnanoporous N-doped graphene for the oxygen reduction reaction. Adv. Mater. **26**, 4145–4150 (2014).
11. Mukherjee, S. & Kaloni, T. Electronic properties of boron- and nitrogen-doped graphene: A first principles study. J. Nanopart. Res. **14**, 1–5 (2012).
12. Xu, H., Ma, L. & Jin, Z. Nitrogen-doped graphene: Synthesis, characterizations and energy applications. J. Energy Chem. **27,** 146–160 (2018).
13. Lu, Y., Huang, Y., Zhang, M. & Chen, Y. Nitrogen-doped graphene materials for supercapacitor applications. J. Nanosci. Nanotechnol. **14,** 1134–1144 (2014).
14. Liu, P. et al. Synthesis of lightweight N-doped graphene foams with open reticular structure for high-efficiency electromagnetic wave absorption. Chem. Eng. J, **368,** 285–298 (2019).
15. Zheng, B., Hermet, P. & Henrard L. Scanning tunneling microscopy simulations of nitrogen- and boron-doped graphene and single-walled carbon nanotubes. ACS nano. **4**, 4165–4173 (2010).
16. Guillaume, S. O., Zheng, B., Charlier, J. C. & Henrard, L. Electronic properties and STM images of doped bilayer graphene. Phys. Rev. B, **85,** 035444 (2012).
17. Wang, Y., Shao, Y., Matson, D. W., Li, J.& Lin, Y. Nitrogen-doped graphene and its application in electrochemical biosensing. ACS nano. **4**, 1790–1798 (2010).
18. Lei, H., Tu, J., Tian, D. & Jiao S. A nitrogen-doped graphene cathode for high-capacitance aluminum-ion hybrid supercapacitors. New J. Chem. **42,** 15684–15691 (2018).
19. Yang, D. et al. Chemical analysis of graphene oxide films after heat and chemical treatments by X-ray photoelectron and Micro-Raman spectroscopy. Carbon. **47**, 145–152 (2009).
20. Lin, Z., Waller, G. H., Liu, Y., Liu, M. & Wong, C. 3D Nitrogen-doped graphene prepared by pyrolysis of graphene oxide with polypyrrole for electrocatalysis of oxygen reduction reaction. Nano Energy **2**, 241–248 (2013).
21. Balaji, S. S., Elavarasan, A. & Sathish, M. High performance supercapacitor using N-doped graphene prepared via supercritical fluid processing with an oxime nitrogen source. Electrochim. Acta. **200,** 37–45 (2016).
22. Lee, K. H., Oh, J., Son, J. G., Kim, H. & Lee, S. S. Nitrogen-doped graphene nanosheets from bulk graphite using microwave irradiation. ACS Appl. Mater. Interfaces **6**, 6361–6368 (2014).
23. Li, X. et al. Superior cycle stability of nitrogen-doped graphene nanosheets as anodes for lithium ion batteries. Electrochem. Commun. **13,** 822–825 (2011).
24. Deng, D. et al. Toward N-doped graphene via solvothermal synthesis. Chem. Mater. **23**, 1188–1193 (2011).

25. Mou, Z., Chen, X., Du, Y. & Wang, X. Forming mechanism of nitrogen doped graphene prepared by thermal solid-state reaction of graphite oxide and urea. Appl. Surf. Sci. **258**, 1704–1710 (2011).
26. Simon, P. & Gogotsi, Y. Materials for electrochemical capacitors. Nanosci. Technol. 320–329 (2010).
27. Conway, B. E. Transition from "supercapacitor" to "battery" behavior in electrochemical energy storage. J. Electrochem. Soc. **138**, 1539 (1991).
28. An, K. H, et al. Electrochemical properties of high-power supercapacitors using single-walled carbon nanotube electrodes. Adv. Funct. Mater. **11**, 387–392 (2001).
29. Wang, Y.et al. Preventing graphene sheets from restacking for high-capacitance performance. J. Phys. Chem. C. **115**, 23192–23197 (2011).
30. Dreyer, D. R., Murali, S., Zhu, Y., Ruoff, R. S. & Bielawski, C. W. Reduction of graphite oxide using alcohols. J. Mater. Chem. **21**, 3443–3447 (2011).
31. Vivekchand, S., Rout, C. S., Subrahmanyam, K., Govindaraj, A. & Rao, C. Graphene-based electrochemical supercapacitors. J Chem Sci. **120**, 9–13 (2008).
32. Jeong, H. M. et al. Nitrogen-doped graphene for high-performance ultracapacitors and the importance of nitrogen-doped sites at basal planes. Nano Lett. **11,** 2472–2477 (2011).
33. Zhang, Y., Cao, B., Zhang, B., Qi, X. & Pan, C. The production of nitrogen-doped graphene from mixed amine plus ethanol flames. Thin Solid Films **520**, 6850–6855 (2012).
34. You, B., Wang, L., Yao, L. & Yang, J. Three dimensional N-doped graphene-CNT networks for supercapacitor. Chem. Comm. **49,** 5016–5018 (2013).
35. Elessawy, N. A., El Nady, J., Wazeer, W. & Kashyout, A. Development of high-performance supercapacitor based on a novel controllable green synthesis for 3D nitrogen doped graphene. Sci. Rep. **9**, 1–10 (2019).
36. Ikram, R. et al. Recent advancements of n-doped graphene for rechargeable batteries: A review. Crystals, **10**, 1080 (2020).
37. Xiong, D. et al. Superior cathode performance of nitrogen-doped graphene frameworks for lithium ion batteries. ACS Appl. Mater. Interfaces. **9**, 10643–10651 (2017).
38. Pan, D. et al. Li storage properties of disordered graphene nanosheets. Chem. Mater. **21**, 3136–3142 (2009).
39. Wu, Z. S., Ren, W., Xu, L., Li, F. & Cheng, H. M. Doped graphene sheets as anode materials with superhigh rate and large capacity for lithium ion batteries. ACS nano. **5**, 5463–5471 (2011).
40. Palnitkar, U. et al. Remarkably low turn-on field emission in undoped, nitrogen-doped, and boron-doped graphene. Appl. Phys. Lett. **97**, 063102 (2010).
41. Sui, Z. Y. et al. A highly nitrogen-doped porous graphene-an anode material for lithium ion batteries. J. Mater. Chem. A. **3**, 18229–18237 (2015).
42. Yang, J. et al. Fabrication of nitrogen-doped porous graphene hybrid nanosheets from metal-organic frameworks for lithium-ion batteries. Nanotechnology **31**, 145402 (2020).
43. Avouris, P., Chen, Z. & Perebeinos, V. Carbon-based electronics. Nanosci. Technol., 174–184 (2010).
44. Sharma, R., Late, D., Joag, D., Govindaraj, A. & Rao, C. Field emission properties of boron and nitrogen doped carbon nanotubes. Chem. Phys. Lett. **428**, 102–108 (2006).
45. Xu. W. et al. N-doped graphene field-effect transistors with enhanced electron mobility and air-stability. Small, **10**, 1999–2005 (2014).
46. Tiwari, D. L. & Sivasankaran, K. Nitrogen-doped NDR behavior of double gate graphene field effect transistor. Superlattice Microst. **136**, 106308 (2019).
47. Kashid, R. V. et al. Field emission characteristics of pristine and N-doped graphene measured by in-situ transmission electron microscopy. J. Appl. Phys. **113**, 214311 (2013).
48. Li, M., Zhang, L., Xu, Q., Niu, J. & Xia, Z. N-doped graphene as catalysts for oxygen reduction and oxygen evolution reactions: Theoretical considerations. J. Catal. **314,** 66–72 (2014).
49. Choi, C. H., Chung, M. W., Kwon, H. C., Park, S. H. & Woo, S. I. B, N-and P, N-doped graphene as highly active catalysts for oxygen reduction reactions in acidic media. J. Mater. Chem. A. **1**, 3694–3699 (2013).

50. Rajasekar, R., Nayak, G.C., Malas, A., Das, C.K.M.D. Development of compatibilized SBR and EPR nanocomposites containing dual filler system. **35**,878–885 (2012).
51. Nayak, G.C., Rajasekar, R., Bose, S., Das, C.K.J.N.T. Effect of MWNTs and SiC-coated MWNTs on properties of PEEK/LCP blend. (2009).
52. Nayak, G.C., Sahoo, S., Rajasekar, R., Das, C.K.C.P.A.A.S.M. Novel approach for the selective dispersion of MWCNTs in the Nylon/SAN blend system. **43**(8), 1242–1251 (2012).

Graphene-Based Polymer Composites: Physical and Chemical Properties

Srikanta Moharana, Bibhuti B. Sahu, Lipsa Singh, and Ram Naresh Mahaling

Abstract The graphene-based polymer composites are of immense interest for their end-use applicability in the field of electromagnetic interference shielding devices, tissue engineering, sensor, power storage, supercapacitors, and energy storage devices. Graphene oxide is one of the finest nanomaterials with outstanding physical and chemical properties for the choice of scientific and engineering applications. The present chapter is focused mainly on two categories. In the first category synthesis technique is based on electrospinning for the fabrication of graphene-reinforced polymeric composites. In the second one, we have primarily emphasized graphene-based composites with many organic and polymeric materials including polyvinyl alcohol (PVA), poly(vinylidene fluoride) (PVDF), epoxy, polystyrene (PS), polypropylene (PP), polyimide (PI), polyurethane (PU), polyaniline (PANI), polypyrrole, and polythiophene in more detail. In addition, the thermal, mechanical, and electrical properties of these graphene-based polymeric composites have been discussed in a lucid manner. The concluding section of this current chapter throws light on the current challenges and opens the path for these new promising composite materials for their technological applications as per the contemporary demands.

Keywords Graphene · Polymer · Composites · Thermal properties · Dielectric properties · Energy storage

S. Moharana (✉) · L. Singh
School of Applied Sciences, Centurion University of Technology and Management, Bhubaneswar, Odisha, India

B. B. Sahu
Department of Physics, Veer Surendra Sai University of Technology, Burla, Odisha, India

R. N. Mahaling (✉)
Laboratory of Polymeric and Materials Chemistry, School of Chemistry, Sambalpur University, Jyoti Vihar, Burla, Odisha 768019, India
e-mail: rnmahaling@suniv.ac.in

S. Sahoo et al. (eds.), *Defect Engineering of Carbon Nanostructures*, Advances in Material Research and Technology, https://doi.org/10.1007/978-3-030-94375-2_7

1 Introduction

During the past decade, carbon-based polymer composite materials (graphene oxide) are of great importance for several strategic technologically advanced applications like electronic devices, electromagnetic interference shielding devices, tissue engineering, sensor, power storage, supercapacitors, etc. [1–5]. Recently, academic researchers have examined materials with enhanced properties of composite systems that are dimensionally more appropriate in the areas of modern science and technology and found carbon-based polymeric materials are of a wide range of potential applicability due to their structural characteristics. Typical carbon-based materials such as graphite, carbon nanotubes (CNTs), and graphene, in particular, have tremendous attention due to their excellent thermal, electrical, and mechanical performance. Graphite is considered to be the most crystalline form of the element in which carbon atoms are arranged in a hexagonal crystal structure [5–11]. However, the single-phase graphite (act as filler) cannot accomplish the high energy storage density while maintaining satisfactory dielectric and mechanical breakdown strength. As a result, small quantities of fillers are incorporated into the polymer matrix to constitute the graphite-based polymer composites [11–13]. These polymeric composites comprising graphene-based fillers have opened a scope of promising fields for material science and engineering due to their lightweight, easy processing, high specific strength, and exceptional dielectric and electrical performance [14, 15].

Graphene has attracted immense interest due to its extraordinary properties with respect to the combination of physics, chemistry, and materials science. Graphene is the most significant among the different members of the carbon family having sp^2 hybridized two-dimensional structures with honeycomb crystal lattice [16–18]. For instance, Geim et al. [17, 19] have successfully isolated single-layered transferable graphene nanosheets by using a peeling process involving the scotch tape method. These graphene-based composites have exceptional properties, including high current density, chemical inertness, high thermal conductivity, excellent optical transmittance, high carrier mobility, and high mechanical strength with superior Young's modulus. Initially, graphene was taken out from graphite using the micromechanical cleavage technique. This technique is usually used for easy production of high-quality graphene and to promote more experimental activities [19]. Further, the combination of graphene with polymer matrix composite is an excellent candidate for better electrical performance used as a sensor because of its 2D structure, which makes it very efficient to detect absorbed molecules. Besides, the high electrical conductivity and optical transparency of graphene support as superior candidates for end-use practical applicability in the field of conducting electrodes, touch screens, organic light-emitting diodes (OLEDs), and organic photovoltaic cells [20, 21].

Similarly, the Toyota research group [22] reported on the discovery of polymer-based composite, which gave a new dimension in the field of material science. Specifically, the inorganic filler-based composite materials (inorganic polymer) have attracted increasing attention due to their distinctive properties and wide potential applications in aerospace, automotive, construction, and electrical and electronic industries [3–5, 20, 21]. Consequently, most academic and industrial researchers have developed polymer nanocomposites based on layered materials such as montmorillonite type [layered double hydroxide (LDH)]. The thermal and electrical conductivity of clay minerals is reasonably poor [23, 24]. Thus, to overcome these limitations, carbon-based materials such as carbon nanotubes (CNTs), carbon nanofibers (CNF), carbon black (CB), etc. have been incorporated into polymer matrix for preparing polymer composites [25–28]. In the previous literatures, it has been reported that carbon nanotubes (CNTs) are especially used as conductive fillers (nanofillers) but their production cost is high [25–30]. So it is very difficult for mass production of CNT-based functional composite materials. Interestingly, graphene has been chosen as alternative nanofillers to overcome the limitations of conventional nanofillers (Na-MMT, LDH, CNT, CNF, etc.) due to high surface area, aspect ratio, good flexibility, thermal and electrical conductivity, and low-cost production [23–31].

The significant performances of graphene compared to the neat polymer are remarkable when it becomes polymer-filled graphene composites. Polymer-based graphene nanocomposites exhibit better mechanical, electrical, gas barrier, thermal, and flame-retardant properties with respect to pristine polymer matrix. However, it has been reported that the enhancement of electrical and mechanical behavior of graphene-based polymer composites is much superior to that of other carbon filler-based polymer nanocomposites [6–11, 32]. These enhancements in the behavior of resultant composites depend on the distribution of graphene layers in the polymer matrix as well as interfacial bonding between the graphene layers and polymer matrices. This interfacial interaction between graphene and pristine polymer may give rise to the final properties of the graphene-reinforced polymer nanocomposites. However, pure graphene is not compatible with organic polymers and also exhibits in-homogeneity in the composites. On the other hand, graphene oxide surface shows various kinds of oxygen-containing functional groups (including hydroxyl, epoxy, carboxyl, etc.) that alter the van der Waals interaction considerably, which makes good compatibility with organic polymeric materials [33–36]. In addition, there are some additional carbonyl and carboxyl functional groups placed in the sheets, resulting in strong hydrophilic graphene oxide, which readily swells and disperses in water. Therefore, graphene oxide has drawn considerable interest as nanofillers for polymer nanocomposites [21]. The graphene oxide sheets are commonly dispersed in aqueous media, which are incompatible with the organic polymers. Thus, graphene and graphene oxides are electrically insulating, which are incompatible with the fabrication of conducting polymer-based composites [37]. Further, surface modification is the crucial aspect for the improvement of the physical and chemical properties as well as a molecular level of distribution, which have a positive influence on the application of graphene reinforcing polymers. As a result, it is of enormous significance to fabricate polymer–graphene nanocomposites with improved strength and toughness simultaneously.

Because of scientific interest and potential applications of graphene-based polymer composite, research has been increased to a surprising dimension, opening new challenges and opportunities for better performance as well as end-use practical applicability in the various technological fields. This typical chapter is mainly divided into two categories. The first one is particularly focused on the electrospinning technique of graphene–polymer composites; on the other hand, the second one covers properties and its various polymer-based graphene composites for energy storage and various important applications. Finally, a conclusion has been drawn on this new and exciting field of research of graphene-based polymer composites, followed by summary and perspective.

2 Synthesis of Graphene-Based Polymer Composites

2.1 Electrospinning Technique

Electrospinning technique is a simple, straightforward, and versatile method to synthesize one-dimensional (1D nanofibers) and two-dimensional (2D, nanowoven fabric) nanostructured materials [38–43]. Moreover, it is a multipurpose synthesis method for the fabrication of polymer nanofibers with precise diameter in the region of nanometers to micrometers [44]. There are various electrospun functional polymer nanofibers that can be reinforced with carbon-based materials, including CNTs, graphenes, nanodiamonds, nanodots, etc., which can significantly improve the properties (thermal, mechanical, dielectric, and electrical) of the composites and also leads to the potential applications, especially in the field of biology and sensors [38–44]. Among various methods of synthesis, electrospinning is a well-established technique for the fabrication of nanosized polymer fiber-based materials [44]. This technique has earned popularity due to its simplicity and implementation in diverse areas of tissue engineering, chemical and biological sensing, and energy sectors. The concept of electrospinning was observed from the study of electrospraying technique by Rayleigh and Zeleny in the last century [45]. Before a couple of decades, the term electrospinning was coined, due to its versatility and easy processability; this technique was rapidly established in different fields. Moreover, with due course of time, it is upgraded and modified for effective mass production of nanosized fibers. Scientists have tried to improve the physicochemical properties of electrospun material. Poly(sulfone amide) (PSA) has been used as a spinnable polymer recognized for its exceptional thermal properties (heat resistance, flame retardant, thermal stability, etc.). Moreover, it is also used for developing products for protective use in aerospace, high-temperature environment, and civil fields [46, 47]. The electrospun PSA nanofiber is unique for retaining superior performances in terms of crystallization, thermal and mechanical properties, and its nanostructure [48]. There are various advanced techniques relating to electrospinning, like coaxial electrospinning, mixed and multilayer electrospinning, forced air-assisted electrospinning, and air gap electrospinning, which have emerged for enhancing the properties of electrospun-nanosized fibers [49]. The main components of electrospinning processing setup comprise a high-voltage power supply, spinneret (commonly blunt

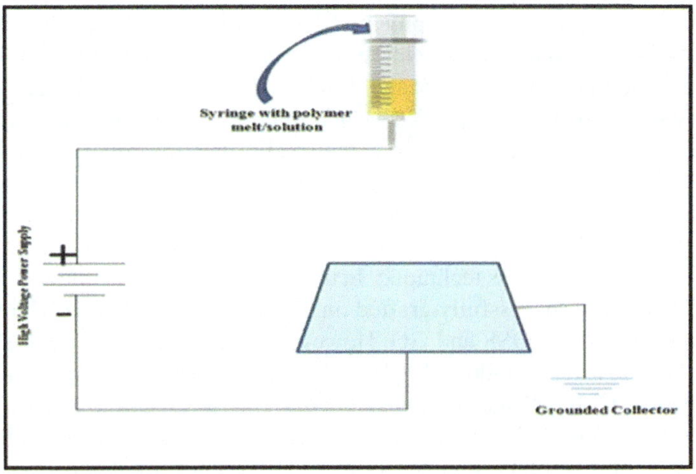

Fig. 1 Electrospinning experimental setup

needled syringe), and grounded collector. This unique technique for manufacturing continuous fiber from polymer melt/solution follows the principle of surface tension acting on polymer drops. The basic process of production is first initiated by placing the solution into the syringe held by a retort stand or by a syringe pump. Then the tip of the needle of the syringe is connected with a positive terminal of the voltage supply. Out of which, the polymer liquid is pumped by the needle to the collector (basically grounded) below the syringe [50]. The experimental setup is shown in Fig. 1.

There are various studies available based on graphene-based polymer composites by using electrospinning technique reported by several researchers. For instance, Bao et al. [51] have successfully prepared electrospun nanocomposites comprising polyvinyl acetate and graphene oxide (GO) and significant enhancement in optical properties was observed. Das et al. [52] have reported on loading of electrospun nanofibers on functionalized graphene for increasing crystallinity and superior thermal stability. Kim et al. [53] have been successful in synthesizing silica-encapsulated carbon nanofiber composites from a graphene–polyacrylonitrile (PAN) solution by the application of electrospinning technique. Further, using electrospinning nylon-6 shows interesting modified physical properties as well as fiber morphology because of polyelectrolytic and polymorphic nature [53]. Correa et al. [54] have fabricated rGO-polycaprolactone (PCL) composites with various weight percentages of rGO contents (0.25, 0.5, 0.75, and 1 wt%) by using electrospinning technique. In this composite system, two different voltage setups (10 and 15 kV) and distance of 10 cm are commonly used for electrospinning. The thermal, mechanical, microstructure, electrical, porosity, and absorption water analysis of the rGO-PCL composites were also made to the scaffold. However, a nanocomposite with electrospun at 10 kV exhibits enhancement of mechanical performance with an increase in the 190% of Young's modulus than that of the nanocomposites without rGO. On the other hand, the samples of electrospun at 15 kV signify deterioration with the incorporation of rGO but simultaneously increase the electrical conductivity and porosity. Besides, the incorporation of 0.75 and 1 wt% of rGO-reinforced PCL nanocomposite shows detriment on properties, which may be attributed to the formation of aggregation of particles into the polymer matrix. Further, the effect of voltage on the electrospinning technique plays a significant contribution toward the final properties of the rGO-reinforced PCL nanocomposites. Guo et al. [55] reported that the thermally conductive CMG-polyimide nanocomposites by using in situ polymerization and electrospinning hot press technique. In these nanocomposites, the NH_2-modified POSS molecules are successfully grafted on the surface of GO and CMG through a reaction between NH_2-POSS and GO. However, the coefficient of thermal conductivity (I), heat resistance index (THRI), and glass transition temperature (Tg) of the synthesized CMG-PI nanocomposites with 5 wt% of CMG was significantly enhanced to 1.05 Wm^{-1} K^{-1}, which is about four times larger than that of the neat PI matrix (0.28 Wm^{-1} K^{-1}). Further, the Tg and THRI values simultaneously improved to 213.0 °C and 282.31 °C, respectively. An enhanced thermal conductivity model was also suggested that predicted the value of coefficient of thermal conductivity in the resultant nanocomposites, which is more accurately calculated than that of the typical Maxwell, Russell, and Bruggemen classical models. Hou et al. [56] have prepared homogeneous polyamide 6 (PA6)-graphene-based nanocomposite fibers using melt spinning and drawing process. It is observed that the resultant nanocomposites, graphene sheets are well dispersed within the matrix uniformly and also increased physio-mechanical performance of the nanocomposites. The graphene was oxidized to form graphene oxide, which is then reacted with amine compound to form the graphene bonding with amine functional groups and simultaneously synthesized nanocomposite fibers. The presence of PA6 on the surface of the graphene sheet is confirmed by FTIR, TGA, and AFM analysis. Moreover, 0.1 wt% of graphene content in the nanocomposites exhibited enhanced tensile strength with an increase in the graphene loading, which is much higher than that of the pristine polymer matrix. Roostaie et al. [57] synthesized polyamide–graphene oxide nanocomposites via electrospinning technique through stainless steel wire as a robust substrate. The nanocomposites with high porosity and diameter in the range of 100–250 nm were produced, which is confirmed by using scanning electron microscopy. However, the polyamide–graphene nanocomposite is fast and sensitive that offers sub (2–5 mg L^{-1}) and low ppt detection limits (0.7–1 mg L^{-1}) at an equilibrium time of about 15 min. This technique was applied to real water samples and relative recoveries in the range of 89–101% are obtained, which is no matrix effect. Similarly, Karumuthil et al. [58] have reported electrospun poly(vinylidene fluoride-trifluoroethylene)-based zinc oxide and exfoliated graphene oxide nanocomposite fibers for the applicability in the field of piezoelectric nanogenerators via electrospinning technique. These nanocomposites show superior energy generation efficiency and potential material for energy harvesting applications. Further, the synthesized nanocomposites were characterized by scanning electron microscopy, Fourier-transform infrared spectroscopy, and X-ray diffraction technique.

2.2 Dip Coating Technique

Dip coating method is commonly utilized for the preparation of graphene-based composites when a liquid polymer is used as a matrix. In this technique, initially neat graphene foam is incorporated into the polymer solution with varying dipping time which determines the quality and formation of coating and composites. However, the curing of polymer–graphene system takes place under suitable time and temperature conditions after completing the dip coating. Various researchers have fabricated graphene-based polymer composites via dip coating technique. Neito et al. [59] synthesized graphene foam (GrF)-doped polylactic acid-poly-e-caprolactone copolymer (PCL) hybrid scaffold with an attractive performance of graphene including excellent structural properties and enables retention of the porous 3D networked structure. However, the outstanding wettability of graphene foam-doped PLC scaffold showed improved strength ($\approx$3700%) and increased ductility ($\approx$3100%). Samad et al. [60] have fabricated free-standing graphene foam (GF) using three different processing techniques, including vacuum-assisted dip-coating of nickel foam (Ni-F) with graphene oxide (GO), reduction of GO to reduced graphene oxide (rGO), and etching out the nickel scaffold. It is observed that neat GF was examined by using microstructure, chemistry, and mechanical integrity. The microstructure of the GF mimics with Ni-F, as individual bones of GF were hollow, due to the entire removal of nickel. Moreover, PDMS-GF composites were characterized by their ability to sense both compressive and bending strains with varied electrical resistance. Sun et al. [61] have prepared reduced graphene oxide (rGO) through dip coating method by using graphene oxide suspension followed by thermal annealing. The growth of GO in this technique was tested through UV–Vis, microstructure, and the structure of the neat GO films via AFM, SEM, and XPS. They revealed that GO films can grow uniform thickness on the surface of the quartz slides and a controlled microstructure of the resultant films by coating cycles and content of solid GO suspension. The best quality of rGO film was achieved when the content of solid GO suspension (0.3 mg mL^{-1}) with five coating cycles followed by annealing at 600 °C in Ar/H_2 atmosphere. The sheet resistance (60 kV%21) and the transmittance (81%) at 550 nm may be a talented approach for synthesizing neat rGO films or composite films. Cheng et al. [62] have reported highly conductive multifunctional graphene-coated glass fiber with elevated electrical conductivity via sol–gel and dip coating technique. These graphene-based glass fibers have better electrical conductivity (24.9 S/cm) than that of the nanocarbon-coated fiber and commercial carbon fiber. The properties of wettability and electrical conductivity of the coated glass fibers are strongly dependent on the dip coating times and coating thickness in association with the coverage degree and compact structure of the graphene coatings. This high conductivity of graphene-coated glass fibers has an enormous interest in the field of flexible conductive wires, highly sensitive sensors, and multifunctional fibers. However, dip coating method is effective for producing very thin membrane and useful for some properties like high adhesion as well as reproducibility. Chatterjee et al. [63] have reported that cotton-woven and knitted fabrics have electrical conductivity along with the various concentration of graphene oxide as well as the number of coating cycles. The increase in the number of reduced graphene oxide sheets is due to an increase in coats. In a reduced graphene oxide mixture, silk fabric was dipped frequently [64]. From some studies, it has been reported that graphene composites and electrical heat elements were dip-coated on non-woven fabrics [65, 66]. Preparation of high-quality graphene oxide nanocomposites was occurred by vacuum dip coating method exhibiting flux of molecular sieves mechanism, which was dependent upon temperature and pressure. To get various concentrations of graphene oxide sol on the modified γ-alumina support, the graphene oxide membrane layer was prepared by vacuum dip coating method. Dip coating procedure is exhibited in four steps. The first step is the immersion speed of the sample holder. To prevent the substrate blade from removing the graphene on the dispersion surface, it is necessary to control the immersion speed. In the second step, it exhibits the substrate's immersion time inside the dispersion. In other words, the adherence of some graphene occurred on the substrate surface. The third step is the removal speed of the substrate which is a parameter to control the deposition of graphene on the top layer of the multilayered graphene.

2.3 *Solution Casting Technique*

Solution casting method is the fabrication method commonly used for the preparation of graphene-based composites. In this method, graphene foam is completely infiltrated with the solution of polymer. The mold containing graphene is filled with polymer. In mold comprising graphene, the polymer is poured. The nodes and branches of the graphene are coated by the polymer which goes through the pore deeply. In the presence of heat, polymerization of polymer containing the embedded graphene occur and in this way, graphene-based polymer composites are formed. Various researchers have fabricated graphene-based polymer composites via solution casting technique. Thomas et al. [67] reported the effect of neat graphene which is incorporated into nitrile rubber matrix by green approach as well as two different routes of fabrication such as latex casting and dry rubber mixing. By planetary ball milling process, accomplishments of pristine few layers graphene occurred and by using X-ray diffraction, Raman spectroscopy, and transmission electron microscopy analysis, it was characterized. Then the incorporation of that produced graphene as well as that multiwalled carbon nanotubes occurred into nitrile rubber latex separately. Bahrami et al. [68] reported that by using electrospinning and solvent casting method, polyurethane-graphene is fabricated as membranes. Characterization of such membrane occurs due to some properties like electrical, mechanical, physiochemical as well as biological properties. The electrospun mats have significantly higher electro conductivity than the casting films. Electro conductivity of the composites is improved due to the graphene which acts as an electrical bridge. The mats have higher mechanical properties as compared to that of film and it is improved by increasing the concentration of graphene up to 5 wt% and then reduction occurred. Dispersion of all exfoliated graphene occurred in polyvinyl alcohol by using the solution casting technique. Polyvinyl alcohol is highly soluble in water due to the presence of hydroxyl groups in it [69]. The polymer electrolytes are prepared by the simple solution casting method. For better solubility of graphene oxide, acetone as well as 1-methyl-2-pyrrolidinone (NMP) act as a solvent in the ratio (40:60) [70, 71]. In this solvent, poly(vinylidene fluoride-co-hexafluoropropylene) PVdF-HFP having 50 wt% was dissolved and stirred for 12 h. Later, the addition of the plasticizer ethylene carbonate having 35 wt% has been done and stirred again for 2 h. Finally, the addition of ionic salt $LiClO_4$ (15 wt%) was performed to that above mixed solution and stirred for about 3 h. Then, on a glass petri dish, the final solution was casted and kept at 80 °C in the hot air oven for 24 h. Then, the membrane is exfoliated from the petri dish after keeping the membrane in a vacuum oven at a pressure of 400 mbar and 80 °C for 2 h. Yang et al. [72] reported that by solution casting method, 1 wt% graphene-oxide-doped poly(vinylidene fluoride-co-hexafluoropropylene)/EMIMBF4 ion gel polymer electrolyte was prepared for supercapacitor application. Fattah et al. [73] reported that by using solution casting method, one can prepare 2–8 wt% of graphite-oxide-doped poly(vinylidene fluoride-co-hexafluoropropylene)-EMI-BTI for the application of electron double-layer capacitor.

2.4 Spray Deposition Technique

Spray deposition method is used for the preparation of graphene-based composites in which electrostatic spray is used sometimes for the deposition of polymer matrix in the form of powder on graphene foam and thus, graphene foam-based polymer composites are formed. Various researchers have fabricated graphene-based composites via spray deposition method. Mohammed et al. [74] reported that polyaniline-graphite nanofiber nanocomposites can be prepared by spray-coated single-mode fiber sensor at room temperature. The ratio of polyaniline/graphite nanofiber nanocomposites polymer film is about 2:1. By differing the diameter of waist single-mode fiber which is spray-coated with polyaniline/graphite nanofiber, a comprehensive investigation in which polyaniline/graphite nanofiber nanocomposites-coated single-mode fiber, as well as etched-tapered single-mode fiber sensor, was performed at room temperature. Soltani et al. [75] reported that by using conventional as well as substrate vibration-assisted ultrasonic spray coating, a highly conductive transparent graphene-doped PEDOT:PSS composites thin film can be fabricated. Liquid droplets of a precursor can be controlled by electrostatic spray deposition method by applying electrostatic force at bank as well as microscale droplet size [76, 77]. Deposition parameters controlled droplet properties such as size and charges as well as the flow rate of the precursor and the voltage applied between the nozzle and substrate. Deposition method is more efficient as compared to uncharged droplet method. Due to more effectiveness, in the deposition method, the electric field controlled the motion of the droplet [76]. The standard of the film was influenced by the size of the particles as well as the dispersal mechanism [76, 77]. The electrostatic spray deposition method plays a major role in various applications. Preparation of the reduced graphite oxide thin films was done via the electrostatic spray deposition method. This approach consists of a syringe pump used to feed the solution of precursors through a nozzle and thus fine droplets were produced after applying DC high-voltage power which was used for the biasing of the precursor solution. Adelowo et al. [78] have reported the applications of lithium-ion capacitor by using reduced graphene oxide which is sprayed as electrodes. In the electrostatic spray deposition method, reduced graphene oxide films were synthesized with the substrate's area of about 50×50 mm^2. The pros of this method are to synthesize thin film at ambient temperature which is a simple process rather. Moreover, the time of deposition controls the thickness of the film and in the presence of substrate temperature, graphene oxide can be reduced. Thus, the electrostatic spray deposition method is the candidate for the fabrication of reduced graphite oxide thin films in a large area. By using electrostatic spray deposition technique, fabrication of acetone gas sensor was done which is based on reduced graphene oxide film in order to signify the preparation of electronic devices.

2.5 *Melt Mixing Technique*

Melt mixing process is utilized for the preparation of graphene-based composites in which nanofiller is combined with polymer melt physically. This method is cost-effective, more versatile as well as environment-friendly. Thermoplastic polymers are widely utilized in this technique at elevated temperatures. In this method, there is no requirement of any solvent. Mixing of polymer matrix with graphene or its derivatives occurred in the molten state. By using a conventional method such as extrusion as well as injection molding, mixing of graphene or its derivative occurred mechanically with thermoplastic polymer at elevated temperature. Due to the cost-effectiveness, there can be some possibilities of producing large quantities of materials. In this technique, there is no involvement of any additive, compounds, or chemicals within the mixing unit. Various researchers have fabricated graphene-based polymer composites via the melt mixing process. Adak et al. [79] reported that polyurethane-functionalized graphene nanocomposites films can be synthesized using solution master batching and melt mixing process and then by compression molding. In polyurethane matrix, graphene sheets can be exfoliated partially and it is uniformly dispersed due to differing in the concentration of graphene which is about 0–3 wt%. Consequently, with an increase in concentration of graphene, there can be some improvement in the helium gas barrier of nanocomposites film, which shows the reduction in gas permeability of about 30% at 3 wt% graphene loading. The increase in the concentration of graphene significantly increases the stiffness as well as the tensile strength of the nanocomposites. To accelerate artificial weathering conditions up to 300 h, the prepared functionalized graphene nanocomposites film was exhibited. Nordin et al. [80] have reported that poly(lactic acid)/thermoplastic polyurethane graphene composites can be synthesized by using melt mixing process. By using a resistance meter, electrical conductivity was tested and it is observed that in the presence of graphene nanoplatelets the resistivity of composite started to percolate. The threshold change in percolation is due to a change in blend composition, and by this, it is shown that the lowest threshold per location was observed at poly(lactic acid) 90/thermoplastic polyurethane 10 (PLA 90/TPU 10) blends. By calculating wetting coefficient along with Owen as well as Wendt equation, it was predicted that graphene nanoplatelet can be localized in poly(lactic acid)/thermoplastic polyurethane blend, and according to the prediction, it is observed that graphene nanoplatelet is present preferentially in thermoplastic polyurethane phase. The increase in thermoplastic polyurethane content leads to an increase in the elongation at break of the composites. The blending of elongation at break occurred after the addition of graphene nanoplatelet in poly(lactic acid) 50/thermoplastic polyurethane 50 blend. Noorunnisa et al. [81] reported that the linear low-density polyethylene and graphene nanoplatelet composites can be fabricated by using twin-screw extruder under various extrusion conditions. The electrical, mechanical as well as thermal properties may alter the screw speed, feeder speed as well as graphene nanoplatelet content. Thermal stability as well as conductivity improved by 2.7% and 43%, respectively, which is caused by the inclusion of graphene nanoplatelets in the matrix. Due to high thermal stability of the graphene nanoplatelets as well as the production of photon and charge carrier network in the matrix of polymer, some improvement in case of electrical conductivity from 10^{-11} to 10^{-5} S/m at 150 rpm may be observed.

3 Defects of Graphene

Recently, in nanotechnology, graphene and graphene-based polymeric composite materials are the most promising candidates. Although they have immense potential applications in the field of energy storage, electronics and optoelectronic devices, biosensors and gas sensors, etc., still it is very difficult to fabricate defect-free materials. So, it is very crucial for understanding and to differentiate between various defects types as well as the influence in the alteration in basic electronic properties of these composite systems. There are natural imperfections, and defects due to growth are generally observed in the structural defects [82–84] that deteriorate graphene-based device performances. Moreover, defects are introduced for getting better ideas about phenomena related to graphene. Whenever crystalline order is disturbed without the presence of any alien atom the defects in three-dimensional crystals are known as intrinsic defects. Various intrinsic defects sometimes lead to interesting effects and potential applications as well, which are observed like point defects, cluster defects, and boundaries/edges defects responsible for reducing the crystal symmetry [85–87]. However, extrinsic defects deal with the introduction of impurities or foreign atoms into the crystalline order. Stone et al. reported on point defects (Stone Wales defect) that the graphene lattice has a tendency to reconstruct the non-hexagonal rings without involving the removed or added atoms produce two pentagons and heptagons each with the rotation of any one of the C–C bonds through

90° [88]. Meyer et al. [89] reported on the single vacancies in graphene which is the simplest defect and had been observed also in the TEM image and the defect of the missed lattice point atom can be shown in Fig. 2. The single vacancies undergo the Jahn–Teller distortion due to the saturation of two of the three sagging bonds in the direction of the missing atom which results in the formation of a five and nine-membered ring.

Moreover, double vacancies occur by either combination of two SVs or by the removal of two neighboring atoms. There is no sagging bond present in a fully reconstructed DV so that two pentagons and one octagon appeared instead of four hexagons in perfect graphene, as shown in Fig. 3 [90]. Also, the atomic network becomes coherent with slight agitation in the bond lengths around the defect [89].

Defects at the boundaries/edges of graphene result due to local changes, i.e., continuous removal of carbon atoms from the boundaries which can be done by sputtering edge atoms with electrons with energies lower than the threshold energy for displacing atoms from a perfect graphene. Under these conditions, armchair edges

Fig. 2 Single-vacancy (SVs) TEM image of the defect in graphene (reprinted with permission from Ref. [89])

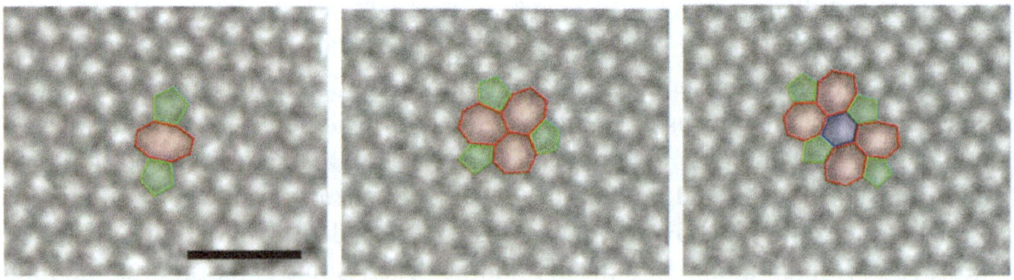

Fig. 3 Double vacancy (DV) TEM image of the defect in graphene (reprinted with permission from Ref. [90])

can be transformed to crisscross edges. An intermediate structure can be considered as a defective edge [91]. The removal of one carbon atom from a crisscross boundary leads to one pentagon in the middle of a row of hexagons at the edge. On the other hand, edge reconstructions result in different combinations of pentagons and heptagons at the boundaries. Banhart has reported on Foreign Adatom that if a non-carbon atom is there in graphene, it influences the properties of the graphene which is purely dependent on bonding with the foreign atom [92]. If the corresponding bond is weaker then due to van der Waals interaction physisorption occurs but the existence of stronger covalent bond results in chemisorptions.

3.1 Techniques for Creation of Defects

Generally, there are three techniques by which the non-equilibrium defects in graphene can be resulted such as particle irradiation, crystal growth, and chemical methods.

3.1.1 Particle Irradiation

The point defects can be obtained by irradiation of graphene, maybe with electrons/ions, as a result of ballistic expulsion of carbon atom [93], and threshold energy (~18–20 eV) has to be provided to abscond from its lattice site. Moreover, the atom may be then engrossed on the graphene expanse and drift on the surface as Adatom or stammer away from graphene. Ion irradiation is another physical method basically employed for producing defects in graphene [94], where several SVs and DVs are graphene-irradiated by using different noble gas ions, and due to this, graphene becomes fundamentally transparent.

3.1.2 Crystal Growth

Growth of graphene in the form of a layer on a huge scale usually do not occur, but defect is essentially required, as in the case of chemical vapor deposition (CVD), defects take place naturally for growing the graphene layers. However, high-temperature growth makes possible by the relaxation and leads to thermal equilibrium so that defects can anneal in a rapid manner. Line defects are created when simultaneous nucleation of graphene layers occurs at different regions over a substrate.

3.1.3 Chemical Methods

When carbon atoms react with the non-carbon species in a graphene layer there are losses of atoms which result in defects. In this type of reactions oxidation is the most common one, and due to this, oxygen, hydroxyl, or carboxyl groups are attached to the graphene. In these chemical treatments highly defective surface-functionalized graphene is obtained by an oxygen group called graphene oxide [95] which are covered uniformly with hydroxyl or carboxylic groups. Moreover, at room temperature very few numbers of reactions are allowed because graphene is highly inert except for its boundaries.

3.2 Properties of Graphene with Defects

The chemical and electronic properties of graphene are immensely affected by defects. Moreover, from various simulations, it has been observed that the functional groups such as hydroxyl and carboxyl can have better attachment in the surface of graphene incorporated with vacancies defects [96]. Similarly, in the case of hydrogen saturated boundaries of graphene, the local reactivity is also enhanced substantially. The theoretical equations, i.e., Schrodinger equation in relation to electron will be replaced by Dirac equation for graphene induced with defects that directly influence the electronic structure. In almost all defective graphene scattering of the electron waves takes place and orientation of electron changes [96, 97].

4 Properties of Graphene-Based Polymer Composites

The researchers have focused on different kinds of polymeric composites comprising carbon-based nanomaterials, including CNTs, CF, graphene, graphene oxide, etc., and studied electrical and mechanical properties in the field of energy storage devices. In this section, we have discussed especially the properties of graphene as well as graphene-oxide-based composites using different polymer matrices. The graphene-based polymer composites might be useful for academic and industrial researchers for the development of various kinds of new graphene-filled polymer composite systems.

4.1 Poly(Vinyl Alcohol) (PVA)–Graphene Composites

Santos et al. [98] have reported polyvinyl alcohol-doped (PVA) ternary sulfonated graphene oxide (SGO) with graphite into polyethylene oxide (PEO) and poly(vinyl pyrrolidone) (PVP) via solution casting technique. It is observed that the swelling ratio is decreased by 17 wt% of SGO composite membrane and lowers two orders of magnitude (0.18×10^{-6} cm^2 s^{-1}) of the permeability as compared to borohydride anion. These composite systems are characterized by scanning electron microscopy (SEM), transmission electron microscopy (TEM), X-ray diffraction (XRD), Fourier-transform infrared spectroscopy (FTIR), and Raman spectroscopy. The direct borohydride fuel cell (DBFC) is used due to its ease of processability, low energy consumption, simple to use and sustainable, as well as cost-effective, and also water is used as a solvent. It is also revealed that the interaction in acidic SGO and cross-linked polymer improved the membrane's mechanical resistance, ionic conductivity, and oxidative stability. In addition, the SGO-encapsulated PVA-PEO-PVP ternary membrane exhibits lower permeability of boro-hydride than that of the Nafion 117, which gives a higher selectivity. Cobos et al. [99] have successfully synthesized polyvinyl-alcohol (PVA)-silver (Ag)-graphene oxide (GO) (PVA-Ag NPs-GO) nanocomposite via a one-step process with L-ascorbic acid as a reducing agent with a mixture of $AgNO_3$, GO, and aqueous solution of PVA. It is observed that GO sheets grafted with spherical Ag NPs are uniformly dispersed in a polymer matrix. The glass transition temperature, crystallization temperature, mechanical, and water resistance properties of PVA composite are enhanced with the incorporation of Ag NPs-GO filler contents and also show antibacterial activity against *Escherichia coli* and *Staphylococcus aureus*. Similarly, they have also shown the homogeneous distribution of spherical silver nanoparticles over the surface of carbon substitute (10 nm in diameter) by an environmentally friendly process. Moreover, the thermogravimetric measurements signified the thermal stability of PVA composites as well as get enhanced by the incorporation of Ag NPs-GO. The Young's modulus and tensile strength of the composites significantly increased by incorporating filler particles, which is helpful for practical applicability in wound dressing [99–101]. Jain et al. [102] have reported on the effect of chemically reduced graphene oxide (RGO)-PVA matrix composite with various weight percentages of filler contents via a biomimetic approach. The microstructural analysis by scanning electron microscope (SEM) revealed better dispersion of RGO into the polymer matrix, and fibril alignment with high lamellar structure is observed. This incorporation of RGO to polymer matrix improved the tensile strength to 67.21 MPa as compared to that of pristine polyvinyl alcohol (PVA) matrix (33 MPa) (as shown in Fig. 4) [102, 103]. The degree of crystallinity and melting enthalpy of PVA-RGO composites are obtained from thermal analysis. It has been noticed that the incorporation of 1% RGO-PVA composites enhanced Young's modulus (5.53 GPA and 5.03 GPA) from the load penetration depth curve. The RGO shows load transfer to PVA matrix which enhances the modulus and tensile strength of RGO-PVA composite and decreased hardness which is found to be about 0.27 GPA. These RGO-PVA composite systems may be applied in the field of supercapacitance, bone regeneration, biosensors, etc. [103, 104].

Similarly, Khan et al. [105] have reported on sulfonated polyvinyl alcohol-aluminum oxide-graphene-platinum (SPVAAI-GR-PT) film by electroding with platinum using electro less plating technique. It has been observed that the ion exchange capacity (1.8 m eqg^{-1}) of the dry film and water uptake (125% at 45 °C for 10 h of drenching time) is superior as compared to other reported composites. The

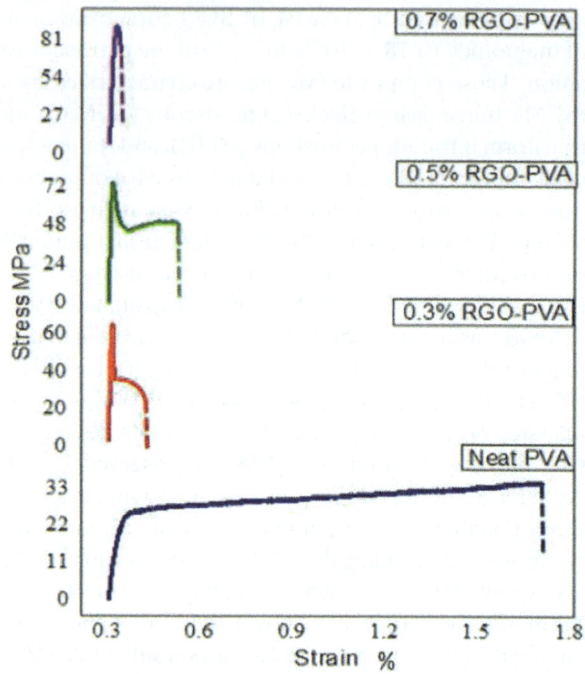

Fig. 4 Variation of stress vs. strain of RGO-incorporated PVA composite. Reprinted with permission from Ref. [102]

electro chemical, electromechanical, structural, morphological properties have been analyzed with various experimental techniques. However, it has good proton conductivity and high temperature stability with better electromechanical properties, which might be a better alternative for conventional polymer-based ionic polymer-metal composite (IPMC) actuator. In addition, the synthesized film has several potential applications in the domain of nano-electronics, actuators, sensors, supercapacitor, and robotics [106, 107].

4.2 Poly(Vinylidene Fluoride) (PVDF)–Graphene Composites

Kang et al. [108] have successfully fabricated MnO_2 doped graphene-based poly(vinylidene fluoride) (PVDF) composite systems for supercapacitor application by using a simple stirring and heating process. It is observed that 5 wt% of CBE shows higher specific capacitance (220 F g^{-1}) than that of the 5% PBE (202 F g^{-1}) at 0.5 Ag^{-1} with identical graphene contents, which is measured from galvanostatic charge–discharge plots (Fig. 5a, b). The cycling stability is found to be 90.07% (Fig. 5c) of the initial specific capacitance of 5% CBE, which is also retained after 1000 cycles. The chemical bond between graphene and PVDF improved electrochemical performance and lead to an effective electron pathway by bridging MnO_2

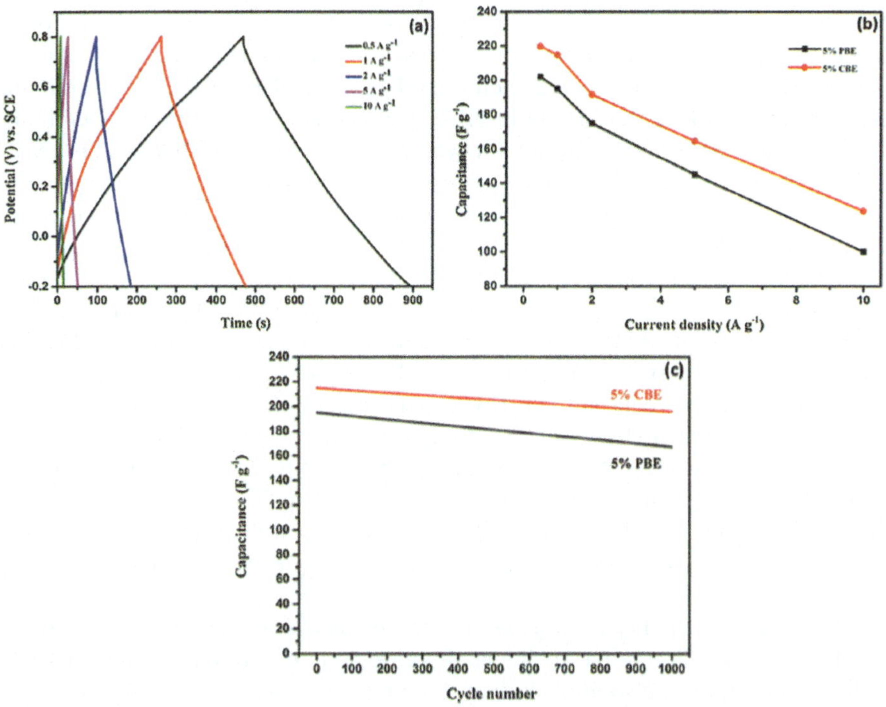

Fig. 5 Variation of **a** charge–discharge curves for 5% of CBE with various current densities, **b** rate capability of the 5% CBE and PBE, and **c** charge–discharge curve of the 5% CBE and PBE. Reprinted with permission from Ref. [108]

particles in the electrodes, which is responsible for the rapid electron transfer and low charge transfer resistance. The MnO_2 supercapacitor incorporated with a composite binder is subjected to cyclic voltammetry (CV) and galvanostatic charge–discharge process (Fig. 5), which shows an excellent electrochemical performance with a maximum specific capacitance of 220 F g^{-1} in 1 ^{mol} L^{-1} Na_2SO_4 electrolyte.

A ternary P(VDF-HFP)-polyaniline (PANI)-graphene oxide (GO) hybrid membrane (10 wt%, 25 wt%, and 40 wt%) was prepared by Hamid et al. [109] by using polymer electrolyte membrane (PEM) for lithium-ion battery. The scanning electron microscopy (SEM), X-ray diffraction (XRD), differential scanning calorimetry (DSC), and Fourier-transform infrared spectroscopy (FTIR) analyses have been performed and the electrochemical stability, porosity, and electrolyte uptake determined the effect of the incorporation of GO into P(VDF-HFP) polymer matrix. However, they revealed very high ionic conductivity ($10^4 \times 10^{-3}$ mS cm^{-1}) of P(VDF-HFP)-PANI membrane with tensile strength increased from 2.8 MPa to 8.8 MPa by the introduction of GO. The ternary P(VDF-HFP)-PANI-GO membrane shows porosity of 89.5% with excellent electrolyte uptake of 367.4%. Moreover, the ionic conductivity gets enhanced by the incorporation of polyaniline (PANI) and the introduction of GO, which resulted in the improved thermal and mechanical properties of the membrane due to their π–π interaction and H-bonding. As an outcome, the electrochemical stability and cycling performance of P(VDF-HFP) membrane were enhanced. The (PVDF-HFP)-PANI-GO ternary polymer electrolyte membrane was found to be more stable even with decomposition voltage (5.6 V) and is nearly 100% Columbic efficiency after 10 cycles. The introduction of diglycidyl ether of bisphenol-A (DGEBA) functionalized reduced graphene oxide (rGO)-based epoxy composites with improved dielectric properties and thermal stability compared to neat graphene oxide (GO) and rGO sheet have been reported by Liao et al. [110]. It is observed that the dielectric constant of epoxy composite filled with 1 wt% of DGEBA-RGO sheets is 32 at room temperature at 1 kHz, which is nine times higher than that of neat epoxy (3.5). On the other hand, the dielectric loss of the resultant composites gets reduced (<1). This enhancement in dielectric properties is due to well-dispersed DGEBA-RGO and strong interaction between filler and polymer matrix, which may be attributed to an induced effective package of grafted DGEBA molecules on the graphene surface. In addition to this, the insulated DGEBA molecular layer affected the sheet contact directly resulting in the suppressed dielectric loss (<1). These high-performance composite films may be applicable in the field of embedded capacitors.

Mahaling et al. [111] have examined the dielectric properties of (Ag)-doped graphene oxide (GO)-poly(vinylidene fluoride-co-hexafluoropropylene) (PVDF-HFP) composites using solution casting technique. It is observed that the composite systems have a high dielectric constant (65) and diminished dielectric loss values (<1) at 10^2 Hz (Fig. 6a, b). However, the Ag nanoparticles as the conductive phase are persistently incorporated on the surface of graphene oxide (GO). The experimental results showed the formations of Ag layers on GO sheet with homogeneous distribution into the P(VDF-HFP) matrix were confirmed by field emission scanning electron microscopy (FESEM). Besides, the percolation threshold of 1.5 vol% of Ag-GO and Ag layer formed on the surface of GO sheet, which influenced with the enhancement of dielectric and electrical performance. These Ag encapsulated GO-P(VDF-HFP) composite systems may have useful application in the field of electronic capacitors.

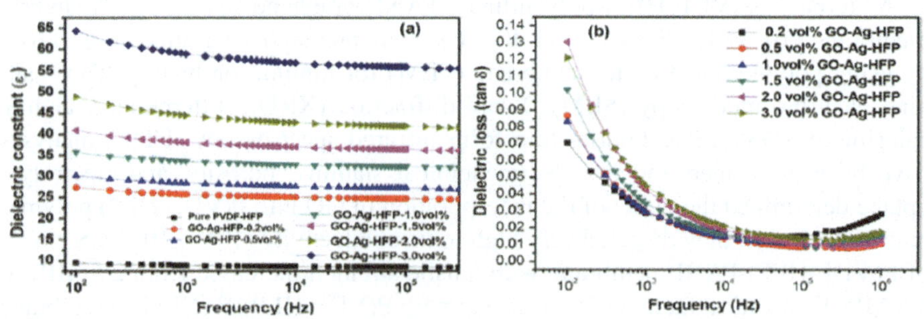

Fig. 6 Frequency dependence of **a** dielectric constant and **b** dielectric loss of Ag-GO-PVDF-HFP composites. Reprinted with permission from Ref. [111]

Chen et al. [112] have prepared a composite of reduced graphene oxide (rGO)-$CaCu_3Ti_4O_{12}$ (CCTO) in polyvinylidene fluoride (PVDF) by solution processing technique. It is observed that the composite has a high dielectric constant (40.4) and a low dielectric loss (0.08) at 1 kHz, when rGO and functionalized CCTO were 0.61 vol% and 12.5 vol%, respectively. In addition to this, the tunneling effects between rGO sheets play an important role in electrical conductivity and dielectric loss of the composite. This composite has application in designing and fabricating polymer composite for nonlinear dielectric and flexible electronic applications. He et al. [113] have developed poly(vinylidene fluoride) (PVDF)-polypyrrole (PPy)@graphene oxide (GO) composite via solution compounding processing. The resulting composite has higher electrical conductivity compared to two-phase PVDF-PPy composites. However, it is observed that the said composites have higher permittivity with a slight reduction in dielectric loss, which may have application in the field of energy storage. Li et al. [114] have fabricated polyvinylidene fluoride (PVDF) composite having high dielectric constant and low dielectric loss, which is obtained by incorporation of reduced graphene oxide(rGO) decorated with magnetic iron oxide (rGO@Fe_3O_4) on PVDF matrix by solution mixing and subsequent melt compression molding technique. This composite has been subjected to X-ray diffraction, Fourier-transform infrared (FTIR), thermogravimetric analysis (TGA), impedance analyzer, and magnetometer. They found that the dielectric constant becomes 1297 and the low dielectric loss is 0.26 at 100_{Hz} with 1.0 wt% RGO@Fe_3O_4 in the PVDF matrix. The excellent dielectric properties may be a consequence of core–shell structure and well dispersion of RGO@Fe_3O_4 in the PVDF matrix. These composites might be applicable in the field of mini capacitors. Fu et al. [115] have reported the development of polydopamine (PDA)-reduced graphene oxide (rGO) by self-polymerization and subsequent chemical reduction, then incorporated into PVDF matrix by solvent blending method. It has been seen that the dielectric constant of PVDF increases to 176 with 0.70 wt% rGO with a reduced tangent loss of 0.337, which is attributed to the reduction of concentration mobility of ionizable carboxyl group by PDA as depicted in Fig. 7a, b. These composites have good flexibility and

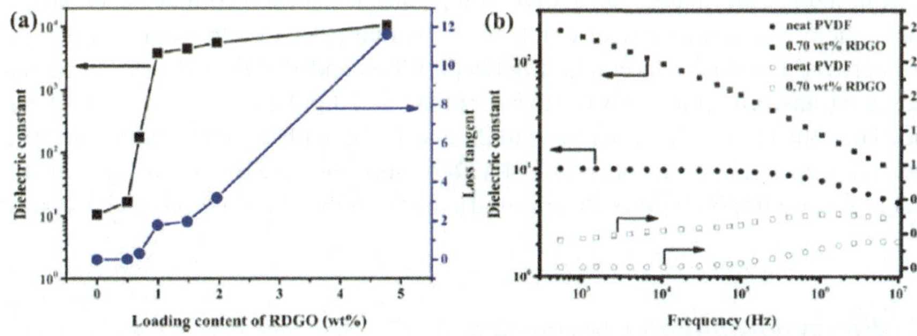

Fig. 7 Variation of **a** dielectric constant and **b** dielectric loss of polydopamine (PDA)-reduced graphene oxide (rGO) composites with 0.70 wt% of rGO contents. Reprinted with permission from Ref. [115]

better dielectric properties suitable for superior application in energy storage devices. The enhancement in dielectric constant is due to duplex interfacial polymerization of graphene–semiconductor interface and semiconductor–insulator interface [116].

Poly(vinylidene fluoride-co-hexafluoropropylene) (PVDF-HFP) incorporated with titanium dioxide-modified rGO (RGO-TiO_2-P(VDF-HFP) [PVDF-HFP] nanocomposite has been prepared by in situ assembling of TiO_2 on GO, which has been reported by Tong et al. [117]. It has been found that the increase in rGO-TiO_2 content increases the dielectric permittivity and low dielectric loss at low frequencies. Further, it is observed that the well-coated TiO_2 dielectric shell acted as an interparticle barrier to prevent direct contact with the graphene sheets. Tong et al. [118] have studied the preparation of poly(vinylidene fluoride-co-hexafluoropropylene) (PVDF-HFP) incorporated polyethylenimine (PEI) nanocomposite by solution casting method. The successful grafting of PEI on GO has been verified by infrared (IR), atomic force microscope (AFM), X-ray photoelectron spectroscopy (XPS), Raman spectroscopy, and thermogravimetric analysis (TGA). Moreover, the dielectric constant of the composite is found to be 67 (1000 Hz) with 8 wt% of rGO-PEI and low dielectric loss (0.12) at specific frequency regions. Li et al. [119] have reported the nanocomposite made from poly (vinylidene fluoride) (PVDF) and poly(vinyl pyrrolidone) (PVP)-anchored rGO (rGO@PVP) nanosheets via solution casting method. Further, the microstructure and dispersion in this nanocomposite have been investigated by atomic force microscope (AFM), X-ray photoelectron spectroscopy (XPS), thermogravimetric analysis (TGA), and Raman spectroscopy. However, the composite has a high dielectric permittivity of 622 and a dielectric loss of 0.2 near the percolation threshold at 100 Hz. This composite with improved dielectric properties and homogeneous dispersion leads to intercalation between rGO and PVDF matrix, and due to the presence of PVP, it resulted in the formation of micro and nano-capacitor structure. The PVDF has the growth of beta form crystal and when induced by the PVP space layer it suppresses the loss near the percolation threshold. These high dielectric constant and low dielectric loss composites may be integrated into electronic energy storage devices. Similarly, Wu et al. [120] have achieved a high dielectric constant of 364 and a low dielectric loss of 0.77 at 1 kHz by synthesizing chlorinated GO-PVDF composite. It is observed that the enhancement in dielectric and electrical conductivity is due to charge transfer complexes. The improved interfacial interaction between fillers and PVDF is due to hydrogen bonds and transformation of PVDF to beta-phase and dipolar interaction. It has been noticed that the tangent loss increases to 2.88 at 1 kHz with 0.4 vol% graphene. The dielectric properties of Cl-doped rGO-PVDF composites are due to the synergistic effect of several important modifications arising from the chlorination of GO sheets.

4.3 Epoxy-Graphene Composites

Prusty et al. [121] have reported on the improvement of flexural and interlaminar properties of carbon fiber-reinforced polymer (CFRP) composite by modification of the surface of carbon fiber using graphene-oxide-based nanofiller (GBN) by electrophoretic deposition (EPD) technique. It has been seen that flexural and interlaminar strength of G-COOH-modified CFRP composite is 9.6% and 22.9% higher than that of control CFRP. The reinforcement mechanism involves esterification between G-COOH and epoxy, which enhanced filtration of epoxy as a result of improved interlocking between epoxy and carbon fibre. However, for a better understanding of the temperature-dependent mechanical behavior and various failure of micromechanism of the composite, they performed dynamic mechanical thermal analysis (DMTA) in the range of 30 °C–180 °C. The scanning electron microscopy (SEM) analysis indicated fiber pulled out to be the most prominent mode of failure in neat CFRP composite which is reduced in the case of G-COOH-modified CFRP composites. Simultaneously, the enhancement in flexural strength for polymer matrix beam reinforced with graphene and carbon filler at nano and micro level is taken place as compared to plain control beam reinforced with graphene. Wan et al. [122] have fabricated epoxy composite filled with graphene oxide (GO) diglycidyl ether of bisphenol-A functionalized GO (DGEBA-f-GO) sheet at various filler loading. It is identified that the presence of DGEBA improves the compatibility and dispersion of GO sheet in an epoxy matrix, while the strong interfacial interaction between sheets and epoxy matrix results in effective load transfer from the matrix to (DGEBA-f-GO). Meanwhile, it is observed that the tensile modulus and strength of epoxy composite with 0.25 wt% DGEBA-f-GO has increased from 0.15 ± 0.11 to 3.56 ± 0.08 GPa (13%) and 52.98 ± 5.82 to 92.94 ± 5.03 MPa (75%) respectively, compared to the neat epoxy resin and fracture toughness (29–41%). The neat GO and DGEBA-f-GO of 0.25 wt% loading produced 26% and 41% improvements in KIC values of epoxy composites, which indicate improved interfacial interaction between DGEBA-f-GO and matrix. The DMTA and TGA analyses indicate increased glass transition temperature and better thermal stability of DGEBA-f-GO-epoxy-based composites compared to two-phase GO-epoxy composite. Surnova et al. [123] have synthesized graphene oxide (GO)-based epoxy composite by homogeneous liquid phase transfer of GO into the epoxy resin resulting in uniform distribution of GO flakes within the epoxy matrix. The study of NPEL-128 epoxy resin curing by DDM showed that the introduction of GO into the epoxy resin exhibits an accelerating effect on the curing reaction. It is investigated that the storage modulus of epoxy matrix gets increased by 25.4% (3035 MPa) by the introduction of 0.2% GO as compared to neat epoxy polymer (2420 MPa). This enhancement is due to the highly exfoliated condition of GO in the matrix and covalent bonding between GO flakes and epoxy matrix at relatively low loading. Feng et al. [124] have synthesized reduced graphene oxide (RGO)@$Ni(OH)_2$, hexagonal boron nitride (h-BN) sheet (lateral size of 4.37 ± 1.68 μm and thickness of 80 ± 21 nm) hybrid in epoxy matrix composites using ultrasonication technique. These composites have high thermal conductivity (Fig. 8a) and flame retardancy properties, as confirmed by peak heat release rate, total heat release, and total smoke production (33.5%, 33.8%, and 43.0%) as compared to the neat epoxy. The RGO@$Ni(OH)_2$ hybrid in a matrix having good dispersion and interfacial interaction suppressed the stacking aggregation behavior of h-BN sheets with induced thermal conductivity (2.01 w/mK).

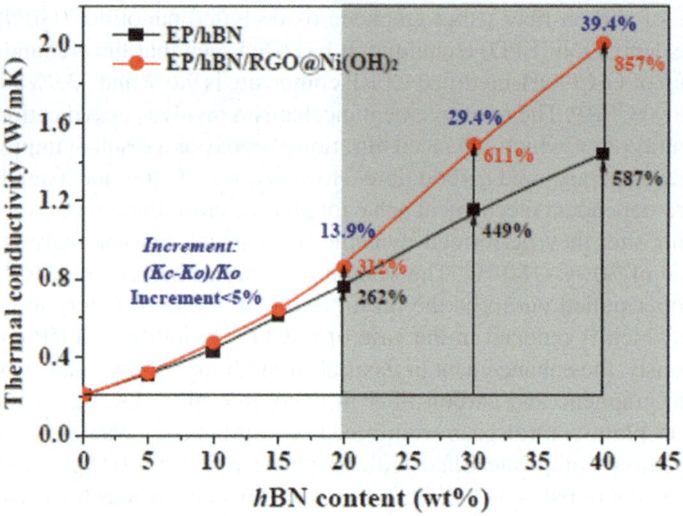

Fig. 8 **a** Thermal conductivity of EP-hBN and EP-hBN-RGO@Ni$(OH)_2$ composites as a function of h-BN contents. Reprinted with permission from Ref. [124]

Besides, improvement in the interfacial adhesive performance of dopamine-modified aramine fiber (AF) grafted with amino-functionalized graphene oxide using the secondary reaction of active hydrogen in polydopamine has been reported by Gong et al. [125]. It was found that this grafting is attributed to the increase in surface roughness and surface-active groups, polarity, and reactivity of polydopamine (PDA-AF surface), which is investigated by Fourier-transform infrared spectroscopy (FTIR), X-ray photoelectron spectroscopy (XPS), and scanning electron microscopy (SEM). The interfacial shear strength of AF-epoxy composite was enhanced by 34% after the incorporation of amino graphene oxide. This reinforcement of AF on GO may be a high-performance fiber, which can be used in electromagnetic interference shielding and environmental remediation.

4.4 Polystyrene (PS)–Graphene Composites

Zhang et al. [126] have developed polystyrene-grafted reduced graphene oxide (RGO-PS) composites via the emulsion polymerization process. It is observed that with an increase in the amount of grafted PS or RGO, the dielectric constant increases whereas dielectric loss remains constant. However, the composites containing RGOs with the highest amount of grafted PS show a low dielectric loss of 0.45 at 100 Hz and conductivity of 4.0410^{-9} S/cm, which is due to improved interfacial polarization induced by the PS layer. On the other hand, grafting of PS also increased the glass transition temperature of composite. Sun et al. [127] have prepared graphene

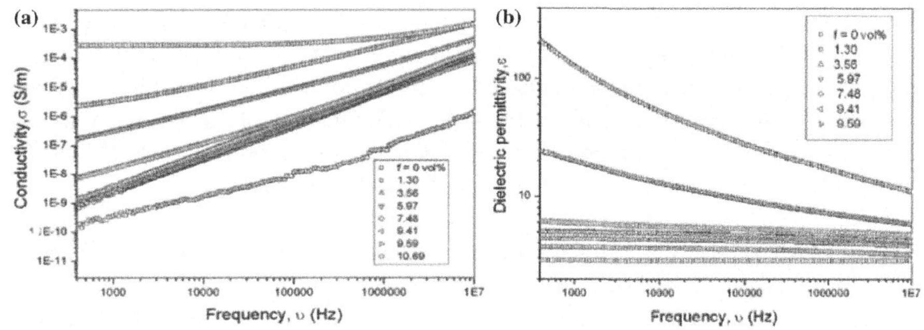

Fig. 9 Frequency dependence of **a** AC conductivity and **b** dielectric constant with different vol% of filler contents. Reprinted with permission from Ref. [128]

oxide (GO) and polystyrene (PS) nanocomposite via pickering emulsion polymerization technique characterized by transmission electron microscopy (TEM), scanning electron microscopy (SEM), and X-ray diffraction (XRD). It was noticed that GO-PS nanocomposite has higher crystallinity and better thermal stability than pure PS chain. These composites have several applications in electromechanical and coating industries. Graphene nanosheet Fe_3O_4(GNS-Fe_3O_4) hybrids have been successfully synthesized by He et al. [128] by one-step solvothermal reduction of iron(III) acetylacetonate [Fe(acac)$_3$] and GO simultaneously, then blended with syndiotactic PS (SPS) by solution blending method. It was investigated that the percolation threshold of GNS-Fe_3O_4 in SPS was 9.41 vol%. At the same time, the composite GNS-Fe_3O_4-SPS has a high dielectric permittivity of 123 at 10^3 Hz (Fig. 9b), which is nearly 42 times higher than pure SPS. Meanwhile, the AC electrical conductivity at 10^3 Hz increased from 3.6×10^{-10} S/m for pure SPS to 2.82×10^{-4} S/m for GNS-Fe_3O_4-SPS composite containing 10.69 vol% of GNS-Fe_3O_4 which showed that an insulator semiconductor transition (Fig. 9a). These composites have various applications in high-charge storage capacitor, electromagnetic interference shielding, and electromagnetic wave absorption.

Tu et al. [129] have developed graphene nanosheet (GN)–polystyrene (PS) latex nanocomposite system using the emulsification process, where the GO sheet is attached to PS particles through electrostatic adsorption. It was shown that the resultant composites have excellent electrical properties with a low percolation threshold as low as 0.054 vol% of GN sheet. Meanwhile, it was observed that the electrical conductivity was 46.32 S/m and thermal conductivity was 0.47 w/mk. This composite can be used in high-energy–density capacitors as dielectric film. Chen et al. [130] have synthesized polystyrene (PS)-Fe_3O_4@Thermally exfoliated and reduced graphene oxide (TGO) composite by solution blending method. The PS-Fe_3O_4-TGO composites are found to be of much higher electrical conductivity and electromagnetic interference (EMI) shielding effectiveness than PS-Fe_3O_4@RGO (reduced graphene oxide) composites due to better reduction of TGO than RGO. It was also revealed that the EMI shielding effectiveness of PS-Fe_3O_4-TGO composites is more than 30 Db in the frequency range of 9.8–12 $_{\text{GHz}}$ with 22.4% of graphene content.

This behavior is due to the combination of electrically conductive TGO sheets and magnetic Fe_3O_4 nanoparticles. Wang et al. [131] have prepared 2-poly(styrene-ran-butylene-b-styrene) (SEBS)-graphene oxide (GO) composites at different styrene segment content using the solution blending method. It is observed that with the addition of 0.5 wt% GO-enhanced tensile strength and modulus of SEBS-30 by 44% and 64%, SEBS-12 was increased by 24% and 39%. Moreover, the micro rack formation by GO increased the toughness and fibrillation of SEBS during the fracture process. Furthermore, the elongation at break and fracture toughness of SEBS-30 was increased by 10% and 64%, respectively.

Graphene oxide (GO)-reinforced polymer composite consisting of poly(4-styrene sulfonic acid) (PSSA) and polyvinyl alcohol (PVA) blend matrix has been synthesized by Deshmukh et al. [132] using the colloidal processing technique. These composites were successfully characterized by Fourier-transform infrared spectroscopy (FTIR), Raman spectroscopy, X-ray diffraction (XRD), UV–Visible spectroscopy, thermogravimetric analysis (TGA), polarized optical microscopy (POM), and scanning electron microscopy (SEM). It is observed that the resultant composites have a higher dielectric constant of 297.91 (50 Hz, 150 °C) with 3 wt% GO loading. Furthermore, the dielectric loss has increased to about 2.64 (50 Hz, 140 °C) with 3 wt% GO loading, which makes it efficient for energy storage application in electronic devices and embedded capacitors.

4.5 Polypropylene (PP)-Based Graphene Composites

Adloo et al. [133] have described the preparation of maleic anhydride grafted polypropylene (PP-MAH)-graphene-based composite via compression molding method, which was characterized by Fourier-transform infrared (FTIR), X-ray diffraction (XRD), atomic force microscopy (AFM) analysis. It is noted that the flexural strength and electrical conductivity is 44.28 MPa and 104.63 S/cm respectively, which is due to the formation of hydrogen bond between graphene and PP-MAH. Wang et al. [134] have reported about reduced graphene oxide-polypropylene (rGO-PP) composite by latex method, which has ultra-low percolation threshold at 0.033 vol% of filler content. However, the composite shows homogeneous dispersion of rGO nanosheet in PP matrix by analysis of scanning electron microscope (SEM). Moreover, there is a transition that occurs from insulator to conductor percolation with an increase in the rGO loading within the polymer matrix. Further, it was observed that the dielectric permittivity of the composite significantly increases as three orders of magnitude. The reduced graphene oxide-polypropylene (rGO-PP) composite with a high dielectric constant might be used as high-energy–density capacitors. Chen et al. [135] have fabricated multiwalled carbon-nanotubes-reinforced polypropylene conductive fibrous membrane via melt electrospinning technique. Initially, CNT is mixed with a small amount of paraffin liquid (PL) and then melt blended with PP matrix via melt electrospinning technique and fibrous membrane with fiber diameter is about 1–3 μm. The microstructural analysis confirms the good orientation and well-dispersed CNTs in PP fiber by scanning and transmission electron microscopy after the addition of PL. These conductive fibrous membranes (CNT-PL-PP) have improved the tensile strength and modulus with superior electrical conductivity and better dielectric constant than that of the CNT-PP fibrous membrane with various percentages of CNT contents, as shown in Fig. 10a–c.

Ding et al. [136] have reported high, thermally conductive, three-dimensional polypropylene–graphene composites via in situ building technique. The composite shows higher thermal conductivity (10.93 Wm^{-1} K^{-1}). This is 55 times greater than that of the pristine PP matrix. Moreover, the polypropylene (PP)–graphene composites have excellent heat dissipation for LED and superior interaction between polypropylene and graphene through hydrogen bonding and π–π conjugate which reduced interfacial thermal resistance. The synthesized composite has wide applicability in the field of heat dissipation of high-power and highly integrated electronic devices. Guo et al. [137] have reported a low weight percentage of graphene nanoplatelets-reinforced polypropylene nanocomposites by using solution mixing

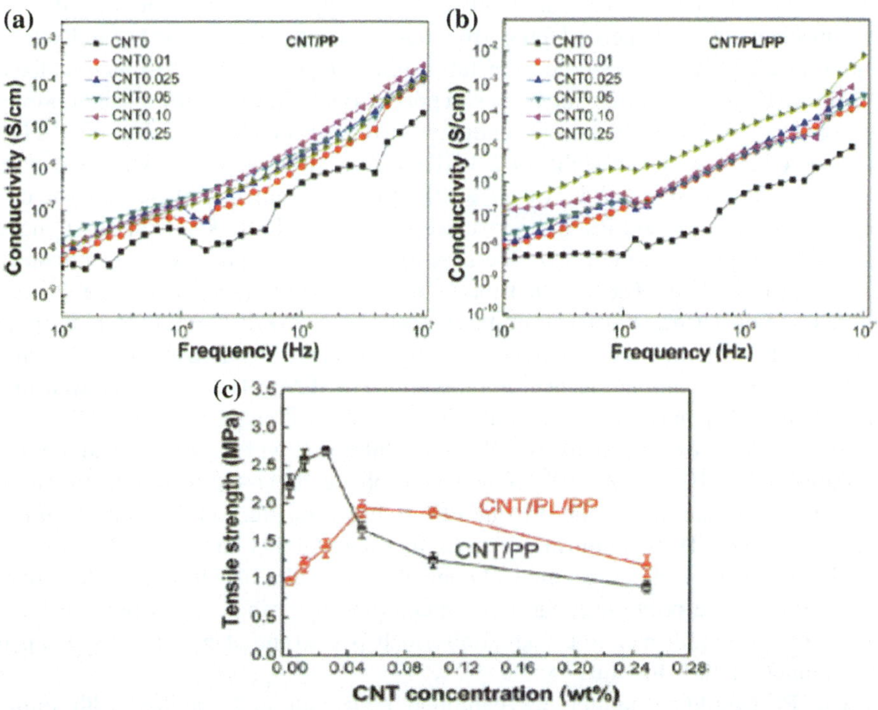

Fig. 10 Frequency dependence of **a**, **b** electrical conductivity and **c** tensile properties of the CNT-PL-PP electrospun fibrous membranes with various concentrations of CNT contents. Reprinted with permission from Ref. [135]

technique and it is widely used in the field of lightweight storage tanks. They studied the effect of physical, mechanical, and morphological performances of polypropylene with various weight percentages of graphene nanoplatelets into the polymer matrix with improved thermal stability and increased graphene contents. The storage modulus provides the elastic modulus of the nanocomposites, while the loss modulus is a measure of frictional losses, which is due to the motion of polymer chains [138]. Moreover, the magnitude of increase in storage modulus is lower than that of Young's modulus, which may be due to the difference in the measurement mode and resolution of instruments.

4.6 Polyimide (PI)-Based Graphene Composites

Chen et al. [139] have synthesized graphene oxide [GO-polyimide (PI)] composite by incorporation of GO into PI resin via in situ polymerization method. It has been observed that this composite has a high tensile strength of 40% for 0.5 wt% GO and tensile elongation of 32% which is three times greater than pure PI film. Further, it has an enhancement in thermal stability. This composite with improving dielectric, mechanical, and thermal properties has application in aerospace industries. Luong et al. [140] have studied the fabrication of polyimide (PI)-functionalized graphene (FGS) nanocomposite by in situ polymerization, which shows an improvement in mechanical and electrical conductivity. It has been observed that the PI/FGS composite has Young's modulus of 2.3 GPa, tensile strength of 131 MPa, and electrical conductivity in the order of 1.7×10^{-5} Sm^{-1}. Meanwhile, the composite can be used in microelectronics and aerospace industries. Kothurkar et al. [141] have fabricated graphene oxide-polyimide-based nanofiber composites via electrospinning technique. It is observed that the GO-PI nanofiber composites were characterized by Fourier-transform infrared spectroscopy, X-ray diffraction, Raman spectroscopy, thermogravimetric analysis, transmission electron microscopy (TEM), and high-resolution scanning electron microscopy (HRSEM). The results showed that GO is bunched up into a bead or spindle-like structure within the nanofiber. However, the dynamic mechanical analysis (DMTA) exhibited 2 wt% GO improved storage modulus (1.4×10^8 to 3.8×10^8 Pa) and also enhanced glass transition temperature (317–323 °C) than that of the pristine PI (as shown in Fig. 11). The thermogravimetric analysis (TGA) confirms superior thermal stability than that of a pristine polyimide matrix with increasing GO content. Ren et al. [142] have reported about high-performance polyimide nanofiber membrane by using electrospinning technique. The polyimide nanofibers have uniform dispersed and fiber diameters obtained in the range of 140–400 nm.

The PI nanofiber membranes exhibited better thermal stability with initial decomposition temperature (T_d) (544.4 °C) and heat resistance temperature (T_{HRI}) (198.8 °C). The mechanical performance of the nanofiber membranes is found with reasonable tensile strength (10.5 MPa) and Young's modulus (927.6 MPa). Further,

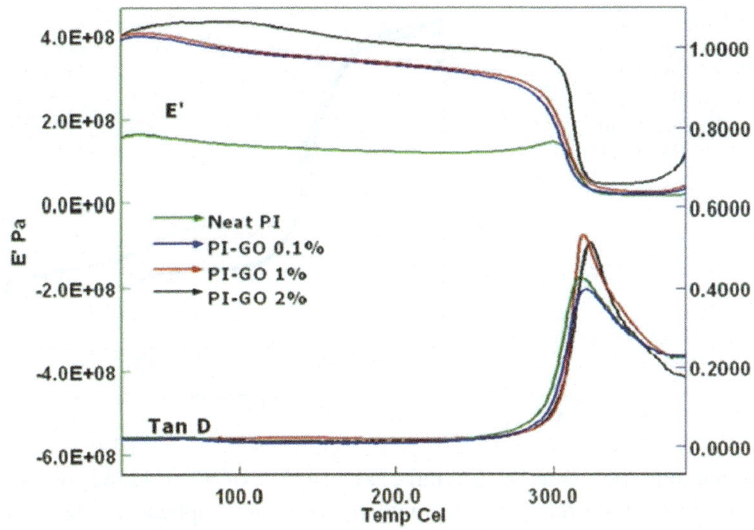

Fig. 11 Dynamic mechanical analysis of PI-GO composite with various weight percentage (0% (neat PI), 0.1%, 1.0%, and 2.0%) of GO contents. Reprinted with permission from Ref. [141]

the filtration performance of the nanofiber membranes showed the best filtration efficiency of 90.4%. These electrospun PI nanofiber membranes could be a promising candidate for hot gas filtration.

4.7 Polyurethane (PU)-Based Graphene Composites

Pokhreal et al. [143] have studied the mechanical properties of graphene nanoplatelets (GNPs), graphene oxide, and functionalized graphene sheets (FGS) incorporated into PU by in situ polymerization. It has been found that due to strong interaction of FGS with PU and 2 wt% of GO-PU the composite has high modulus (25.8 MPa) than GMPs with PU. It has been observed that the FGS/PU composite has high Young's modulus (213%) and high tensile strength (17%) than neat PU (Fig. 12). This enhancement is due to the increased number of hydroxyl groups on the surface of FGS than GO produced with a heavily PU-coated graphene sheet, which resulted in an interface for the stress transfer from PU to FGS. Lu et al. [144] have reported the synthesis of graphene(G)–polymer(P) composite to assess the force separation response of interface between the graphene and polymer matrix with polymers like polyethylene (PE), polyurethane (PU), and polystyrene (PS) by simulation process. However, for comparison a theoretical model proposed by Jiang et al. [145] was suggested, i.e., the interference based on cohesive law model. Meanwhile, it was found that the G/PU system has a weak interface as compared to G/PE and G/PS which helps in determining the mechanical properties, but the interfacial strength of each G/P matrix

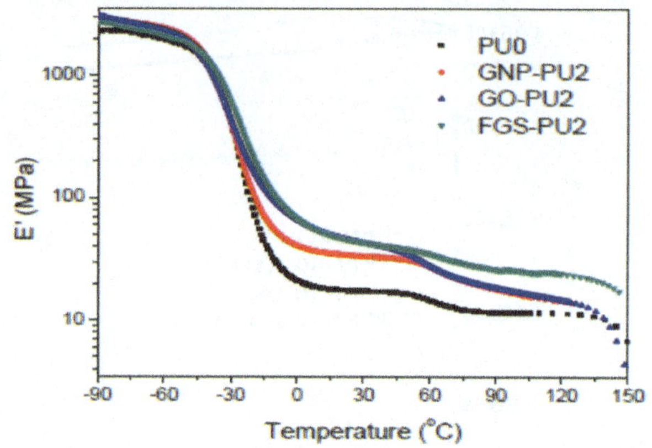

Fig. 12 Dynamic storage modulus with temperature for neat PU and 2 wt% of GO, FGS-based PU composites. Reprinted with permission from Ref. [143]

is greater than pure polymer. Sadasivuni et al. [146] have reported the formation of nanocomposite of polyurethane (PU) with hydrophilic graphene oxide (GO) and hydrophobic-modified GO (m-GO) by the solution mixing method. It has been seen that there is an increase in the dielectric permittivity of PU-mGO nanocomposite as compared to PU/GO, which confirmed more effective dispersion of thin exfoliated sheet of mGO in PU, characterized by scanning electron microscope (SEM) and X-ray diffraction (XRD). Meanwhile, the viscoelastic behaviors are studied by Kraus and Maier and Goritz models. These nanocomposites have application in capacitors. It is found that the stiffness and toughness are increased without any deterioration in the storage modulus.

5 Conducting Polymer-Based Graphene Composites

Conducting-polymer-filled graphene composites have much significant attention in both academic and industrial fields owing to their elevated conductivity, ease of processing, and multifunctional properties. There are several conducting polymer families such as PANI, polypyrrole, and polythiophene generally used as matrices. In this section, the main focus is on conducting polymer-based graphene composites and the subsections are devoted to summarizing their properties and applications in the field of energy storage devices.

5.1 Polyaniline (PANI)-Based Graphene Composites

Gupta et al. [147] have synthesized hybrid film of electrochemically processed graphene nanosheets with electrochemically synthesized (ErGO) conducting polymers polypropylene (ppy) and polyaniline (PANI) by layer-by-layer approach (LBL).

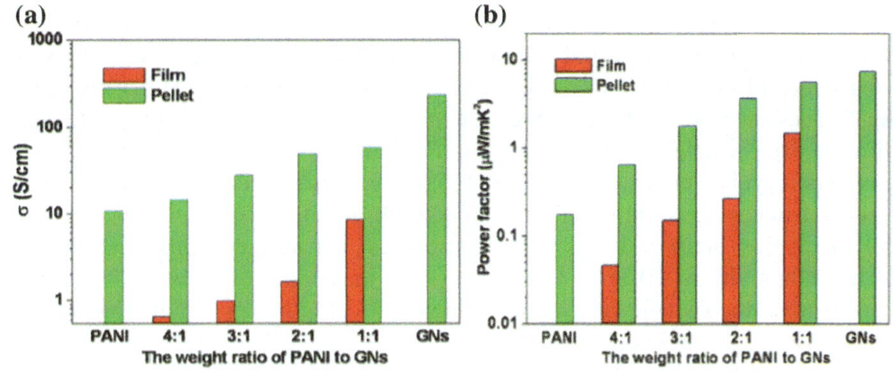

Fig. 13 Variation of **a** electrical conductivity and **c** power factor for PANI-GNs composite with various GNs contents. Reprinted with permission from Ref. [148]

It has been found that there is an increase in the mechanical properties of polymers due to the synergistic effect promoted by nanostructure morphology. The surface topology, chain ordering, residual stress distribution, force curve, and force volume imaging can be done by micro Raman spectroscopy, Raman mapping, atomic force microscopy (AFM), and force spectroscopy. Further, the force curve shows the enhancement in the spring constant (K) for the hybrids. Meanwhile, it was found that the mechanical properties get increased in the given order PANI/ErGO > PPy/ErGO > PANI/GO > PPy/GO > PANI > PPy. However, these hybrids are used in electrolyte-free or dry high energy storage density for dielectric capacitors and might be used in photovoltaic and aerospace device applications. Du et al. [148] have synthesized polyaniline (PANI)-graphene nanosheet (GNS) thermoelectric bulk composite pellets and film in the weight ratio ranging from 4:1 to 1:1, which gives rise to an increase in the power factor (Fig. 13c). It has been investigated that the electrical conductivity of both pellets and films are enhanced from 14.76 to 5889 S/cm and 0.66 to 8.63 S/cm, respectively, which may be due to the larger electrical conductivity of GNs pellet (as shown in Fig. 13a). As a result, this power factor of the pellets and films increased from 0.64 to 5.60 and 0.005 to 1.47 $\mu Wm^{-1} k^{-2}$. Further, the increase in carrier mobility leads to conducting polymer-inorganic composites, which have more usefulness in the field of thermoelectric application.

Haldar et al. [149] have reported the introduction of graphene into the stable matrix of MnO_3O_4-polyaniline composite, which has the applicability in supercapacitors with electrodes having a specific capacitance of 1000 Fg^{-1}. Meanwhile, the composite has good cycling stability of 97%. The initial capacitor at a current density of 25 Ag^{-1} for 3000 cycles is due to the connected percolated conductive path originating from dispersed graphene sheets. It has been found that the energy density and power density were 24.9 Wkg^{-1} and 900 Wkg^{-1}, respectively, at a current density of 1 Ag^{-1}. Gomez et al. [150] have fabricated graphene (G)-polyaniline (PANI) nanocomposite film via wet polymerization method for chemical application, which is characterized by Raman, scanning electron microscopy (SEM), high-resolution transmission electron microscopy (TEM), FTIR, and cyclic voltammetry (CV) at different ratios of graphene and aniline monomer. It has been found that the specific capacitance of G-PANI nanocomposite is 300–500 Fg^{-1} at a current density of 0.1 Ag^{-1}. These composites have application in supercapacitors and packaging to qualification for portable system application. Further, graphene-PANI in the 1:2 ratio gives good cyclic stability. Wang et al. [151] have synthesized a high-performance electrode material of fibrillar PANI doped with graphene oxide (GO) via in situ polymerization method. It has been investigated that the nanocomposites have a high conductivity of 10 Scm^{-1} at 295 K for a mass ratio of aniline/GO (100:1). Furthermore, the specific capacitance is high at 531 F/g in a potential range from 0 to 0.45 V.

The structure and morphology have been investigated by energy-dispersive X-ray technique (EDX), scanning electron microscopy (SEM), and transmission electron microscopy (TEM). This composite can be used as supercapacitor and in other power source system. Jin et al. [152] have designed sulfonated graphene polyaniline nanofiber (SGEPA) composite via oil/water interfacial polymerization whose chemical structure is characterized by scanning electron microscopy (SEM) and high-resolution transmission electron microscopy (TEM). More importantly, the composite shows a better electrochemical performance than pure aniline nanofibers. It has been found that there is a high specific capacitance of 962 f/g at a potential scanning rate of 2 mv/s and high cycle stability of about 78% after 100 cycles. Meanwhile, it also exhibits a high energy density of 68.86 wh/kg at a power density of 102 w/kg. These properties help this composite in the application of supercapacitors. Ansari and co-workers have been successfully reported that highly conductive graphene/polyaniline nanocomposites can be prepared by the in situ oxidative polymerization of aniline in the presence of cetyltrimethylammonium bromide [153]. Because of π–π interactions between PANI and GN, the distribution of GN is uniform in the matrix of PANI and obtained high electrical conductivity. Due to the conducting nature of GN and PANI, high electrical conductivity may be observed. Moreover, they have noticed that at room temperature, pTSA-doped GN-PANI composites exhibited enhanced electrical conductivity, i.e., 26.5 S/cm than PANI doped with inorganic acids [154]. Due to the additive/synergistic effect of GN and PANI, they show high electrical conductivity.

5.2 Polypyrrole (PPy) and Polythiophene (PTh)-Based Graphene Composites

Wang et al. [155] have investigated the polypyrrole (PPy) layers of intercalated graphene sheets by in situ intercalative chemical polymerization with different graphene and polypyrrole ratios, characterized by Fourier-transform infrared spectroscopy (FTIR), X-ray diffraction (XRD), Raman spectroscopy, and transmission electron microscope (TEM). It has been observed that the prepared graphene–PPy composite with mass ratio 1:10 displayed the best electrochemical properties which have high specific capacitance and good cycling stability during charge–discharge processes which are used as supercapacitor electrode due to the homogeneous intercalation of PPy layers on graphene substrate. Further, these supercapacitors have a specific capacitance of 650 fg^{-1} at current density of 0.45 Ag^{-1}, high energy density of 54.0 $whkg^{-1}$ at 1 ^{mA} current, and the highest power density of 778.1 wkg^{-1} at 5 mA current, and the cycling stability is 95% specific capacitance retained after 5000 cycles. Liu et al. [156] have reported the conductivity of polypyrrole fiber-graphene composites to be increased (141 S/cm), which is much better than the pure pyrrole fiber. Ding et al. have successfully reported that graphene/polypyrrole composite fiber is selected as an ideal choice for the fiber-based electrochemical supercapacitor due to its conductivity which is about 137–144 S/m, measured by probe method [157]. Li et al. [158] have fabricated poly [3-(2-(2-(2-(2-(diethanolamino)ethoxy)ethoxy) ethoxy ethoxy) thiophene] (PD4ET)-g-GO nanocomposite by esterification reaction, which is characterized by Fourier-transform infrared (FTIR), X-ray diffraction (XRD), Raman, and X-ray photoelectron spectroscopy (XPS). It has been seen that the specific capacitance of the composite is 971 f/g at a current density of 1 A/g. Meanwhile, the cycling stability shows that the PD4ET-g-GO-based capacitor retained 98% of its initial capacitance within 1000 consecutive charge–discharge cycles, which indicate a good cycling stability. Further, the composites have excellent electrochemical performance, which can be used as a power supplier in red LED diode in series connection. Nayebi et al. [159] have studied the mechanical properties of graphene-polythiophene nanocomposite by molecular dynamic simulation by relax force field. It has been investigated that the mechanical characteristics of tension along the zigzag orientation are higher than any other direction. Furthermore, by increasing the weight concentration of graphene in composite, Young's modulus and breaking stress get enhanced. However, Young's modulus decreases with an increase in temperature and defect in graphene structure. Further, with defect concentration, elastic modulus decreases gradually. Poly (3-hexylthiophene) P3HT with low molecular weight (Mn 6000)/graphene composite has been fabricated by Laguchi et al. [160] using in situ polymerization. It has been investigated by FTIR, atomic force microscopy (AFM), and field-emission scanning electron microscope (FE-SEM) that the P3HT/graphene complex has flat and ultrathin multilayered graphene of approximately <10 nm in thickness. However, the composite has high electrical conductivity. Alabadi et al. [161] have prepared GO-(thiophene-2,5-diyl-co(benzylidene)(TB) composite by in situ polymerization. It has been observed that the composite has good capacitive performance in alkali aqueous electrolytes, less internal resistance, and high specific capacitance (296 fg^{-1}) at a current density of 0.3 Ag^{-1}. Furthermore, over 91.86% of the long-term stability is retained after repeating the galvanostatic charge/discharge over 4000 cycles, which indicates high cycle stability. More importantly, a large energy density of up to 148 $Whkg^{-1}$ at a power density of 41.6 Wkg^{-1} of the GO-TB-based three-electrode device is obtained with alkali aqueous electrolytes.

6 Summary

The graphene-based polymeric composites have become potential candidates for application in various technological fields due to their outstanding physical and chemical properties. Graphene is the most significant among members of carbon family having sp^2 hybridized two-dimensional carbon-based nanofillers with the honeycomb crystal lattice. These composites have significantly improved mechanical, thermal, and electrical performances with apposite applicability in electronic, electromagnetic interference shielding devices, tissue engineering, sensor, power storage, supercapacitors, etc. However, the graphene-based polymer composites have better dispersion and homogeneity due to superior chemical interaction between the particles and polymer matrix, which resulted in improvement in the overall performances of the composite systems. In this chapter, emphasis is especially on the electrospinning processing technique of graphene and the attractive properties of various polymer-based graphene composites. These graphene-based materials have opened a new dimension for the production of low-cost, lightweight, easy processing, and high-performance composites with a wide range of applications for industrial and academic researchers.

References

1. Costa, P. et al. High-performance graphene-based carbon nanofiller/polymer composites for piezoresistive sensor applications. Compos. Sci. Technol. **153**, 241-252 (2017).
2. Sankaran, S., Deshmukh, K., Ahamed, M. B. & Pasha, S. K. K.Recent advances in electromagnetic interference shielding properties of metal and carbon filler reinforced flexible polymer composites: A review.Compos. Part A: Appl. Sci. Manuf. **14**, 49-71 (2018).
3. Lightcap, I.V. & Kamat, P. V.Graphitic design: prospects of graphene-based nanocomposites for solar energy conversion, storage, and sensing. Acc. Chem. Res. **46**, 2235-2243 (2013).
4. Lei, Z., Christov, N. & Zhao, X. S. Intercalation of mesoporous carbon spheres between reduced graphene oxide sheets for preparing high-rate supercapacitor electrodes. Energy Environ. Sci.**4**, 1866-1873 (2011).
5. Evanoff Jr, D. D. & Chumanov, G.Synthesis and optical properties of silver nanoparticles and arrays. Chem. Phys. Chem., **6**, 1221-1231 (2005).
6. Winey, K.I. & Vaia, V.Polymer nanocomposites. MRS Bull.**32**, 314-322 (2007).
7. Yin, Z. et al. Graphene-Based Materials: Synthesis, Characterization, Properties, and Applications.Nano-Micro Small **7**, 1876-1902 (2011).
8. Tong, X., Li, N., Zeng, M. & Wang, Q. Organic phase change materials confined in carbon-based materials for thermal properties enhancement: Recent advancement and challenges. Renew. Sust. Energ. Rev.**108**, 398-422 (2019).
9. Huang, Y., Liang, J. & Chen, Y. An overview of the applications of graphene-based materials in supercapacitors. Small, **8**, 1805-1834 (2012).
10. Khan, M. et al. Graphene based metal and metal oxide nanocomposites: synthesis, properties and their applications.J. Mater. Chem. A **3**, 18753–18808 (2015).
11. Stankovich, S. et al. Graphene-based composite materials. Nature **442**, 282-286 (2006).
12. Xiao, X. C., Xie, T. & Cheng, Y.T. Self-healable graphene polymer composites. J. Mater. Chem. **20**, 3508-3514 (2010).

13. Chen, S., Skordos, A. & Thakur, V. K. Functional nanocomposites for energy storage: Chemistry and new horizons. Mater. Today Chem. **17**, 100304(2020).
14. Sengupta, R., Bhattacharya, M., Bandyopadhyay, S. & Bhowmick, A. K.A review on the mechanical and electrical properties of graphite and modified graphite reinforced polymer composites. Prog. Polym. Sci. **36**, 638–670 (2011).
15. Liang, C. et al. Constructing interconnected spherical hollow conductive networks in silver platelets/reduced graphene oxide foam/epoxy nanocomposites for superior electromagnetic interference shielding effectiveness. Nanoscale **11**, 22590-22598 (2019).
16. Lee, C., Wei, X., Kysar, J. W. & Hone, J. Measurement of the elastic properties and intrinsic strength of monolayer graphene. Science **321**, 385-388 (2008).
17. Kuilla,T. et al. Recent advances in graphene based polymer composites. Prog. Polym. Sci. **35**, 1350–1375(2010).
18. Cui, Y., Kundalwal, S. I. & Kumar, S. Gas barrier performance of graphene/polymer nanocomposites. Carbon **98**, 313-333 (2016).
19. Novoselov, K.S. et al. Electric field effect in atomically thin carbon films. Science **306**, 666-669 (2004).
20. Pang, S., Hernandez, Y., Feng, X. & Mullen, K. Graphene as transparent electrode material for organic electronics. Adv. Mater. **23**, 2779-2795 (2011).
21. Iwan,A. & Chuchmała, A. Perspectives of applied graphene: Polymer solar cells. Prog. Polym. Sci. **37**, 1805-1828 (2012).
22. Usuki, A. et al. Composite material containing a layered silicate. US Pat. 889885 (1989).
23. Garcia, N. J. & Bazan, J. C. Electrical conductivity of montmorillonite as a function of relative humidity: La-montmorillonite. Clay Miner. **44**, 81-88 (2009).
24. Bao, Y.Z.,Cong, L.F.,Huang, Z.M. & Weng, Z.X. Preparation and proton conductivity of poly (vinylidene fluoride)/layered double hydroxide nanocomposite gel electrolytes. J. Mater. Sci. **43**, 390-394 (2008).
25. Ivanov, R. et al. PLA/Graphene/MWCNT composites with improved electrical and thermal properties suitable for FDM 3D printing applications. Appl. Sci. **9**, 1209 (2019).
26. Paszkiewicz, S. et al. A. Enhanced Functional Properties of Low-Density Polyethylene Nanocomposites Containing Hybrid Fillers of Multi-Walled Carbon Nanotubes and Nano Carbon Black. Polym. **12**, 1356 (2020).
27. Achaby, M. El. & Qaiss, A. Processing and properties of polyethylene reinforced by graphene nanosheets and carbon nanotubes. Mater. Design. **44**, 81–89 (2013).
28. Neubauer, E., Kitzmantel, M., Hulman, M. & Angerer, P. Potential and challenges of metal-matrix-composites reinforced with carbon nanofibers and carbon nanotubes. Compos. Sci. Technol. **70**, 2228-2236 (2010).
29. Kim, J. H. et al.Simple and cost-effective method of highly conductive and elastic carbon nanotube/polydimethylsiloxane composite for wearable electronics. Sci. Rep. **8**, 1375 (2018).
30. Wang, Y. et al. Effect of nano-scale Cu particles on the electrical property of CNT/polymer nanocomposites. Compos. Part A: Appl. Sci. Manuf.**143**, 106325 (2021).
31. Balkanloo, P. G., Mahmoudian, M.&Hosseinzadeh, M. T. A comparative study between MMT-Fe_3O_4/PES, MMT-HBE/PES, and MMT-acid activated/PES mixed matrix membranes. Chem. Eng. J.**396**, 125188 (2020).
32. Araby, S. et al. Recent advances in carbon-based nanomaterials for flame retardant polymers and composites. Compos. Part B: Eng.**212**, 108675(2021).
33. Thakur, K. &Kandasubramanian, B. Graphene and graphene oxide-based composites for removal of organic pollutants: A review. J. Chem. Eng. Dat. **64**, 833-867 (2019).
34. Guo, Y., Bao, C., Song, L., Yuan, B. & Hu,Y. In Situ Polymerization of Graphene, Graphite Oxide, and Functionalized Graphite Oxide into Epoxy Resin and Comparison Study of On-the-Flame BehaviorInd. Eng. Chem. Res. **50**, 7772-7783 (2011).
35. Dai, J., Wang, G., Ma, L. & Wu, C. Study on the surface energies and dispersibility of graphene oxide and its derivatives. J. Mater Sci. **50**, 38953907 (2015).
36. Cui, J., Xiong, Z., Qiu, H., LI, J. & Yang, J. Functionalized graphene oxide: Carrier for corrosion inhibitor and barrier in waterborne epoxy coatings. Compos. Part A: Appl. Sci. Manuf. **144**, 106354 (2021).

37. Zong, P. et al. Effect of aminopropylisobutyl polyhedral oligomeric silsesquioxane functionalized graphene on the thermal conductivity and electrical insulation properties of epoxy composites. RSC Adv. **6**, 10498-10506 (2016).
38. Persano, L., Camposeo, A.,Tekmen, C. & Pisignano, D. Industrial upscaling of electrospinning and applications of polymer nanofibers: A review. Macromol. Mater. Eng. **298**, 504-520 (2013).
39. He, X. X. et al. Near-field electrospinning: progress and applications. Phys. Chem. C **121**, 8663-8678 (2017).
40. Zhang, L. F., Aboagye, A. & Kelkar, A. A review: carbon nanofibers from electrospun polyacrylonitrile and their applications. J. Mater. Sci. **49**, 463-480 (2014).
41. Wang, X., Yu, J., Sun, G. & Ding, B. Electrospun nanofibrous materials: a versatile medium for effective oil/water separation. Mater. Today **19**, 403-414 (2015).
42. Kim, J. F.,Kim, J. H.,Lee, Y. M. & Drioli, E. Thermally induced phase separation and electrospinning methods for emerging membrane applications: A review. Alche J. **62**, 461-490 (2016).
43. Mirjalili M. & Zohoori, S. Review for application of electrospinning and electrospun nanofibers technology in textile industry. J. Nanostructures Chem. **6**, 207-213 (2016).
44. Zhang, C. L. & Yu, S. H. Nanoparticles meet electrospinning: recent advances and future prospects. Chem. Soc. Rev. **43**, 4423-4448 (2014).
45. Bhardwaj, N. S. C. Kundu, Electrospinning: a fascinating fiber fabrication technique. Biotechnol. Adv. **28**, 325 (2010).
46. Chen, Z., Xin, B.,Wu, X.,Wang, X. &Du, W. Preparation and characterisation of PSA/CNT composites and fibres. Fibres Text. East. Eur. **94**, 21-25 (2012).
47. Yu, J., Xin, B. & Shen, C. Preparation and characterization of PSA/PEDOT conductive composite yarns. Text. Res. J., **87** (5), 528-541 (2017).
48. Chen, W. J., Xin, B. J. & Wu, X. J. Fabrication and characterization of PSA nanofibers via electrospinning. J. Ind. Text. **44**, 159-179 (2014).
49. Garg, K. & Bowlin, G. L. Electrospinning jets and nanofibrous structures.Biomicrofluidics**5**, 013403 (2011).
50. I. F. Wahab et al. Electrospun graphene oxide-based nanofibers. Adv. Carbon Nanostructures, 101–120 (2016).
51. Bao, Q. et al. Atomic-layer graphene as a saturable absorber for ultrafast pulsed lasers. Adv. Funct. Mater. **19**, 3077-3083 (2009).
52. Das, S. et al .Electrospinning of polymer nanofibers loaded with noncovalently functionalized graphene. J. Appl. Polym. Sci. **128**, 4040-4046 (2013).
53. Kim, S. Y., Kim, B. H.,Yang, K. S.& Kim, K. Y. The formation of silica nanoparticles on the polyacrylonitrile-based carbon nanofibers by graphene via electrospinning. Mater. Lett. **71**, 74-77 (2012).
54. Correa, E., Moncada, M .E.,Gutiérrez,O. D.,Vargas, C. A. & Zapata, V. H. Characterization of polycaprolactone/rGO nanocomposite scaffolds obtained by electrospinning. Mater. Sci. Eng. C Mater. Biol. Appl. **103**, 109773(2019).
55. Guo, Y. et al. Significantly enhanced and precisely modeled thermal conductivity in polyimide nanocomposites with chemically modified graphene via in situ polymerization and electrospinning-hot press technology. J. Mater. Chem. C **6**, 3004-3015 (2018).
56. Hou, W. et al. Preparation and physico-mechanical properties of amine-functionalized graphene/polyamide 6 nanocomposite fiber as a high performance material. RSC Adv. **4**, 4848-4855 (2014).
57. Ehteshami, S., Feizbakhsh, A., Sarrafi, A. H. M., Panahi, H. A. & Roostaie, A. An electrospun polyamide/graphene oxide nanocomposite as a novel fiber coating. Anal. Methods **10**, 2123-2128 (2018).
58. Karumuthil, S. C., Rajeev, S. P., Valiyaneerilakkal, U., Athiyanathil, S. & Varghese, S. Electrospun Poly(vinylidene fluoride-trifluoroethylene)-Based Polymer Nanocomposite Fibers for Piezoelectric Nanogenerators. ACS Appl. Mater. Interfaces **11**, 40180-40188 (2019).

59. Nieto, A., Dua, R., Zhang, C., Boesl, B.& Ramaswamy, S.Three dimensional graphene foam/polymer hybrid as a high strength biocompatible scaffold. Adv. Functional Mater. **25**, 3916-3924 (2015).
60. Samad, Y. A. Li, Y. Alhassan, S. M. & Liao, K. Novel graphene foam composite with adjustable sensitivity for sensor applications.ACS Appl. Mater. Interf. **7,** 9196–9202 (2015).
61. Sun, H. B., Yang, J., Zhou, Y. Z., Zhao, N. & Li, D. Preparation of reduced graphene oxide films by dip coating technique and their electrical conductivity. Mater. Technol. **29,** 14-20 (2014).
62. Fang, M. et al. Preparation of highly conductive graphene-coated glass fibers by sol-gel and dip-coating method. J. Mater. Sci. Technol., **3**, 1989-1995 (2019).
63. Chatterjee, A., Kumar, M. N. & Maity, S. Influence of graphene oxide concentration and dipping cycles on electrical conductivity of coated cotton textiles. J. Text. Inst. **108**, 1910-1916 (2017).
64. Berendjchi, A., Khajavi, R., Yousefi, A. A. & Yazdanshenas, M. E. Surface characteristics of coated polyester fabric with reduced graphene oxide and polypyrrole. Appl. Surf. Sci. **367**, 36-42 (2016).
65. Liu, X. et al. Fabricating conductive poly(ethylene terephthalate) nonwoven fabrics using an aqueous dispersion of reduced graphene oxide as a sheet dyestuff. RSC Adv. **4**, 23869-23875 (2014).
66. Kongahge, D., Foroughi, J., Gambhir, S., Spinks, G. M. & Wallace, G. G. Fabrication of a graphene coated nonwoven textile for industrial applications. RSC Adv. **6**, 73203-73209 (2016).
67. B. Thomas, H. J. Maria, G. George, S. Thomas, N.V. Unnikrishnan, K. Joseph, A novel green approach for the preparation of high performance nitrile butadiene rubber-pristine graphene nanocomposites. Compos. Part B Eng. **175**, 107174 (2019).
68. Bahrami, S., Solouk, A., Mirzadeh, H. & Seifalian, A. M. Electroconductive polyurethane/graphene nanocomposite forbiomedicalapplication. Compos. Part B Eng. **168**, 421-431 (2019).
69. Fan, J. et al. Gum arabic assisted exfoliation and fabrication of Ag–graphene-based hybrids. J. Mater. Chem. **22**, 13764-13772 (2012).
70. Johnson, D. W., Dobson, B. P.& Coleman, K. S. A manufacturing perspective on graphene dispersions. Curr. Opin. Colloid Interface Sci. **20**, 367-382 (2015).
71. Konios, D., Stylianakis, M. M., Stratakis, E. & Kymakis, E. Dispersion behaviour of graphene oxide and reduced graphene oxide. J. Colloid Interface Sci. **430**, 108-112 (2014).
72. Yang, X. et al. A high-performance graphene oxide-doped ion gel as gel polymer electrolyte for all-solid-state supercapacitor applications. Adv. Func. Mater. **23**, 3353 (2013)
73. Fattah, N. et al. An approach to solid-state electrical double layer capacitors fabricated with graphene oxide-doped, ionic liquid-based solid copolymer electrolytes. Mater. **9**, 450 (2016).
74. Mohammed, H. A. et al. Fabrication and characterizations of a novel etched-tapered single mode optical fiber ammonia sensors integrating PANI/GNF nanocomposite.Sens. Actuators B: Chem. **287**, 71-77 (2019).
75. Soltani-kordshuli, F., Zabihi, F. & Eslamian, M. Graphene-doped PEDOT: PSS nanocomposite thin films fabricated by conventional and substrate vibration-assisted spray coating (SVASC). Eng. Sci. Technol. Int. J. **19**, 1216-1223 (2016).
76. Jaworek, A. Electrospray droplet sources for thin film deposition. J. Mater. Sci. **42**, 266-297 (2007).
77. Eslamian, M. Spray-on thin film PV solar cells: advances, potentials and challenges. Coatings **4**, 60-84 (2014).
78. Adelowo, E., Baboukani, A. R., Chen, C. & Wang, C. Electrostatically sprayed reduced graphene oxide-carbon nanotubes electrodes for lithium-ion capacitors. J. Carbon Res. **4**, 31 (2018).
79. Adak, B., Joshi, M. & Butola, B. S. Polyurethane/functionalized-graphene nanocomposite films with enhanced weather resistance and gas barrier properties. Compos. Part B: Eng. **176**, 107303 (2019).

80. Nordin, N. M., Buys, Y. F., Anuar, H., Ani, M. H. & Pang, M. M. Development of conductive polymer composites from pla/tpu blends filled with graphene nanoplatelets. Today Proceeding **17**, 500 (2019).
81. Khanam, P. N. et al. Melt processing and properties of linear low density polyethylene-graphene nanoplatelet composites. Vacuum **130**, 63-71 (2016).
82. Panchakarla, L. S.,Govindaraj, A. & Rao, C. N. R. Nitrogen- and boron-doped double-walled carbon nanotubes. ACS Nano **1**, 494 (2007).
83. Subrahmanyam, K. S., Govindaraj, P. L. & Rao, C. N. R. Simple method of preparing graphene flakes by an arc-discharge method. J. Phys. Chem C **113**, 4257-4259 (2009).
84. Goldberg, D. & Goldberget, D. The tribological properties of solid lubrication graphite coatings prepared by a sol–gel method. Carbon **38**, 2017-2020 (2005).
85. Meyer, J. C., Simon, S.K., Park, J.H., Skakalova, V. & Künze, D. Experimental analysis of charge redistribution due to chemical bonding by high-resolution transmission electron microscopy. Nat. Mater. **10**, 209-215 (2011).
86. Banhart, F., Kotakoski, J. &Krasheninnikov,A. V. Structural defects in graphene. ACS Nano **5**, 26-41 (2011).
87. Ruiz-Vargas, C.S. et al. Softened elastic response and unzipping in chemical vapor deposition graphene membranes. Nano Lett., **11**, 2259-2263 (2011).
88. Stone, A. J. & Wales, D. J. Theoretical studies of icosahedral C_{60} and some related species. Chem. Phys. Lett. **128**,501-503 (1986).
89. Meyer, J. C. et al. Direct imaging of lattice atoms and topological defects in graphene membranes. Nano Lett. **8**, 3582-3586 (2008).
90. Banhart, F., Kotakoski, J. & Krasheninnikov, A. V. Structural defects in graphene. ACS nano **5**, 26-41 (2011).
91. Girit, C.O. et al. Graphene at the edge: stability and dynamics, Science **323**, 1705-1708 (2009).
92. Banhart, F. Interactions between metals and carbon nanotubes: at the interface between old and new materials. Nanoscale **1**, 201-213 (2009).
93. Krasheninnikov, A. & Banhart, F. Engineering of nanostructured carbon materials with electron or ion beams. Nat. Mater. **6**, 723-733 (2007).
94. Tapaszto, L. et al. Tuning the electronic structure of graphene by ion irradiation. Phys. Rev. B **78**, 233407 (2008).
95. Bagri, A. et al. Structural evolution during the reduction of chemically derived graphene oxide. Nat. Chem., **2**, 581-587 (2010).
96. Cortijo, A. & Vozmediano, M.A.H. Effects of topological defects and local curvature on the electronic properties of planar graphene. Nucl. Phys. B **763**, 293-308 (2007).
97. Li, J., Tang, T., Luo, L. & Yao, J. Enhancement and modulation of photonic spin Hall effect by defect modes in photonic crystals with graphene. Carbon **134**, 293-300 (2018).
98. Gouda, M.H. et al. Poly(vinyl alcohol)-based crosslinked ternary polymer blend doped with sulfonated graphene oxide as a sustainable composite membrane for direct borohydride fuel cells. J. Power Sources **432**, 92-101 (2019).
99. Cobos, M., Pinta, I. D. L.,Quindos, G., Fernandeza, M. J. & Fernandeza, M.D. One-step eco-friendly synthesized silver-graphene oxide/poly (vinyl alcohol) antibacterial nanocomposites. Carbon **150**, 101-116 (2019).
100. Feng, L. & Liu, Z. Graphene in biomedicine: opportunities and challenges. Nanomedicine **6**,317-324 (2011).
101. Liu, Z., Robinson, J. T., Tabakman, S. M., Yang, K. &Dai, H. Carbon materials for drug delivery & cancer therapy. Mater. Today **14**, 316-323 (2011).
102. Sharma, B., Shekhar, S., Gautam, S., Sarkar, A. & Jain, P. Nanomechanical analysis of chemically reduced graphene oxide reinforced poly (vinyl alcohol) nanocomposite thin films. Polym. Testing **70**, 458-466 (2018)
103. Malik, P., Bhasha, B. & Jain, P. Influence of surface modified graphene oxide on mechanical and thermal properties of epoxy resin. Orient J. Chem. **34**, 1597-1603 (2018).
104. Wang, Z. et al. Ultrahigh dielectric constant and low loss of highly-aligned graphene aerogel/poly (vinyl alcohol) composites with insulating barriers. Carbon **123**, 385-394 (2017).

105. Khan, A., Jain, R. K., Luqman, M. & Asiri,A. M. Development of sulfonated poly (vinyl alcohol)/aluminium oxide/graphene based ionic polymer-metal composite (IPMC) actuator. Sens. Actuator A Phys. **280**, 114-124 (2018).
106. Huang, Y., Liang, J. &Chen, Y. The application of graphene based materials for actuators. J. Mater. Chem. **22**, 3671-3679 (2012).
107. Chattaraj, R., Bhaumik, S., Khan, S. &Chatterjee, D. Soft wearable ionic polymer sensors for palpatory pulse-rate extraction. Sens. Actuators A: Phys. **270**, 65-71 (2018).
108. Dong, J., Wang, Z. & X. Kang. The synthesis of graphene/PVDF composite binder and its application in high performance MnO_2 supercapacitors. Colloids and Surfaces A: Physicochem. Eng. Aspects **489**, 282–288 (2016).
109. Ahmad, A. L., Farooqui, U. R. & Hamid, N. A. Synthesis and characterization of porous poly(vinylidene fluoride-co-hexafluoro propylene) (PVDF-co-HFP)/poly(aniline) (PANI)/graphene oxide (GO) ternary hybrid polymer electrolyte membrane. Electrochim. Acta **283**, 842-849(2018).
110. Wan, Y. J., Yang, W. H., Yu, S. H., Sun, R., Wong, C. P. & Liao, W. H. Covalent polymer functionalization of graphene for improved dielectric properties and thermal stability of epoxy composites. Compos. Sci. Technol. **122**, 27-35 (2016).
111. Moharana, S. & Mahaling, R.N. Silver (Ag)-Graphene oxide (GO) - Poly (vinylidene fluoride-co-hexafluoropropylene) (PVDF-HFP) nanostructured composites with high dielectric constant and low dielectric loss. Chem. Phys. Lett. **680**, 31-36 (2017).
112. Chen, Z., Liu, Y., Fang, L., Jiang, P. & Huang, X. Role of reduced graphene oxide in dielectric enhancement of ferroelectric polymers composites. Appl. Surf. Sci. **470**, 348-359 (2019).
113. He, Z. Z. et al. Compos. Part ALargely enhanced dielectric properties of poly (vinylidene fluoride) composites achieved by adding polypyrrole-decorated graphene oxide. Appl. Sci. Manuf. **104**, 89–100 (2018).
114. Li, Y. et al. Fe_3O_4 decorated graphene/poly (vinylidene fluoride) nanocomposites with high dielectric constant and low dielectric loss. Compos. Sci. Technol. **171**, 152-161 (2019).
115. Li, Y. et al. Polydopamine coating layer on graphene for suppressing loss tangent and enhancing dielectric constant of poly (vinylidene fluoride)/graphene composites. Compos. Part A Appl. Sci. Manuf. **73**, 85-92 (2015).
116. Zhang, Y., Wang, Y., Deng, Y., Li, M. & Bai, J. Enhanced dielectric properties of ferroelectric polymer composites induced by metal-semiconductor Zn-ZnO core–shell structure. ACS Appl. Mater. Interfaces **4**, 65-68 (2012).
117. Tong, W. et al. Amorphous TiO_2-coated reduced graphene oxide hybrid nanostructures for polymer composites with low dielectric loss. Chem. Phys. Lett. **638**, 43-46 (2015).
118. Tong, W. et al. Achieving significantly enhanced dielectric performance of reduced graphene oxide/polymer composite by covalent modification of graphene oxide surface. Carbon **94**, 590-598 (2015).
119. Li, H. et al. Poly (vinyl pyrrolidone)-coated graphene/poly (vinylidene fluoride) composite films with high dielectric permittivity and low loss. Compos. Sci. Technol.**121**, 49-55 (2015).
120. Wu, Y. et al. Exceptional dielectric properties of chlorine-doped graphene oxide/poly (vinylidene fluoride) nanocomposites. Carbon **89**, 102-112 (2015).
121. Gangineni,P. K. Mechanical behavior of Graphene decorated carbon fiber reinforced polymer composites: An assessment of the influence of functional groups. Compos. Part A: Appl. Sci. Manuf. **122**, 36-44 (2019).
122. Wan, Y. J. et al. Grafting of epoxy chains onto graphene oxide for epoxy composites with improved mechanical and thermal properties. Carbon **69**, 467-480 (2014).
123. Surnova, A., Balkaev, D., Musin, D. Amirov, R. & Dimiev, A. M. Fully exfoliated graphene oxide accelerates epoxy resin curing, and results in dramatic improvement of the polymer mechanical properties. Compos. Part B. Eng. **162**, 685-691 (2019).
124. Feng, Y. et al. Multiple synergistic effects of graphene-based hybrid and hexagonal born nitride in enhancing thermal conductivity and flame retardancy of epoxy. Chem. Eng. J. **379**, 122402 (2020).

125. Gong, X. et al. Amino graphene oxide/dopamine modified aramid fibers: Preparation, epoxy nanocomposites and property analysis. Polym. **168**, 131-137 (2019).
126. Zhang, T. et al. Grafting of polystyrene onto reduced graphene oxide by emulsion polymerization for dielectric polymer composites: High dielectric constant and low dielectric loss tuned by varied grafting amount of polystyrene. Eur. Polym. J. **94**, 196-207 (2017).
127. Liu, Y. et al. Polystyrene/graphene oxide nanocomposites synthesized via Pickering polymerization. Prog. Org. Coat. **99**, 23-31 (2016).
128. He, F. et al. Fabrication of graphene nanosheet (GNS)–Fe_3O_4 hybrids and GNS–Fe_3O_4/syndiotactic polystyrene composites with high dielectric permittivity. Carbon **58**, 175 (2 013)
129. Tu, Z. et al. A facile approach for preparation of polystyrene/graphene nanocomposites with ultra-low percolation threshold through an electrostatic assembly process. Compos. Sci. Technol. **134**, 49-56 (2016).
130. Chen, Y. et al. Enhanced electromagnetic interference shielding efficiency of polystyrene/graphene composites with magnetic Fe_3O_4 nanoparticles. Carbon **82**, 67-76 (2015).
131. Wang, J., Jin, X., Zhang, X., Wu, H. & Guo, S. Effect of tunable styrene content on achieving high-performance poly (styrene-b-ethylene-ran-butylene-b-styrene)/graphene oxide nanocomposites. Compos. Sci. Technol. **164**, 229-237 (2018).
132. Deshmukh, K. et al.Graphene oxide reinforced poly (4-styrenesulfonic acid)/polyvinyl alcohol blend composites with enhanced dielectric properties for portable and flexible electronics. Mater. Chem.Phys. **186**, 188-201 (2017).
133. Adloo, A., Sadeghi, M., Masoomi, M. & Pazhooh, H. N. High performance polymeric bipolar plate based on polypropylene/graphite/graphene/nano-carbon black composites for PEM fuel cells. Renew. Energy **99**, 867-874 (2016).
134. Wang, D. et al. Dielectric properties of reduced graphene oxide/polypropylene composites with ultralow percolation threshold. Polym. **54**, 1916-1922 (2013).
135. Cao, L., Su, D., Su, Z. & Chen, X. Fabrication of multiwalled carbon nanotube/polypropylene conductive fibrous membranes by melt electrospinning. Ind. Eng. Chem. Res. **53**, 2308-2317 (2014).
136. Song, N. et al. Highly thermally conductive polypropylene/graphene composites for thermal management. Compos. Part A: Appl. Sci. Manufact. **135**, 105912 (2020).
137. Bafana, A.P. et al., Polypropylene nanocomposites reinforced with low weight percent graphene nanoplatelets. Compos. Part B: Eng. **109**, 101-107 (2017).
138. Song, P. et al. Fabrication of exfoliated graphene-based polypropylene nanocomposites with enhanced mechanical and thermal properties. Polym. **52**, 4001-4010 (2011).
139. Chen, M. et al. Dielectric and mechanical properties and thermal stability of polyimide–graphene oxide composite films. Thin Solid Films **584**, 232-237 (2015).
140. Luong, N. D. et al. Enhanced mechanical and electrical properties of polyimide film by graphene sheets via in situ polymerization, Polym. **52**, 5237-5242 (2011).
141. Ramakrishnan, S., Dhakshnamoorthy, M., Jelmy, E. J., Vasanthakumari, R. & Kothurkar, N. K. Synthesis and characterization of graphene oxide–polyimide nanofiber composites. RSC Adv. **4**, 9743-9749 (2014).
142. Yi, B., Zhao, Y., Tian, E., Li, J. & Ren, Y. High-performance polyimide nanofiber membranes prepared by electrospinning. High Performance Polym. **31**, 438-448 (2019).
143. Pokharel, P. et al. Effects of functional groups on the graphene sheet for improving the thermomechanical properties of polyurethane nanocomposites. Compos. Part B **78**, 192-201 (2015).
144. Yuan, Z., Lu, Z., Yang, Z., Sun, J. & Xie, F. A criterion for the normal properties of graphene/polymer interface. Comput. Mater. Sci. **120**, 13-20 (2016).
145. Jiang, L.Y. et al. A cohesive law for carbon nanotube/polymer interfaces based on the van der Waals force. J. Mech. Phys. Solids **54**, 2436-2452 (2006).
146. Sadasivuni, K. K. et al. Dielectric properties of modified graphene oxide filled polyurethane nanocomposites and its correlation with rheology. Compos. Sci. Technol. **104**, 18-25 (2014).

147. Gupta, S., McDonald, B., Carrizosa, S. B. & Price, C. Microstructure, residual stress, and intermolecular force distribution maps of graphene/polymer hybrid composites: Nanoscale morphology-promoted synergistic effects. Compos. Part B: Eng. **92**, 175-192 (2016).
148. Du, Y. et al. Simultaneous increase in conductivity and Seebeck coefficient in a polyaniline/graphene nanosheets thermoelectric nanocomposite. Synt. Met. **161**, 2688 -2692 (2012).
149. Haldar, P., Biswas, S., Sharma, V., Chowdhury, A. & Chandra, A. Mn_3O_4-polyaniline-graphene as distinctive composite for use in high-performance supercapacitors. Appl. Surf. Sci. **491**, 171-179 (2019).
150. Gómez, H. et al. Graphene-conducting polymer nanocomposite as novel electrode for supercapacitors. J. Power Sources, **196**, 4102-4108 (2011).
151. Wang, H., Hao, Q., Yang, X., Lu, L. & Wang, X. Graphene Oxide Doped Polyaniline for Supercapacitors. Electrochem. commun. **11**, 1158-1161(2009).
152. Jin, Y., Huang, S., Zhang, M. & Jia, M. Preparation of sulfonated graphene–polyaniline nanofiber composites by oil/water interfacial polymerization and their application for supercapacitors. Synth. Met., **168**, 58-64 (2013).
153. Ansari, M.O. et al. pTSA doped conducting graphene/polyaniline nanocomposite fibers: Thermoelectric behavior and electrode analysis. Chem. Eng. J. **242**, 155-161 (2014).
154. Luo, Y. et al. Electricity generation from indole and microbial community analysis in the microbial fuel cell. J. Hazard Mater. **176**, 759-764 (2010).
155. Liu, Y. et al. Graphene/polypyrrole intercalating nanocomposites as supercapacitors electrode Electrochim. Acta **112**, 44-52 (2013).
156. Li, S. et al. One-step synthesis of graphene/polypyrrole nanofiber composites as cathode material for a biocompatible zinc/polymer battery. ACS Appl. Mater. Interfaces **6**, 16679-16686 (2014).
157. Ding, X., Spinning fabrication of graphene/polypyrrole composite fibers for all-solid-state, flexible fibriform supercapacitors. J. Mater. Chem. A **2**, 12355-12360 (2014).
158. Li, Y. et al. Remarkably enhanced performances of novel polythiophene-grafting-graphene oxide composite via long alkoxy linkage for supercapacitor application. Carbon, **147**, 519-531 (2019).
159. Nayebi, P. & Zaminpayma, E. A molecular dynamic simulation study of mechanical properties of graphene–polythiophene composite with Reax force field. Phys. Lett. A **380**, 628-633 (2016).
160. Iguchi, H. Et al. Preparation of uncurled and planar multilayered graphene using polythiophene derivatives via liquid-phase exfoliation of graphite. Flat Chem. **8**, 31-39 (2018).
161. Alabadi, A., Razzaque, S., Dong, Z., Wang, W. & Tan, B. Graphene oxide-polythiophene derivative hybrid nanosheet for enhancing performance of supercapacitor. J. Power Sources **306**, 241-247 (2016).

Defect in Carbon Nanostructures Through Electrospinning: Status and Prospect

Raunak Pandey, Prabhav Thapa, Magdalena Bogdan, K. C. Nayak, Nannan Wang, Michał Bystrzejewski, and Santosh K. Tiwari

Abstract A novel technique for the fabrication of nanofibers through electrospinning has proven effective in the field of nanotechnology. Nanofibers of polymers and composites produced by drawing charged threads from polymer solutions or polymer melts by the effective use of electric force have applications in daily use of the twenty-first century. But the use of metals, ceramics, carbohydrates, crystalline polysaccharides, and cellulose as a solvent for electrospinning is also gaining a popular trend now. From the development of electric force way back in the 80s to the application of these forces to draw polymers from electric charges, electrospinning is modified. The modification is so vast that structural change from needle to needleless electrospinning, from single spinneret to multiple jets is developed. Structural changes using carbon nanostructures with the change in polymer solution from inorganic to organic and from organic to herbal are also examined for several prototype applications. These modifications in characteristics leading to improved hydrophilicity, tensile strength, electrical properties and permeability prove electrospinning a worthy candidate for nanotechnology in terms of energy storage, tissue engineering and biomedical applications. These applications are proof that electrospinning has the capability of shaping technology with mass industrial productions.

R. Pandey · P. Thapa
Department of Chemical Science and Engineering, Kathmandu University, Dhulikhel, Nepal

M. Bogdan
Department of Social Medicine and Public Health, Medical University of Warsaw, Wigury 61, 02-091 Warsaw, Poland

K. C. Nayak
Air and Thermal Systems Specialist, Rolls-Royce, Bangalore, India

N. Wang · S. K. Tiwari
Key Laboratory of New Processing Technology for Nonferrous Metals and Materials, Ministry of Education, School of Resources, Environment and Materials, Guangxi University, Nanning, China

M. Bystrzejewski · S. K. Tiwari (✉)
Faculty of Chemistry, Warsaw University, 1 Pasteur Str., 02-093 Warsaw, Poland

M. Bystrzejewski
e-mail: mibys@chem.uw.edu.pl

S. Sahoo et al. (eds.), *Defect Engineering of Carbon Nanostructures*, Advances in Material Research and Technology, https://doi.org/10.1007/978-3-030-94375-2_8

Referring to the applications from past to present, electrospinning has been developed by designing the machine to be more versatile, tailoring the collectors and spinnerets, using different voltages for different applications and using different polymer solutions for enhanced properties, which happen to create a benchmark assembly for developing artificial intelligence and smart materials used as an advancement in science and technology. In this chapter, the historical prospect of electrospinning, its development and modifications for the effective production of nanofibers consisting of carbon nanomaterials are discussed with suitable examples. In addition, selected engineering applications of nanofibers modified with carbon nanomaterials for biomedical, energy, environmental, sensory, agricultural, optoelectrical, and food packaging are briefly discussed.

Keywords Electrospinning · Nanotechnology · Smart materials · Historical prospect · Applications

1 Introduction

Applications in science and technology from the scale of a nanometer or the point of view of a nanometer can be termed as nanotechnology. Among various techniques of nanotechnology, electrospinning is also a novel approach for fabricating nanofibers to be used in applications of science and technology. Polymer fibers and wires comprising of natural, synthetic, or both properties are influenced under electrical forces providing electrostatic fiber with a diameter ranging from two nanometers to several micrometers can be referred to as electrospinning [1]. This initial discovery of the electrostatic force was later applied to fluid to study its influence providing enough reference for the development of electrospinning [2]. Research on electrospinning, pre-modification, and post-modification of the nanofibers offers controlled morphological structures of the produced fibers, enhanced characteristics helping to obtain the desired properties for the particular applications in the field of diverse subject of science and technology [3]. These researches also provided the development of polymer nanofibers in the form of natural and synthetic, which is friendlier towards that particular applications. The use of synthetic polymers started when electrospinning was patented in the 1900s [4] and was used extensively then. But nowadays, synthetic polymers are being slowly replaced by natural polymers although the processing of the solution is difficult. Researchers are focusing more on the natural polymer solution for the fabrication of nanofibers because it is more beneficial in terms of biocompatibility, cell attachment, and cell proliferation [5]. The use of synthetic, natural nanofibers is diverse, and all these nanofibers can be surface modified after electrospinning or while electrospinning by modifications of the device itself to produce hollow and core–shell nanofibers. These hollow and core–shell nanofibers have biologically active material incorporating a controlled release profile and excellent characteristics of large surface area to volume ratio for diverse applications [6].

The development of polymer-based nanofibers is speeding up fast but simultaneously, carbohydrate-based nanofibers, native crystalline polysaccharide-based nanofibers, metal and metal oxide-based nanofibers, and cellulose-based nanofibers are also being developed, which may limit the use of polymer-based nanofibers in the future. The polymer-based nanofibers obtained from electrospinning have diverse applications in environment, energy, and other fields of nanotechnology. They also play an important role in the widely understood health field, including biomedical engineering and pharmaceutical sector. For tissue engineering and regenerative applications, scaffold porosity, pore size, and morphology of biodegradation profile and biocompatibility have more significance [7]. Tailoring the characteristics of nanofibers to ensure enough crystallinity, electrical conductivity, energy-storing, and supply is developed for applications in the energy sector. Research on piezoelectric devices had gained much attention for artificial intelligence and smart material system having the characteristics of wide bandwidth, fast electromechanical response, low power requirements, and high generative forces [8]. These applications make the energy sector an enthusiastic field for research. Electrospun nanofibers producing a high surface area to volume ratio is important for the efficient interaction of nanoparticles with the desired substrate. Hence, the released kinetics of incorporated nanoparticles plays an important role in offering a huge application of enzyme inhibition in food industries [9, 10]. Electrospun fibers can be produced for the controlled release of the incorporated particles, which is applicable in the agricultural sector as well [11]. On the other hand, electrospun fibers acting as membrane offer less permeability of the materials passing through it, which makes it a perfect candidate in filtration systems. These membrane-based fibers incorporate antimicrobial properties for enhanced filtration for environmental protection [12, 13]. With less permeability of nanofibers and the good orientation of fibers, nanofibers offer applications in clothing in extreme conditions, windproof clothing, personal protective equipment (PPEs), and so on. However, optimal manufacturing conditions are required [14, 15]. Thus, nanofibers have been playing an important role in various applications of daily use and will be playing a crucial role in the future upcoming.

In this review, a brief introduction of nanotechnology, electrospinning, and its nanofibers is discussed along with historical prospects of electrospinning in terms of applications. Various applications of nanotechnology using electrospinning techniques are also discussed briefly, which will be a basis for future applications (Fig. 1).

2 Nanotechnology, Electrospinning and Its Importance

Research tools of substantial value are provided by nanotechnology to shape the future of Earth. Nanotechnology refers to the technique of producing materials on a nanoscale exhibiting a unique blend of properties. Engineering and innovating the technological levels in nanometer scale have always been a great strength of this field of science, and it will take over the world in about a decade. Improvement or even

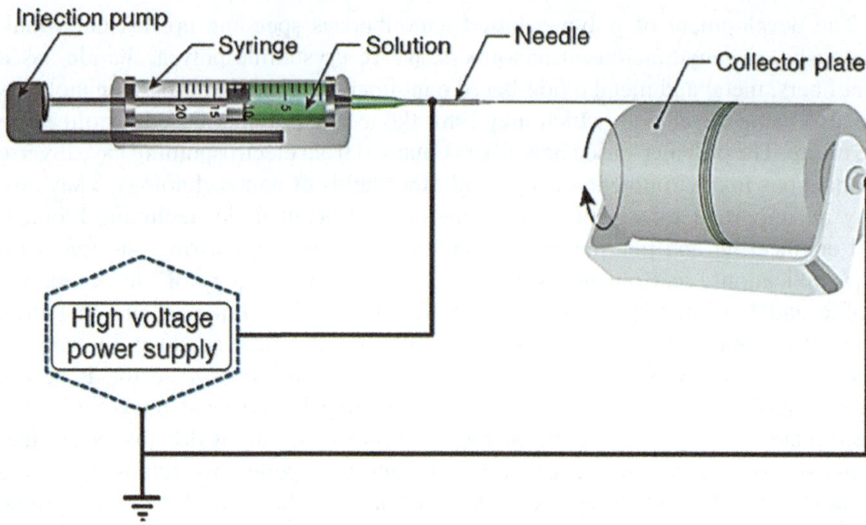

Fig. 1 The Electrospinning apparatus (Reproduced with permission from Ref. [16])

revolution in the technological field is provided by nanotechnology, which, of course, is important for the point of view in different areas of science. The incorporation of characteristics such as enhanced strength, durability, reactivity, conductivity, sieve-like behavior has provided a resistive foundation for nanotechnology to enter into the scientific era. The reduction of the size but the enhancement of properties of different appliances can be observed today, all thanks to the use of nanomaterial, the creation of nanotechnology. The manufacturing of nanomaterials can be done via two approaches: top-down approach and bottom-up approach. In a bottom-up approach, nanomaterials are produced from its elemental level, while the top-down approach is its reverse [16, 17]. Some of the manufacturing techniques are lithography, sol–gel method, self-assembly, chemical synthesis, electrospinning, and so on. Out of all existing approaches of nanotechnology, electrospinning is one of the most promising techniques.

Being a diverse field, nanotechnology's subdivision of electrospinning provides a novel technique of fabrication of nanofibers. Production of very thin fibers by the action of electric force in nanometers scales with large surface areas, ease of processing and scale-up for industrial scale, modification, and functionalization, superior mechanical properties make this field an important field for scientific research of nanotechnology [5]. Electrospinning is the technique of drawing continuous fibers from the required polymer solution. The present-day electrospinning apparatus involves a syringe pump with a needle, high voltage supply, and a collector [18]. The polymer solution is drawn in a controlled manner from the syringe pump. A high DC voltage is placed across the grounded collector plate and tip of the needle. During the release of solutions from the tip of the needle, the charge at the collector overcomes the surface tension of the polymer solution. This results in the conical structure of the fiber, which is called the Taylor cone. The reason for the attraction of the drawn polymer solution to the collector is due to the difference in polarity between needle and collector [19]. The main reason for the use of electrospun fibers is due to their nano-scale diameters and high surface-to-volume ratio. Similarly, nanoscaffolds resemble good morphology and high porosity, which enhances the good attachment of nanoparticles [20, 21]. The ease of modifications of morphology in electrospun nanofibers has intensified its popularity. The ability to functionalize the surface of nanofibers is one of the reasons, for its applications in numerous fields [15, 22].

Utilization of nanotechnological field with electrospinning in agriculture, biomedical, energy, and other related fields have been proven effective. The contribution of electrospinning in nanofiber fabrication due to its enhanced characteristics related to morphology has been crucial in these fields. Tissue engineering, drug delivery, and regenerative medicines are very relevant examples to be viewed for the importance of electrospun nanofibers. The enhanced properties of electrospun nanofibers provide important help in tissue engineering for bone, cartilage, cardiovascular, and other important parts of human bodies. Along with this example, the importance of nanofibers in energy storage through piezoelectric characteristics, battery applications, pressure characteristics, super capacitance characteristics is more than enough to shape up the concept of renewable energy, which is in dire need. Environmental issues of air and water pollution have been dealt with the use of nanofibrous membranes, smart materials in textiles, energy applications have also gained a lot of interest by the use of nanofibers produced from electrospinning. All these examples of research are trying to create a bridge to visualize smart materials and artificial intelligence, which can be also called super materials. Thus, these applications provide enough examples of the importance of electrospinning and ultimately of nanotechnology (Fig. 2).

3 Historical Prospect

Blending science and engineering of the biological, chemical, and physical world provided a platform for manufacturing nanofibers through electrospinning. These fibers can be drawn via electrostatic force, which is applicable in various fields [23]. The origin of the electrospinning technique dates back to the sixteenth century. The development of electrostatic force to the discovery of the Taylor cone has a long history. The first recorded electrostatic force was in 1628 AD, which was well explained by William Gilbert. In 1882 AD, Sir Rayleigh proposed a theory on how much charge a liquid droplet could carry, which was later considered as 'Rayleigh Limit'. Later in 1900 AD, J. F. Cooley first personalized the apparatus of electrospinning with three types of spinning heads. He analyzed the agglomeration of solution in the auxiliary electrode [24]. He inspected that needle spinning might be a better approach for the production of fibers and patented it [24, 25]. His postulates were

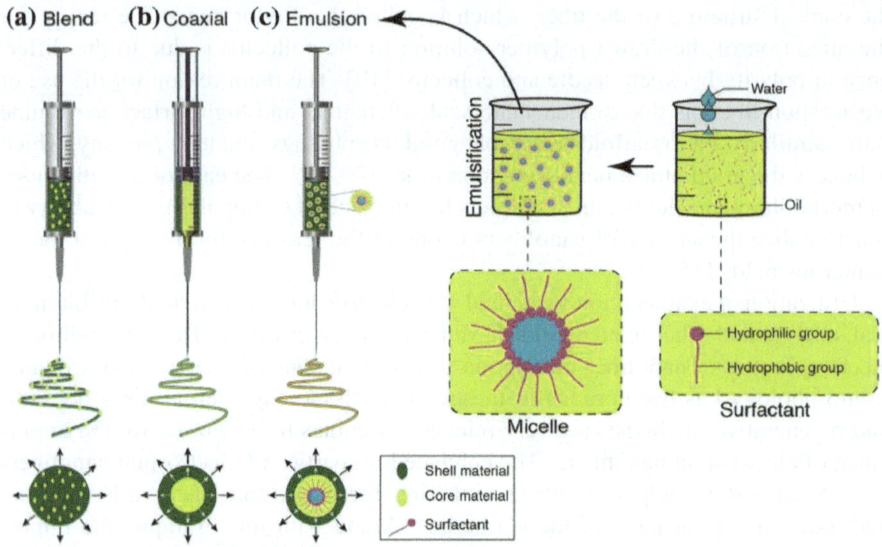

Fig. 2 Schematic displays of the spinneret loaded with a bioactive agent for **a** blend, **b** coaxial, and **c** emulsion electrospinning (Reproduced with permission from Ref. [16])

later studied by John Zeleny who discovered the behavior of fluids under the influence of electrostatic force [2]. He concluded that the discharge stress of the liquid is higher on the sharp ends than cylindrical ends. He also discovered that positive polarity discharge liquids oxidizes the tip of needles more than negative polarity [24]. Later, the first distortion of the liquid drop was developed by Sir Geoffrey Ingram Taylor.

The arrival of modern electrospinning apparatus seems a boon in the history of nanotechnology, however, there is a lot of room for improvements [2, 23]. The mimicking of electrospun fibers was focused on the development of gas masks under the name of "Petryanov Filters", which were resistant to smoke and gases. In the 1950s, it was known that electrospinning can draw thin fibers of viscous liquids like cellulose acetate; however, it was yet to be commercialized [26]. It was discovered that a corona discharge can be produced from the pointed electrode with high voltage. When a liquid with the same charge is placed near the electrode, it will repel the molecules in an orderly manner. This technique used for the spraying will give homogenous fabrication of liquid solution over the substrate [26]. Later, it was Sir Geoffrey Ingram Taylor who did a detailed study of a fluid droplet under the electrostatic field, which is the theoretical study of electrospinning, and found out the features of droplets, finally leading to the formation of the Taylor cone [25, 27]. The technique is known for quite a long, however, its applications were seen only in recent years. It was in 1993 where the applications of electrospun fibers were presented at a conference [25].

The application of electrospinning was noticed in the 1970s, where textiles from electrospun nanofibers started to appear. Development of non-woven e-textiles for the use of sensors and actuators that held perfect piezoelectric properties was observed then [28]. In the early 1980s, development of glucose biosensors for biomedical application was observed widely [29]. By the time, it was the late 1980s, a wide range of electrospinning polymers along with their suitable solvents were already discovered [30]. It was from 1990 AD where applications of electrospinning started booming in numerous fields. The development of DNA fibers via electrospinning was reported [31]. Along with that, poly(ethylene oxide) (PEO) and ultrathin membrane nanofibers started to develop on large scale [32, 33]. Since then, a lot of applications of electrospun nanofibers have been developed but with time, there have been certain modifications in the electrospinning process. For instance, organic solvents evaporate when it travels from the spinneret to the collector, which can be harmful to the environment and living beings nearby [32]. Along with that, the use of low viscosity polymer solutions will produce non-homogeneous fibers in collectors [34]. To omit this problem, melt electrospinning was introduced, which restricted the use of organic solvents. This proved to be a promising approach towards safe electrospinning in tissue engineering, food, sensors, textiles, and many other applications [35, 36]. However, the use of melt electrospinning can be done for certain polymers only [36]. It avoids the use of heat-sensitive materials/chemicals as it involves a heating device right before the spinneret, which made researchers to emphasize on the safety issues: the use of heating devices and electric voltage can intervene in each other's path [35]. To overcome this, many researchers find green electrospinning as a good technique. It uses diluted solvents and has no heating issues associated with other melt procedures. However, the use of organic solvents in a small fraction can be lethal when it comes to medical applications. Further, the concept of suspension electrospinning was employed. The use of this approach still prohibits the use of all the polymers [37]. Still, there is a lot of research going on to develop an efficient, safe and feasible electrospinning process (Fig. 3).

With time, not only process, but there were also certain changes in the electrospinning apparatus as well. For instance, at the initial stage, the use of only a single spinneret was implied. The production from a single capillary was not sufficient to produce nanofibers scaffolds on a large scale. To increase the efficiency, multi-jet electrospinning was introduced: this system consisted of more than one ejection system [39]. However, using multiple jets still appeared to be a non-feasible process. In 2004, the development of needless electrospinning was introduced. It used cylindrical, circular, disc, and spiral coil as a spinneret. From a single roller, multiple jets of fibers were produced [40]. Along with that, there were changes in collectors as well. The initial collector consisted of plate and drum collectors in a stationary position. Later, drum collectors with the rotating disc were introduced. Similarly, other collectors such as mandrel, wire drum, annular, conductive wires are explored [41–43]. Going back to history, it can be observed, the size of the electrospinning apparatus was huge and had many errors. The dimensions of these apparatus are smaller in size and are portable. In recent years, portable electrospinning apparatus has been reported. The performance of these apparatus is remarkable [44]. To sum

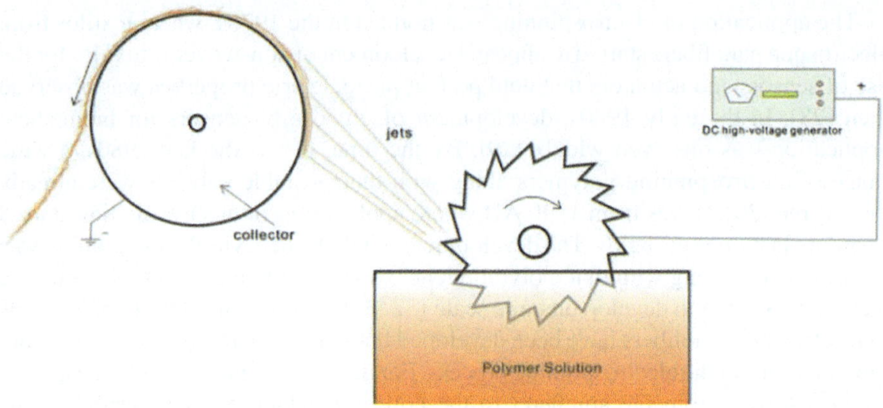

Fig. 3 Schematic drawing of apparatus for the production of artificial fibers by Formhals. (Reproduced with permission from Ref. [38])

up the major events in electrospinning, a lot of advancements can be seen. However, there are a lot of improvements to be done in the process. Numerous research is going on, to find the safe, low cost and feasible electrospinning process for applications in nanotechnology, not only on a laboratory scale but industrial scale too (Table 1).

4 Applications

Historical prospects suggest that different transition phases have occurred in the modifications of electrospinning for different applications of nanotechnology. Earlier, the research was more and more focused on tissue engineering and textiles but nowadays, researchers are trying to devise a way to store energy efficiently. Along with energy, research on effective agricultural techniques and food packaging industries is also being done. The focus on newer applications does not mean researchers have forgotten the fields of tissue engineering and textiles. Further innovation in these fields is also taking place. The transition from big structural applications to nanostructural applications for the same basis has made the research world competitive, and scientists are trying different ways to accomplish it. The different applications of nanotechnology driving the world of the twenty-first century are discussed below (Fig. 4).

4.1 Biomedical Applications

The biomedical application of nanotechnology is vast. Electrospinning is employed by nanotechnology to fabricate surface-modified and functionalized nanofibers

Table 1 Major Discoveries relating to the electrospinning device [2, 23, 24, 38, 44–53, 54, 55]

S. N.	Name	Date	Key investigation and innovation
1	Josue Heilmann [45]	1828	Josue Heilman developed a commercial machine fiber combing. He later highlighted the significance of fiber drawing and its profitability. Finally, the machine was patented in 1829 in England.
2	Joshua Wordsworth [2]	1840	Joshua Wordsworth used the patent and collaborated with Louis Schwabe to counterfeit the embroidered goods. Louis tried to develop a process for fiber spinning. He tried to fabricate the glass and was in search of a suitable spinning solution for the production of fibers.
3	Christian Friedrich Schoben [2, 46]	1846	Christian Friedrich Schoben developed nitrocellulose, which helped Audemars to develop a long thread of nitrocellulose. The solution was mixed with the solvent and passed towards the collector through the use of draught air as a solvent carrier. He later discovered that the solvent was evaporated till it reached the collector.
4	Charles Vernon [2]	1888	Charles Vernon used electricity for the first time in spinning the fibers. Vernon used the insulated dish with a high voltage electrical supply and did the spinning of beeswax, shellac, gutta-percha, and sealing wax.
5	Hilaire De Chardonnet [47]	1889	Hilaire De Chardonnet demonstrated the work to develop artificial silk. Chardonnet used cellulose derivatives to produce the fiber. Hilaire passed the solution of the polymer through the small holes which are called spinneret in the present-day scenario. His work later generated the concept of using needles as a spinneret during the electrospinning of solutions.
6	John F. Cooley [24, 45]	1900	John F. Cooley patented the first theoretical electrospinning device, with three types of spinning heads. He also described the coaxial spinning, air-assisted electrospinning.
7	W. J. Morton [23]	1902	W. J. Morton proposed the use of dispersion of fluid with assisted indirect and direct current, which later was used in most electrospinning methods. This discovery helped Hagiwara to use the charged liquid drops for manufacturing nanofibers. Finally, the use of electrostatic force and air for the production of artificial fibers in 1929 was done by Hagiwara.

(continued)

Table 1 (continued)

S. N.	Name	Date	Key investigation and innovation
8	Anton Formhals [48, 38]	1934	Formhals developed the first electrospinning apparatus. He is considered the originator of electrospinning. Formals presented the systematic procedure along with demonstration. The apparatus included, spinner (Sawtooth) dipped in a polymer reservoir. The potential was developed between the disc and collector. Nanofibers with electrostatic force, assisted by centripetal force, was used for the transfer of polymer. Formhals later developed the multi-head spinnerets, which invented the concept of multijet electrospinning.
9	Gladding, Ernest K [49]	1939	Gladding, Ernest K. Modified the apparatus and initiated the use of charged electrode collector. They placed the electrode behind the collector, passing polymer solution with multiple spinnerets for the development of discrete fibers.
10	Charles Norton [50, 51]	1936	The concept of the first melt electrospinning was patented by Charles Norton in the Massachusetts Institute of Technology. Norton discovered and found out the way to eradicate the polymer solution by melting the fibers before spinning. This concept later developed the use of a heater in the tip of the spinneret that enhanced the safe electrospinning process.
11	Henry R. Child [52]	1944	H. R. Child suggested the use of non-conducting spinneret. For that, he proposed the use of glass as a material of a spinneret.
12	W. C. Heubner [53]	1953	Heubner proposed the preheating of spinning dope for the faster evaporation of the solvent. Likewise, he discovered the use of a teardrop-shaped electrode inside collector for the easy fiber removal after the spinning is complete.
13	J. E. Owens [51]	1970	J. E. Owens and S. P. Scheinberg experimented by heating the polymer above the boiling point. and near the critical pressure of the solvent. This was done for the rapid removal of solvent.

(continued)

Table 1 (continued)

S. N.	Name	Date	Key investigation and innovation
14	G. E. Martin [54]	1977	The method developed by G. E. martin led to the use of grounded spinneret and the use of Vaan de Gaaf triboelectric generator. The layered fibers produced by this method were identical to the human tissues
15	Simm, Gosling C, Bonart R, Falkai BV [52]	1979	Bonart and his team performed the first demonstration of Needless electrospinning. He used the ring as a spinneret in place of needles and patented the apparatus.
16	Larrondo and St John Manley [50]	1981	Larrondo and St. John performed the first scientific publication on melt electrospinning. The paper presented the trilogy on the topic. They experimented to draw a fiber from the polymer melt. They created the pendant molten droplet to draw the fibers electrostatically, which became the fundamental approach of producing the fibers from the melt.
17	Thien.V. How [55]	1985	T.V. How studied the effect of fiber orientation with the speed of collector. This work done by T.V. led to the development of the tuning of the speed of the collector and the controlled flow rate of the polymer solution.
18	Coffee, Pirrie, Reneker's group (University of Akron) [44]	1998	Coffee and Pirrie developed the battery-powered handheld electrospraying devices, which later helped Reneker's group from the University Of Akron, to develop the portable electrospinning apparatus. Till now, numerous applications of portable electrospinning have been recorded.
19	Renker and Rangkupan [49, 53]	2001	Renker and Rangkupan performed the first melt electrospinning of PE, PET under the vacuum. The electrospinning apparatus was the same except it was placed in the vacuum. This experiment concluded that even the smallest electric field can draw the fiber from polymer melts.
20	Oldrich J. Fillip, S David L, Vadav K, Lenka M [52]	2005	Oldrich and the team used the rotating roller as the spinneret for the manufacturing of the fibers. This technique filled the void of minute efficiency of the conventional electrospinning approach. Later, this method was used for the production of fibers on a large scale.

Fig. 4 Application potential of 1D nanomaterials (Reproduced with permission from Ref. [44])

for applications in the field of bio-medics. Tissue engineering, wound healing, wound dressing are some of the topics of the biomedical field that has been influenced by electrospinning and nanotechnology. Electrospinning and modification/functionalization of the surface of the nanofibers tend to exhibit the characteristics of wound healing and wound dressing applications discussed below.

4.1.1 Antimicrobial

While wound dressings, the presence of bacteria and microbes compromises the wound, infecting it. So, antimicrobial agents need to be used during wound dressing. The use of silver nanoparticles as an antimicrobial agent is common where the DNA of bacteria is destroyed [56]. Large surface to mass ratio, higher interfacial adhesion between polymer and filler provides durable, non-leachable, biocompatible, load sharing, and improved nano-composite used in antimicrobial applications [57, 58].

Evidence of applications provides silver particles loaded poly(ethylene terephthalate) (PET) used as an antimicrobial agent because of characteristics of reduction of cytotoxic effects, anti-inflammatory properties, which have potential applications in wound dressings [59]. Polycaprolactone (PCL) nanofibers modified with tin oxide (TiO_2) and silver (Ag) nanoparticles have applications in antimicrobial providing inhibitory effects to Escherichia coli and Staphylococcus aureus. Inhibitory effects are seen by disrupting the cell wall, releasing toxic ions, damaging protons efflux bombs, modifying the membranes, and producing reactive oxygen to disrupt the cell wall [60]. Polyhydroxyalkanoate (PHA)/graphene silver (Ag) nanoparticles (NPs) synthesized from electrospinning provide excellent applications in anti-microbial activity because of the enhanced biodegradability and flexibility of PHA. These characteristics support wound healing and distribution of Ag-NPs on a high surface area of reduced graphene oxide (rGO) enabling direct contact with bacterial cells [61]. Synthesis of polyvinylpyrrolidone (PVP)/ciprofloxacin (CIP)/ethyl cellulose (EC)-Ag nanoparticles was performed with electrospinning providing fast release of CIP and Ag NP for antibacterial activity. They have a low frequency of spontaneous resistance used for wound dressings [56]. The cross-linked electrospun *N*-halamine fibers exhibited excellent antibacterial properties with E-coli [57]. Characteristics of good biocompatibility, high mechanical strength, wettability, hydrophilicity by electrospinning polyurethane, and soy protein nanofibers with zinc oxide (ZnO) nanoparticle and CIPHCl have potential applications in wound dressing [62]. Wide application of silver nanoparticles has provided researchers to review the negative aspects and in some cases, cytotoxic effects on normal cells and accumulation in vital organs have been found [59]. But silver nanoparticles linked in nanofibers have been known to provide antimicrobial effects.

4.2 Tissue Engineering and Wound Healing

The concept of tissue engineering and wound healing includes the development of biological substitutes used for restoring, maintaining, or improving the function of tissue to organs using various engineering and life science applications and principles [63]. The mechanism of wound healing and tissue engineering looks into the biological pathways acting as cues for bio-molecular signal transduction that can be controlled by nanofiber utilization [64]. Tissue engineering and wound healing applications have a similar working mechanism so, intrinsic properties need to be optimized for both electrospinning and polymer matrix such as polymer ratio, molecular weight, solvent composition, and other characteristics for applications in the biomedical field [65]. Properties of large surface to volume ratio, porosity, cell adhesion, strength, flexibility, protein adsorption, cell spreading, hydrophilicity, biocompatibility, mechanical support, cell proliferation needs to be modified for effective use in tissue engineering and wound healing applications [6, 7, 66–71]. For these characteristics of nanofibers, different modifications such as electrospinning techniques, use of solvents, and so on need to be performed. Possible substitution from needle electrospinning to needleless electrospinning has also been introduced as modifications for an excellent result. Earlier use of inorganic materials was prevalent in tissue engineering and wound healing applications. But nowadays, scientists have been looking and studying ways for regeneration application with herbal and plant extracts. Use of aloe vera extract [6], soy protein [62], propolis [65], keratin [70], ayurvedic-indhulekha oil [72], orange spiny oyster shells [73], lavender oil [74], pulp-dentin [75], and use of other herbal, natural extracts can be seen. Nanotechnology can also aid in helping to cure Parkinson's disease and other neurological diseases, which are not addressed properly by the medical diagnosis [76].

For applications in tissue engineering, organic, inorganic, and herbal additives have been used. Inorganic additives of gold embedded co-electrospun or blend electrospun polyaniline (PAN) solution provided flexibility in the fabrication of nanofibers by showing non-toxicity, possessing biocompatibility, promoting cell growth, cell attachment, cell spreading, and proliferation of nanofibers. These modifications helped in suitable applications of bone tissue engineering [77]. Tin oxide (TiO_2) additives in TiO_2/polyvinyl alcohol (PVA)/chitosan (PC) electrospun nanofibers have potential applications of tissue engineering showing biocompatibility, cell viability, high mechanical and physical support, and enhanced surface area for the better reaction of the cells [78]. Calcium carbonate ($CaCO_3$)/TiO_2 additives in PVA/$CaCO_3$/TiO_2 nanofibers were fabricated by electrospinning along with hydrolysis from hydrochloric acid (HCl) where Ca^{++} ions showed the capability for bone regeneration. Characteristics of cell biocompatibility, osseointegration, bio-mineralization, and cell proliferation confirm these nanofibers can be used in bone tissue engineering [79]. Electrospun PCL was added with ZnO and europium hydroxide nanorods (EHNs) showing excellent properties of hydrophilicity, cell adhesion, migration and proliferation, non-toxicity, and mechanical properties. With the view of these properties, applications in cardiovascular and ischemic diseases wound healing, and tissue regeneration by promoting vascularization are effective [64]. Electrospun poly (butylene adipate-co-terephthalate) (PBAT)/polypyrrole (PPy) nanofibers show changes in hydrophilicity after PPy was added to the nanofibers providing osteoblasts behavior and alkaline phosphatase activity (ALP) activity used in bone tissue engineering [80]. Polyurethane and graphene oxide were loaded with retinoic acid during electrospinning to obtain nanofibrous exhibiting good mechanical strength, deceased super-hydrophobicity, enhanced biodegradation, improved wettability, support in cell adhesion, good biocompatibility, and increased cell viability for application in osteogenic expression in bone tissue engineering [67]. Electrospinning polyvinylpyrrolidone and Eudragitv RS100 and adding dextran or poly (ethylene oxide) have potential applications in the therapeutic application and drug delivery for treating oral mucosal lesions. Electrospinning gave rise to mucoadhesive properties minimizing material toxicity and immune reactions, increased hydrophilicity to form gel-like formation, better cell attachment and adhesion [81]. Polycaprolactone (PCL) composite was loaded with magnesium oxide aiding in the application of bone tissue engineering due to its improved characteristics of tensile properties, cohesion, cell responses, bone cell proliferation and attachment, differentiation, and hydrophilicity [82]. Properties of increased tensile strength and mechanical properties, increased bioactivity, cell attachment, adhesion, and proliferation when poly(3-hydroxybutyrate)/carbon nanotubes were fabricated using electrospinning technique were seen. Potential applications in the synthesis of alkaline phosphatase and concentration of calcium in the extracellular matrix in tissue engineering are seen [83]. Graphene oxide was surface grafted with Poly(ethylene glycol) (PEG) for reinforcement of Poly(lactic acid) (PLA) showed improvement in hydrophilicity, thermal stability, mechanical properties of tensile strength, adhesion, proliferation, attachment and retained cytocompatibility for potential application in tissue engineering [66].

Organic applications of tissue engineering in the regeneration of meniscus are done by electrospinning gelatin and PLGA nanofibers. The characteristics of increasing bioactivity, non-declination of porosity and pore size, desired mechanical and biological properties are seen aiding in meniscus tissue engineering [84]. Electrospun regenerated silk fibroin mats along with laminin have excellent capability in the improvement of neurological disorders such as Parkinson's disease, Huntington's disease, amyotrophic lateral sclerosis, traumatic spinal cord injuries, and peripheral nerve injuries. Characteristics of cell proliferation, differentiation, adhesion, decreased cytotoxicity, good biocompatibility through the neural progenitor cell are seen, which provides enough base for treatment of neurological disorders [7]. PCL/chitosan nanofibers showcased better tensile strength and mechanical properties, crystallinity, decreased cytotoxicity, adequate support for porcine tracheobronchial epithelial (PTBE) cell growth and proliferation, and cell attachment, which has application in respiratory tissue engineering [71]. Good compatibility, hydrophobicity, adhesion are the characteristics that are obtained when cilostazol-loaded poly(ε-Caprolactone) is electrospun aiding in device implantation and vascular injury beneficial for reendothelialization [85].

The addition of herbal additives of Indhulekha oil consisting of karpura, amalaki, and draksha in polyurethane nanofibers acts as antioxidant enhancing cell adhesion, the proliferation of fibroblast cells, biocompatibility, hydrophobicity, thermal stability, and delayed clotting time. These properties help the nanofibers for use in tissue engineering applications [72]. Polylactic acid (PLA)/sodium alginate (SA)/tricalcium phosphate (TCP) nanofibers were fabricated by electrospinning, which consists of properties for bone formation, increased mechanical strength, biocompatibility, and cytocompatibility used for bone tissue engineering. TCP was extracted from orange spiny oyster shells used as a solution for electrospinning [73]. PCL/keratin/chitosan was prepared via electrospinning and aloe vera was encapsulated for enhancing the properties of good mechanical strength, aiding cell proliferation, migration, adhesion, and attachment, decreasing cytocompatibility and toxicity. These modified nanofibers are used in skin tissue engineering, and wound dressings [6].

The characteristics and properties needed for wound healing applications and tissue engineering are similar. Low toxicity, better proliferation, excellent hydrophilicity are some of the characteristics required for better working of the nanofibers. PVA nanoscaffolds as a prototype for wound healing and compromised healing has potential in tissue engineering exhibiting characteristics of excellent biocompatibility, low toxicity, good mechanical properties, flexibility, chemical, and mechanical stability. Hybrid PVA using hydrogel PEG, gelatin with propolis extract showed the characteristics of increased in vitro cell viability, low cytotoxicity, stimulatory effects on macrophage proliferation, reduction of healing time, accelerated regeneration process, and recovery of damaged tissues by anti-microbial and anti-inflammatory actions. These characters enable the interaction of biomaterials with epithelial cells inducing tissue regeneration [65]. Gelatin and poly-e-caprolactone (PCL) nanofibers produced from needle-less electrospinning have higher porosity and surface area indicating the formation of fibroblasts and keratinocytes. They were capable of growth, proliferation, and non-cytotoxic effects having applications in limbal and mesenchymal cells in treatment for eyes, ulcers, skin, and other tissues. Collagen deposits in granulation tissue (GT) were not increased, which does not suggest the formation of pathological fibrosis or scarring. But the promotion of cell adhesion indicated needless electrospinning a good candidate for wound healing and tissue engineering [86]. Hyaluronic acid (HA) nanofibers along with keratin were used for the application in wound healing of burns and related ulcers. Electrospun HA and keratin exhibited the characteristics of good hydrophilicity, improved mechanical properties, cell attachment, cell proliferation, adhesion, and minimal cytotoxicity necessary for this application [70]. PCL nanofibers synthesized with titanium dioxide nanorods (TNR) are used in wound healing applications by promoting cell adhesion and proliferation, cell viability, cell migration, angiogenesis, and good mechanical properties [58]. Pulp-dentin and periodontal regeneration from electrospun PVA/hydroxyapatite/carbon nanotube are prevalent due to the characteristics of biocompatibility, enhanced mechanical properties, formation of both soft and hard tissue by osteogenesis and vascularization, proliferation, viability, cell attachment and differentiation along with apatite formation [75]. Electrospinning of ibuprofen and acetylsalicylic acid-infused PVP-dextran nanofiber mats to create analgesic, anti-inflammatory, degradable scaffolds were seen. Enhanced hydrophilicity, low antibacterial activity, and cytotoxicity were exhibited as having possible wound healing applications [87]. Gelatin-glycerol nanofibrous membranes have applications in medical applications of wound healing as they showed improved characteristics of hydrophilicity, attachment, and proliferation [88]. No cytotoxic effect, increase in cell proliferation and viability, increased cell growth are some notable characteristics when starch/PCL nanofibers are electrospun for application in biomedical fields of science [89]. Hydro-distillate of lavender oil and Ag nanoparticles were added to electrospun polyurethane nanofibers for enhanced characteristics of effective antibacterial activity, improved hydrophilicity, improved rigidity for easy diffusion and permeation, improved growth, attachment, and proliferation. These characteristics provide an enhanced and suitable environment for fibroblastic cell growth for application in wound healing and dressing [74]. Thus, biomedical fields have been developed. Earlier, development was slow and limited to only certain types of polymers and additives but nowadays organic, inorganic, or herbal additives have been used in the applications for better results of nanofiber properties.

4.3 Energy

Energy harvesting has become a subject of interest in today's world. Identification of energy sources, their utilization, and storage needs enough research in nanotechnology. Compact, easy to handle, and easy to segregate should be the characteristics of these energy harvesting materials, which help to provide a strong structure towards creating potential renewable energy applications.

4.3.1 Secondary Battery

Battery construction provides electrolytes, electrodes, and separators present in them. The battery needs to be robust, user-friendly, compact and should have the capability to store charge effectively. Different modifications in construction can provide many enhanced properties for charge storage applications. For instance, electrolytes can be both solid and liquid but solid electrolytes enhance safety and stability, show good contact with lithium electrode and increase the polarization characteristics than liquid lithium ion batteries. Free-standing flexible electrodes showing excellent electrode conductivity, corrosion, and fatigue resistance for flexible energy storage for smart clothing and wearable electronics are fabricated [90]. Carbon nanofibers and carbon nanotubes can be effective for good conduction of electricity and provide excellent mechanical strength used in energy storage [91]. New modifications on lithium-ion batteries by adding sulfur have proven effective. But doping with carbon, metallic impurity, metal oxides, metal carbides and metal fluorides must be done. Doping provides the formation of a strong chemical bond with sulfur or polysulfide during cycling [90]. Cost-effectiveness and high theoretical capacity are seen but challenges of poor Coulombic efficiency, high solubility, large volume change, and fast capacity degradation can also be found. Viewing the applications for lithium-sulfur batteries, carbon nanomaterials as multichannel (LRC) nanofiber has added a remedy for these challenges. LRC fulfills a host for sulfur, enabling effective ion diffusion, fast electron transport, good accommodation of volume change when absorbed to polysulfide PAN. These carbon–sulfur cathodes have good electrochemical capacity and electrical conductivity. Also, polyvinylidene fluoride (PVDF)-based separators for lithium metal batteries after heat treatment showed enhanced physical properties, high ionic conductivity for battery operations. Work on battery stability at high temperature is also found where electrospinning mixture of PAN and ammonium polyphosphate (APP) showed the enhanced capabilities for battery applications [90]. For other applications, $Li_{0.33}La_{0.557}TiO_3$ (LLTO) nanofibers when used as ceramic filler for poly(ethylene oxide)-lithium bis(trifluoromethanesulfonyl) imide (PEO-LiTFSI) electrolyte provided enhanced ionic conductivity and electrochemical stability along with improvements in mechanical properties and fast lithium-ion conduction. Successful suppression of lithium dendrite was also observed. $Li_{1.3}Al_{0.3}Ti_{1.7}(PO_4)_3$ (LATP) nanocomposite immersed in PVdF-HFP solution can also be used in lithium battery applications as a separator in aqueous lithium-air batteries for its high ionic conductivity [92]. Additives of molybdenum dioxide–carbon nanofibers in the lithium-sulfur battery have provided better electrochemical performance and charge transfer than sulfur cathode during the charging and discharging process used in battery applications [93].

4.4 Solar and Thermal Energy Harvesting

The applications of thermal and solar energy harvesting also need much attention. Applications for storing these alternative sources of energy will certainly be fruitful for future generations. Solar applications using cellulose acetate fibers electrospun with photosensitive cadmium selenide (CdSe) nanoparticles can be used in solar thermal conversion devices. They have greater advantages of enhanced mechanical properties, excellent fiber-forming ability, biodegradability, water stability, and cost-effectiveness. Loading CdSe can increase the absorption capacity of cellulose acetate (CA) fibers [94]. The use of zinc oxide and zinc acetate in PVA while electrospinning synthesizes solar cells for potential application in the solar thermal sector. Improvement in the energy conversion efficiency due to enhanced surface activity was seen with zinc oxide and acetate [95]. For thermal applications, the hybrid inorganic and organic structure of copper sulfide (CuS)/polyvinylpyrrolidone (PVP) on flexible polymers of polyethylene naphthalate (PEN) were developed for use as thermal shielding film as it showed enhanced thermal shielding efficiency [96].

4.5 Piezoelectric Devices and Nanogenerators

Piezoelectric response using stress applications can be linked to energy harvesting as increasing strain produces mechanical energy for piezoelectric devices. These devices are developed into nanogenerators for energy production. Crystallinity, electroactive polar phase formation, and planar orientation are important characteristics to perform piezoelectric response [97, 98]. Because of their wide bandwidth, fast electromechanical response, relatively low power requirements, and high generative forces, researchers are more and more excited about researching piezoelectric materials [8]. Piezoelectric nanogenerators have created interest among researchers but presently, acoustic nanogenerators have been developed as it demonstrates ultra-high sensitivity to acoustical vibration. Applications in speech recognition due to vibration and use as ultrafast capacitors by sensing acoustic signals for mechanical energy harvesting for nanogenerators have provided a need for newer and sophisticated research [99].

The application of piezoelectric devices and nanogenerators needs the properties of higher β phase content, stacking effect, smaller effective radius curvature during bending, greater applied strain, and a higher fraction of contribution for PVDF nanofibers enhancing output piezoelectric current and voltage [100, 101].

Poly(vinylidene fluoride-co-hexafluoroproplyene) (PVDF-HFP) and nickel ferrite ($NiFe_2O_4$) nanocomposites are developed by electrospinning for mechanical energy harvesting and charge storage due to good crystallinity. Low dielectric loss values and homogeneous fiber distribution influence the electronic environment making the nanocomposites suitable for energy storage applications and output voltage generation. Magnetic $NiFe_2O_4$ demonstrates good piezoelectric properties showing good response due to its good crystalline behavior and the filler polymer interfacial interaction [97]. Addition of ZnO and EGO in P(VDF-TrFE)-based polymer nanocomposite enhanced piezoelectric properties of self-poled fibers. The alignment and realignment leading to increase and decrease of potential confirm the generation of electrical energy at the output of the device leading to piezoelectric applications [102]. Methylammonium lead bromide ($CH_3NH_3PbBr_3$) doped PVDF nanofibers (PMNF) performed well in a piezoelectric application by inducing the piezoelectric β-phase content by altering the properties for the overall development of crystallinity and piezoelectricity. This nanofiber showed well-improved properties for piezoelectric nanogenerators by the incorporation of semiconducting perovskite. But acoustic nanogenerators can also be fabricated using this nanofiber exhibiting superior mechanical property, enhanced crystalline electroactive phase content instead of piezoelectric nanogenerators by induction charges. Ultimately, piezoelectric nanogenerators and acoustic nanogenerators are used in energy harvesting applications [99]. For nanogenerators, high-performance triboelectric nanogenerators are fabricated with PVDF-silver nanowire and nylon nanofiber acting as triboelectric layers. Charge trapping capability and surface charge potential are enhanced by adding metals, which has potential applications in self-powering commercial liquid crystal display [98]. EuGNF nanofibers were produced when Eu^{3+} was doped in P(VDF-HFP)/graphene composite while electrospinning has applications in first piezoelectricity and second in biosensors. Embedded graphene sheets inside the EuGNFs improves the sensitivity and conductivity of the nanofibers. The improved degree of crystallinity and complete conversion of piezoelectric β-phase leads to better piezoelectric sensitivity and surface conductivity. These characteristics influence the output power of nanogenerators aiding in applications for self-powered portable electronic devices. These devices have the capability of harvesting energy from respiration, environmental air flow and wrist movements [103]. PVDF and lead zirconate titanate nanowires generate high output voltage to convert tiny mechanical strain into electricity showing excellent piezoelectric characteristics and stable robustness. Commercial LEDs were light up by triboelectric nanogenerators (TENGs) promoting self-powered human–machine interfaces using these nanowires [104]. Thus, applications of nanofibers in energy harvesting have provided a base for researches in unturned corners of energy applications such as piezoelectric generators, pressure generators, and so on.

4.6 Supercapacitors

Supercapacitors/pseudocapacitors are known to be a relevant topic for discussion for the concept of power management. The use of metal oxides proves effective in battery and pseudocapacitors/supercapacitor applications. Research on their characteristics of synthesis and growth mechanisms to improve electrochemical performance, energy density, and lifecycle stability are important [91]. Applications of nanofibers as supercapacitors/pseudocapacitors are effective as it provides manganese dioxide (MnO_2) to exhibit excellent ecofriendly properties for these applications. By combining MnO_2 and super-aligned electrospun carbon nanofibers (ECNFs), properties of enhanced electrical, electrochemical, and mechanical energy were seen for electrochemical energy storage [91]. Another application of Titanium tetra isopropoxide (TTIP)-poly(vinylpyrolidone) (PVP) fibers when electrospun with TiO_2 and encapsulated with MnO_2 showed excellent super capacitor behavior for storing charges. The characteristics of pseudocapacitive nature, excellent electrochemical nature through high surface area and aspect ratio, better stability, and better ionic conductivity were seen [105]. Gelatin nanofibers including ionic liquid of 1-ethyl-3-methylimidazolium (EMImBF4) exhibit high ionic conductivity and low viscosity, which can be used as electrolytes in electric double-layer capacitors (EDLCs). High thermal stability, high tensile strength, high ionic conductivity make this nanofiber suitable for gel electrolyte applications, which are used for improving device safety without reducing performance [106]. TiO_2 nanoparticles were gathered to carbon nanofibers while electrospinning, thus, enhancing the super capacitance applications of the nanofibers. Improvement of electrochemical performance and the observation of larger area in the nanofibers and quasi rectangular shape of the curves suggest having better capacitive behavior and pseudocapacitive characteristics, respectively [107].

4.7 Environment

With growing issues regarding of earth, conservation of the environment has become mandatory. The concern regarding the preservation of the environment must be overviewed. The deterioration in an environment can be reflected in an increase in the earth's temperature and physical attributes. For many years, researchers are trying to focus on the conservation of the environment. In the same context, electrospun nanofibers can be one of the promising assets for the protection of the environment. For instance, a discharge of dye solution from different industries has caused serious problems in the environment. The applications of electrospun nanofibers can be implied in the removal of particles from an aqueous solution. Out of all, the toxics, Congo red (1-napthalenesulfonic acid, 3,3-(4,4-biphenylene bis (azo) bis (4-amino-) disodium salt) is a serious threat for the formation of carcinogenic mutants. For its removal, the use of electrospun K_xMnO_2 over TiO_2 mats was developed. The result was remarkable as the filtration efficiency and membrane regeneration were high [108, 109]. Similarly, indigo carmine that is used while manufacturing the Denim jeans appeared to cause potential health hazards. To remove this from water sources, a chitin nanowhisker was functionalized over an electrospun PVDF membrane. The high removal efficiency of indigo carmine by PVDF/chitin nanowhisker (ChNws) membrane concludes that it can be a potential membrane for the treatment of wastewater streams [110]. Lately, living bacteria are incorporated in ultrathin fibers, which holds potentiality in cleansing wastewater streams. For the survival of these bacteria, the polymeric substrate of carbon is required to be cyclodextrin as a promising substrate. Biocompatible non-polymeric cyclodextrin fibers were encapsulated with *Lysinibacillus* sp. bacteria. This cyclodextrin fiber (CD/F) biocomposite is applicable in the eradication of heavy metals Ni(II) and Cr(VI) from wastewater resources [111].

Likewise, for the control of air pollution, polyurethane/chitosan nanofibers can be used. The electrospinning of chitosan alone is not feasible due to the high viscosity and lower stability of chitosan. However, its addition with polyurethane can eradicate this problem. It was reported that 15% of the concentration of polyurethane in chitosan/polyurethane solution was electrospun to get nanofibers. The result of the efficiency and quality factor implied that a decrease in air face velocity provided an increase in efficiency of the filtration. The decreasing quality factor was more significant as increased pressure drop at higher face velocities can be seen. These results indicate that this composite can be used for the removal of hazardous gaseous pollutants from an air stream [12]. For the removal of debris from the air, high-efficiency particulate air filters are used. The use of electrospun nanofibers in this filtration system has shown improvement in the high filtration efficiency of nanofibers. Maintenance of low-pressure drop is the key factor for a good filtration process. However, it was reported that the filter fiber with very fine nanofibers becomes compact and leads to a high-pressure drop and low filtration. For omitting this problem, nanoparticles of SiO_2 were placed with fine nanofibers of poly(acrylonitrile) (PAN), which resulted in higher filtration efficiency and low-pressure drop [112]. Likewise, the use of PAN nanofibers prepared by electrospinning was evaluated showing the excellent result of filtration efficiency with optimal manufacturing of PAN nanofibers [113]. Nowadays, the development of antibacterial/antimicrobial filters has been initialized. Incorporation of Ag nanoparticles in PLA nanofibers was done by electrospraying. The evaluation of antimicrobial properties of PLA-Ag NPs was tested against gram-positive and negative bacteria (S. aureus and P. aeruginosa) showing excellent antimicrobial properties [114]. Similarly, in the case with polyurethane (PU)/chitosan (Ch) nanofibers, it was observed that PU nanofibers had less antibacterial activity, while, the PU/Ch nanofibers had a larger inhibition zone indicating antimicrobial properties of PU/Ch nanofibers [12]. For now, the use of electrospun nanofibers in environmental protection is initiated. Still, it is not enough to compensate for the damage done, however, electrospun fibers have played a crucial role to some extent in controlling the pollution of air and water.

4.8 Sensors

Nanotechnology is the single most appropriate tool for smart material harvesting, which will be able to sense different compounds and characteristics as the subject of interest for many applications. Sensors prove effective for gas sensing but not only this, voice recognition tools through acoustic and pressure sensing have also been developed as smart materials. Sensing hydrogen, ethanol, hydrogen peroxide, sugar, and so on have been proven effective by the use of nanomaterials but for this application, gas sensors need the most important parameter of selectivity to identify different gaseous materials [115]. The excellent and unique properties of high aspect ratio, high surface area, high porosity, and selectivity have provided a platform for the design and fabrication of biosensors [116]. Sensing characters are influenced by selectivity but the role of temperature is also significant as sensing characters are dependent on temperature. Chemical interaction should also be considered, as it limits the response of the sensor. Rapid release of adsorbed molecules is prevalent, providing signal recovery of the sensor. Selectivity is controlled by bulk characteristics for larger grains but grain size controls the characteristics for small particles [117].

Studying the applications of nanoparticles in sensors applications, different nanofibers have been used for a variety of sensors. Biotinylated bovine serum albumin (BSA) embedded PLA-PEG nanofibers contain decreased fiber diameter, which helps in the fast immobilization of avidin molecules to fabricate paper-based biosensors. The release of avidin from biotinylated materials can help to fabricate generable biosensors [116]. Tin(IV) oxide/polyaniline/polyhydroxy-3-butyrate biodegradable nanocomposite fibers are electrospun, which has the capability of sensing ethanol gas by decreasing the resistance of sensors in ethanol environment even at the low temperature of 80 °C. The characteristics of large surface-to-volume ratio, accelerating the diffusion of ethanol gas molecules, presence of dopant, enhancement of protonation level of PANI, improving the interaction between ethanol molecules and the electronic properties of p-type PANI and n-type Pd: SnO_2 improve ethanol gas sensing properties for the particular application at low temperature [118]. In another application, Haematite (α-Fe_2O_3) nanoparticles and Co_3O_4 nanoparticles have the capability of ethanol sensing due to the achievement of a balance of large voids [117]. Copper oxide (CuO)/polyvinyl alcohol (PVA) nanofibers were produced from electrospinning, exhibited ethanol, hydrogen, and liquefied petroleum gas (LPG) gas sensing applications. Sensing characters of increasing the sensor resistance with a fast response, a fast and almost complete recovery, low drifts and good stability, and a low limit of detection can be seen [115]. For hydrogen peroxide sensitivity, nanofibers of cerium oxide (CeO_2)/cobalt oxide (Co_3O_4)/poly(3,4-ethylenedioxythiophene) (PEDOT) are prevalent. Peroxidase-like activity is enhanced showing the synergistic effect between multi-components through improved catalytic oxidation of 3,3,5,5-tetramethylbenzidine (TMB) in presence of hydrogen peroxide (H_2O_2). The monitoring of H_2O_2 is mainly done by observing the absorbance changes at 651 nm in UV-visible spectra. Being a good catalyst as a mimetic enzyme, this nanofiber has application in catalysis and catalytic activity along with bio-sensing [119]. Detection of sulfite and ascorbic acid using the colorimetric approach by carbon nanofibers (CNFs)/$MnCo_2O_{4.5}$ nanofibers was developed where nanofibers exhibited the characteristics of high catalytic activity. The formation of oxidized products by oxidase or peroxides mimics typical blue, green, and light yellow colors during the oxidase reaction confirmed by the UV-visible spectra [120]. For glucose sensing and identification in blood, gold particles were blended with electrospun poly(vinyl alcohol) (PVA)/poly(ethyleneimine) (PEI) nanofiber interacting with glucose to form glucose oxidase (GOx) nanofibers [121]. pH-sensitive devices can be fabricated using a tri-axial electrospinning process using pH-sensitive shell and lipid drug [122]. $CH_3NH_3PbBr_3$-doped PVDF nanofibers (PMNF) can be used in pressure sensing and acoustic sensing for noise pollution detection and can be further used as nanogenerators. Characters of improved piezoelectric β-phase ultimately, decreasing fiber diameter and showcasing super flexibility is helpful in high acoustic sensing characteristics. This leads to effective speech recognition tool used in health care monitoring and security sector [99]. Eu^{3+} doped P(VDF-HFP)/graphene composite nanofibers have characteristic application in sensing along with nanogenerators. The graphene sheets of Eu^{3+} doped P(VDF-HFP)/graphene composite nanogenerators (EuGNG) are capable of high output performance with finger touch sensitivity through pressure. Also, acoustic sensors detect acoustic vibrations for speech recognition as well as biomedical sensors and self-powered electronics [103]. Thus, the above applications in gases, smart materials, and medical fields prove nanofibers as a suitable candidate for sensing applications where enough research is required for further optimization.

4.9 Textiles

From a single piece of fabric to a designer's brand, attire aims to protect from heat, cold, toxics, chemicals but primarily worn for comfort. With the advancement in technology, manufacturing techniques of fabrics have changed drastically. Updates in electrospinning and its applications have been increased widely, however, its implication in clothing was observed recently. The improvement in nanofibers production techniques has created manufacturing of textiles through electrospinning easier and efficient. It was found that electrospinning can produce thin fibers having a wide scope in the field of textiles [33]. The main reason to use electrospun fibers is to produce large surface area fibers along with small pore size [123]. The major advantage of these fibers is that they can be directly applied to the substrate, which reduces the cost of fabrication. However, they cannot be used alone because of having low mechanical stability. So, modifications are required [123, 124].

The most important applications were found in protective clothing systems where characteristics of wearability, comfort, breathability, transport properties and water vapor transmission rate need to be considered [124, 125]. Out of all properties, transport property is vital for evaluating the water vapor diffusion rate and air resistance of the fabrics with the help of Dynamic Moisture Vapor Permeation (DMPC) [126].

But, it was reported that the convective flow resistance of electrospun mats was much less than that of conventional clothing [127]. These features need to be addressed, which are all obtained via electrospun fibers. One of the drawbacks noticed was, the mechanical properties of these fibers were found to be poor. The high porosity and random orientation of the fibers can be the reasons; however, the understanding and main cause for the poor mechanical properties remain obscure [33, 126]. When it comes to thermal applications, electrospun nanofibers were found to have good temperature stability. 1,4-phenylenediamine and terephthalic acid 1 (Kevlar) were fabricated with electrospun polyimide, exhibited and found to bear good thermal properties [127]. Similarly, the use of nylon 6,6, polybenzimidazole, polyacrylonitrile, and polyurethane is found to be resistant to aerosols, which is applicable while manufacturing PPE (Personal Protective Equipment) [127, 128]. In terms of military applications, initially, (polytetrafluoroethylene) (PTFE) was widely used. Later, there was a transition from PTFE to polyolefin/polyurethane [33].

Nowadays, electrospun polymeric fibers possess a large specific surface area with lightweight. They can withstand wind and liquid without destroying breathability. For waterproof technology, fibers are laminated in woven or knitted fabric. For instance, polyurethane nanofibrous membrane fabrication resulted in the prevention of water, dust, and enhancement in elastic properties [129]. In the filtration process, glass fibers were replaced by non-woven materials, which were preceded by nanofibers. Nanofibers have a smaller diameter and higher surface area, which attracts more dust and is widely used in filtration clothing. However, to avoid infection, filter masks are incorporated with antimicrobial substances [130]. The membranes and fabrics produced from electrospinning are comparatively better than those produced from the conventional approach. Their main features include breathability, elasticity, and filtration efficiency. Similarly, electrospun fibers do not hinder the flow of water vapors through the pores of fibers [128]. As discussed earlier, these features enhance the transport property of material, which promotes the use of electrospinning in the clothing industry.

4.10 Food Packaging

At present, electrospinning has played an important role while in the production, processing, and packaging of food. In food packaging applications, electrospinning solution has a high concentration, which results in fiber formation while electrospraying solution has a low concentration resulting in the formation of droplets [10]. But, both the process of electrospinning and electrospraying has their importance in food packaging. The low toxicity of electrospun fibers is one of the main reasons for its use in food industries [14].

There are different food packaging processes but the technique of electrospinning resulting in the production of nanocapsules in a dry state is used. Bioactive compound encapsulation is performed during electrospinning. Encapsulation is done to improve the stability, bioavailability, solubility, compatibility, and controlled release of bioactive molecules such as vitamins, antioxidants, living cells acting as preservatives to maintain the odor and taste of the food. Natural biopolymers, proteins, and polysaccharides are used as polymers in electrospinning. Whey protein concentrate, chitosan, zein, alginates, dextran's (nanosponges), and cellulose are electrospun with polymers for food applications [10, 14]. Some studies also suggest that emulsion electrospinning of the matrix can enhance the encapsulation efficiency of the novel compounds as it omits the use of organic solvents and protein inactivation [10, 14, 131]. For instance, PLLA nanofibers were incorporated with PVA nanoparticles resulting from the controlled release of PVA nanoparticles via electrospun fibers [131]. In recent applications, PHAs matrices were filled with CuO nanoparticles by the process of melt mixing. This resulted in a rough cross-section morphology of films; however, there was no agglomeration of nanoparticles, and CuO was uniformly distributed. The transparent, optical nanofibers produced were good enough to be used in food packaging, though it was compromised with melt mixing approaches [132]. Nowadays, in food packaging, less soluble films are also produced via electrospinning. Similarly, in the case of gluten edible films, nanocomposites of chitan/gelatin were placed over it to increase the hydrophilicity of the film, increasing the solubility of film. However, low solubility of the nanocomposites can also be used to stop the moisture exchange between the food and environment [133]. Innovation of electrospun hydrophobic membranes can also be used in moist food packaging with low solubility nanofibers. Considering this aspect, poly(3-hydroxybutyrate) (PHB) fibers were developed, exhibiting good thermal and optical properties to have application in food packaging [134].

The use of smart materials in the food packaging industries has gained much interest nowadays. Smart food packaging includes designing smart tags, antibacterial electrospun membrane coatings, analyzing the thermal fluctuation of food, and so on. For instance, the development of poly(ethylene-*co*-vinyl alcohol) (EVOH)/graphene nanoplatelets (GNPs) nanofibers provides the properties of the quality level such as maintaining freshness and temperature and decreasing the contaminants of the food. The use of phase change materials can be an indicator for determining the change in thermal variations within the food. These materials will change their phase, which indicates the change in temperature of food [135]. Similarly, the combination of electrospinning and electrospraying is also used for the incorporation of nanoparticles. Initially, PLLA nanofibers were incorporated with PVA nanoparticles. This made the controlled release of PC nanoparticles more due to electrospinning [131]. Nowadays, enzymes can also be incorporated into the food by electrospinning. Enzymes will enhance the biological activity within food but the high cost of production and sensitivity to a certain change in external conditions have made it's use limited. Similarly, the use of small nanocarriers will result in a smaller surface area of food and higher reactivity of enzymes [10, 14]. In food packaging, the use of electrospun PLA matrix is incorporated with enzymes, metals, and peptides, which are mostly used to avoid the increment of bacteria [135]. These bacteria if exposed to contaminates or open environments will damage the food quality and give rise to the fetid smell. For instance, PLA was encapsulated with cinnamon essential oil, which showed that there was no hindrance in the performance in the bioactivity of material along with controlled proliferation [14, 135, 136]. To sum up the idea, the electrospun film exhibits the properties that can be used in food industries. However, a lot of improvements are to be done for a better approach in food manufacturing.

4.11 Agriculture

The electrospinning application in the field of agriculture is evolving. Most of the applications in agriculture include pre and post-crop growing procedures. Some features of electrospun fibers required for application in agriculture include their uniformity, high surface area, and smaller diameter, which is important when it comes to releasing kinetics of any incorporated materials inside the nanoparticles. One of the key problems in agriculture is the use of fertilizers. Fertilizers catalyze the growth of plants and crops. From an economic viewpoint, fertilizers are very expensive. Mostly due to high temperature and humidity, the loss of fertilizers by leaching and volatilization is a common problem. To prevent this, electrospun fibers of wheat gluten, biodegradable cellulose mats can be incorporated with urea. It was reported that controlled release of urea via nanofibers was achieved. At the same time, the thermal stability of the system was up to 117 °C [11, 137–139]. Similar is the case with nitrogen phosphorous potassium (NPK) fertilizer where the distribution of fertilizer over chitosan nanoparticles was accomplished. Results showed uniform distribution of nanoparticles controlling the release of the fertilizer by the outer shell into plants [140]. The morphological structures of fertilizers are in beads or circular structures. Lately, the production of nature-friendly organic fertilizers is also introduced. It was noticed that PCL fibers became thinner while loading into liquid fertilizers, and the surface area of PCL nanofibers was also increased as the loading of liquid increased [141].

While trading the fruit internationally, Mediterranean fruit fly (*Ceratitis Capitata Wied*) has created a major problem. In California, Florida, and Texas of Uinted States of America (USA), lot of *Ceratitis* extermination technique is carried on. To control these pests, a pheromone is used, however, its effect lasts for a short time, and they are highly volatile. For the controlled release of pheromone, nanoparticles as pheromone dispensers can be used [142]. Similarly, in fabric making process, 25% of the total cotton is recovered as waste. The recovery of 90% cellulose from waste cotton is done by dissolving it in ethylene diamine. However, electrospun cellulose with thin fibers was produced, which can be used in air filtration, PPE and nanocomposites. An alternative approach is the production of biodegradable cellulose mats, which can be used as a shell for holding pesticides and fertilizers as discussed earlier [138]. Germination of seed is an initial step during the growth of the plant. Usually, seeds are placed below the ground and nourished till they grow. Application of nanotechnology in rice seeds where it was coated with fungicide carboxy-thiram and cabendazim by electrospinning provides evidence that the germination in treated seeds was higher than non-treated seeds [143]. Nowadays, green electrospinning in agriculture is also a subject of interest for researchers. The use of organic solvents while preparing nanoparticles can be harmful when it comes to direct applications on a living organism. To omit this problem, nanocarriers developed from green electrospinning can be a better approach [36]. Similarly, water-based nanostructures can be used in agricultural applications. Water-based PVA nanofibers loaded with pheromone were produced to protect the fruits from various insects [144]. To conclude, there are recent advancements in agricultural applications of electrospinning. However, a lot of exploration is yet to be done.

4.12 Others

Electrospun nanofibers have applications in polarization, optoelectronics, and photocatalysis too. Research on the use of nanofibers in applications of polarization and optoelectronics is very recent. Modifying electrospun nanofibers enhances the characteristics such as the degree of polarization of electrospun nanofibers providing enough efficiency for the applications in polarization and optics. Fabrication of $CsPbBr_3$ nanowires with polymeric fiber assembly produced polarization higher than fiber dispersion being simple, flexible, and customizable with potential applications in optoelectronics [145]. Electrospinning polyvinyl acetate (PVA) and polymethyl methacrylate (PMMA) nanofibers have excellent properties of polarization as they showed excellent efficiency of the degree of polarization, mechanical strength, thermal conductivity, and electrical conductivity having potential applications in optical industries [146]. PMMA/P3HT nanofibers exhibit excellent electrical and optical properties showing response to certain stimuli by color change because of the twisting of the polymer backbone. Characteristics of large surface areas are advantageous for analyte adsorption and fluorescence having potential application in optical sensors [147]. These references prove that the applications of nanofibers in polarization are growing.

Photocatalytic applications are also a new subject for optical applications of electrospun nanofibers. Photocatalytic activity and its application in optoelectronics can be viewed where activity is performed by the mechanisms of photo-absorption, the generation of electron–hole pairs, charge carrier transfer, and charge carrier utilization. Enhancement of these activities can be due to specific surface area, crystallinity, light absorption capacity, and separation efficiency of electron–hole pairs [148]. Fabrication of MCM-41 nanofibers by electrospinning of polyvinyl alcohol (PVA), cetyltrimethylammonium bromide (CTAB), and tetraethyl orthosilicate (TEOS) has potential application in photocatalysis. Harvesting of visible light, adsorption, and catalytic degradation process of MB dye can be done but silver bromide (AgBr) nanoparticles should also be added during electrospinning [149]. Mesoporous $ZnFe_2O_4@TiO_2$ nanofibers were fabricated using electrospinning, promoting the high surface area, adsorption, desorption, and diffusion of reactants and products with potential applications seen in photocatalytic activity [148]. ZnO/Al_2O_3 nanolaminate has properties of enhanced thickness, grain size, several bilayers, defect concentration aiding in recombination of photo-generated carriers for application in optical sectors [150]. Modification in electrospinning through these additives in polarizers and photocatalysts has proven effective in optoelectronics applications. Researchers are searching for new techniques, new polymers for effective application aiding enhancement in optical and electronic efficiency for better performance. Thus, these applications are evident in optoelectronics field of nanotechnology, which is a backbone for different energy applications.

5 Conclusion

With the development of nanotechnology, it's applications will also diversify. This review has presented an overview of the historical prospect of electrospinning and its applications in nanotechnology for diverse fields of this generation. Later on, the fields will broaden and applications will also increase at a very demanding and conclusive rate. Along with the diversification, the transition from a research approach to an industrial approach is vital. Technological advances through nanotechnology are very prone so, the industrial approach is very important. For instance, the use of electrospun and modified nanofibers for wound healing, tissue engineering, and wound dressing is proven effective and has provided a basic framework for regenerative medicines. In the case of energy applications, the use of piezoelectric nanofibers has proven effective in fabricating energy storage devices and its use in supercapacitors and nanogenerators. Along with this, optical applications for catalytic and polarization activity have created a base for optoelectrical aspects. Environmental applications for fabricating various nanofibers to be used in filtration, purification of air and water, agricultural applications in fertilizers and seed germination, sensor applications for identifying various gases, chemicals have also proven to be effective by the use of nanotechnology. These examples provided above are researches conducted in the laboratory, and these researches need to be addressed at an industrial level as well. Mass production for the use in applications of the daily life of human beings is important. As the mass production of computer chips for has provided ease of work in various sectors, the researches presented above should also be recommended for mass production. From history till date, these modifications provide various frameworks for different applications of nanotechnology, which will surely help to fabricate the technological advances of the future world.

Acknowledgement SKT would like to thank NAWA and the University of Warsaw, Poland for the research grant (PPN/ULM/2019/1/00008/DEC/2) and scientific cooperation.

References

1. Bhardwaj, N. & Kundu, S. C. Electrospinning: A fascinating fiber fabrication technique.

Biotechnol Adv. **28**, 325–347(2010).
2. Tucker, N. et al. The history of the science and technology of electrospinning from 1600 to 1995. J. Eng. Fibers Fabr. **7**, 63-73 (1995).
3. Pillai, C. K. S. & Sharma, C. P. Electrospinning of Chitin and Chitosan Nanofibres. Trends Biomater. Artif. Organs. .**22**, 25–28 (2009).
4. Morton. W. J. Method of dispersing fluids. US Pat 705691, **28**,131–134, (1902).
5. Agarwal, S., Wendorff, J. H. & Greiner, A. Use of electrospinning technique for biomedical applications. Polymer (Guildf) **49**, 5603–5621(2008).
6. Shoba, E., Lakra, R., Kiran, M.S. and Korrapati, P.S., 2017. Fabrication of core–shell nanofibers for controlled delivery of bromelain and salvianolic acid B for skin regeneration in wound therapeutics. Biomedical Materials, 12(3), p.035005.
7. Li, G. et al. Laminin-coated electrospun regenerated silk fibroin mats promote neural progenitor cell proliferation , differentiation , and survival in vitro. Front. Bioeng. Biotechnol.**7**, 190 (2019).
8. Mokhtari, F., Latifi, M. & Shamshirsaz, M. Electrospinning / electrospray of polyvinylidene fluoride (PVDF): piezoelectric nanofibers. J. Text. Inst. **107**, 1037-1055 (2016).
9. Kai, D., Shy, S. & Jun, X. Biodegradable polymers for electrospinning : Towards biomedical applications. Mater. Sci. Eng. C **45**, 659–670(2014).
10. Bhushani, J. A. Electrospinning and electrospraying techniques : Potential food based applications. Trends Food Sci. Technol. **38**, 21-33 (2014)
11. Sun, J. et al. Nanofibers by Green Electrospinning of Aqueous Suspensions of Biodegradable Block Copolyesters for Applications in Medicine , Pharmacy and Agriculture. Macromol. Rapid Commun. **31**, 2077–2083 (2010).
12. Konwarh, R., Karak, N. and Misra, M., 2013. Electrospun cellulose acetate nanofibers: the present status and gamut of biotechnological applications. Biotechnology advances, 31(4), pp.421-437.
13. Yang, Z., Peng, H., Wang, W. & Liu, T. Crystallization behavior of poly(ε-caprolactone)/layered double hydroxide nanocomposites. J. Appl. Polym. Sci. **116**, 2658–2667 (2010).
14. Zhang, C., Feng, F. & Zhang, H. Emulsion electrospinning: Fundamentals, food applications and prospects. Trends Food Sci. Technol. **80**, 175-186 (2018).
15. Frenot, A. & Chronakis, I.S. Polymer nanofibers assembled by electrospinning. **Curr** Opin. Colloid. Interface Sci**.** **8**, 64–75 (2003).
16. Nikmaram, N., Roohinejad, S., Hashemi, S., Koubaa, M., Barba, F.J., Abbaspourrad, A. and Greiner, R.,. Emulsion-based systems for fabrication of electrospun nanofibers: Food, pharmaceutical and biomedical applications. RSC advances **7**,28951-28964 (2017).
17. Bhushan, B. Introduction to nanotechnology: history, status, and importance of nanoscience and nanotechnology education. Glob. Perspect. Nanosci. Edu., 1–31 (2016).
18. Agarwal, B. S., Wendorff, J. H. & Greiner, A. Progress in the field of electrospinning for tissue engineering applications. Adv. Mater. **21**, 3343–3351(2009).
19. Dong, Z., Kennedy, S. J. & Wu, Y. Electrospinning materials for energy-related applications and devices. J. Power. Sources **196**, 4886–4904(2011).
20. Sill, T. J. & Von Recum, H. A. Electrospinning: Applications in drug delivery and tissue engineering. Biomaterials **29**, 1989–2006 (2008).
21. Wade, R. J. & Burdick, J. A. Advances in nanofibrous scaffolds for biomedical applications : From electrospinning to self-assembly. Nano Today **9**, 722–742(2014).
22. Ji, X., Xu, Y., Zhang, W., Cui, L. and Liu, J., 2016. Review of functionalization, structure and properties of graphene/polymer composite fibers. Composites Part A: Applied Science and Manufacturing, 87, pp.29-45.
23. Navarro-Pardo, F., Martinez-Hernandez, A.L. and Velasco-Santos, C., 2016. Carbon nanotube and graphene based polyamide electrospun nanocomposites: a review. Journal of Nanomaterials, 2016.
24. Kleivaite, V. & Milašius, R. Electrospinning - 100 years of investigations and still open questions of web structure estimination. Autex Res. J. **18**, 398-404 (2018).

25. Figen, A. K. History, basics, and parameters of electrospinning technique. Electrospun. Mater. Their Allied Appl., 53–69 (2020).
26. Ralston, O. C. Practical applications of electrostatic phenomena to particulate matter. Trans. Am. Inst. Electr. Eng. Part I Commun. Electron. **75**, 155–159 (2013).
27. Teo, W. E. & Ramakrishna, S. A review on electrospinning design and nanofibre assemblies. Nanotechnology **17**, R 89 (2006).
28. Dawson, D.M., Hiremath, S., Mahmoodi, N., Saeidpourazar, R. and Aphale, S., 1974. Distributed Sensors and Actuators via Electronic-Textiles. Appl. Phys, 45(3768).
29. Sarmah, S. & Tamuli, R.P. Conducting Polymers: Biomedical Engineering Applications. Encycl. Biomed. Polym. Polym. Biomater., 1982–1996 (2016).
30. Asmatulu, R. & Khan, W. S. Synthesis and applications of electrospun nanofibers. Synth Appl. Electrospun. Nanofibers, 1–306 (2018).
31. Detecting the Rouse and sub-Rouse modes in poly (butyl acrylate) and poly (ethyl acrylate) through two-dimensional dynamic mechanical spectra. Journal of Macromolecular Science, Part B, 53(10), pp.1642–1653.
32. Doshi, J. and Reneker, D.H., 1995. Electrospinning process and applications of electrospun fibers. Journal of electrostatics, 35(2-3), pp.151-160.
33. Gibson, P., Schreuder-Gibson, H. & Pentheny, C. Electrospinning technology: Direct application of tailorable ultrathin membranes. J. Ind. Text. **28**, 63–72 (1998).
34. Brown, T. D., Dalton, P. D. & Hutmacher, D. W. Melt electrospinning today: An opportune time for an emerging polymer process. Prog. Polym. Sci. **56**,116–166 (2016).
35. Zhang, L. H. Recent advances in melt electrospinning. RSC Adv. **6**, 53400–53414 (2016).
36. Agarwal, S. & Greiner A. On the way to clean and safe electrospinning-green electrospinning: Emulsion and suspension electrospinning. Polym. Adv. Technol. **22**, 372–378 (2011).
37. Liverani, L., Vester, L. & Boccaccini, A. R. Electrospun biomaterials and related technologies. electrospun. Biomater. Relat. Technol., 149–168 (2017).
38. He, J. H., Liu, Y. & Xu, L. Apparatus for preparing electrospun nanofibres: A comparative review. Mater. Sci. Technol. **26**, 1275–1287 (2010).
39. Varesano, A., Carletto, R. A. & Mazzuchetti, G . Experimental investigations on the multi-jet electrospinning process. J. Mater. Process. Technol. **209**, 5178–5185 (2009).
40. Niu, H. & Lin, T. Fiber generators in needleless electrospinning. J Nanomater, 2012 (2012).
41. Liu, C. K. et al. Preparation of short submicron-fiber yarn by an annular collector through electrospinning. Mater. Lett. **62**, 4467–4469(2008).
42. Katta, P., Alessandro, M., Ramsier, R. D. & Chase, G. G. Continuous electrospinning of aligned polymer nanofibers onto a wire drum collector. Nano Lett. **4**, 2215–2218(2004).
43. Alfaro De Prá, M. A., Ribeiro-do-Valle, R. M., Maraschin, M. & Veleirinho, B. Effect of collector design on the morphological properties of polycaprolactone electrospun fibers. Mater. Lett. **193**, 154–157(2017).
44. Babar, A. A., Iqbal, N., Wang, X., Yu, J. & Ding, B. Introduction and historical overview. In Electrospinning: Nanofabrication and applications. William Andrew Publishing, 3–20 (2019).
45. Ghosal, K., Agatemor, C. Tucker, N. & Kny, E. Electrical spinning to electrospinning: A brief history, in Electrospinning: From Basic Research to Commercialization, 1–23 (2018).
46. Cook, J. G. Handbook of Textile Fibres, Volume 1: Natural Fibres. Elsevier (1984).
47. Parsons, J. L. Count Hilaire De Chardonnet, Scientist and Inventor. Ind. Eng. Chem. **17**, 754–755 (1925).
48. Carothers, W.H. and Hill, J.W., 1932. Studies of polymerization and ring formation. XV. Artificial fibers from synthetic linear condensation superpolymers. Journal of the American Chemical Society, 54(4), pp.1579–1587.
49. Hutmacher, D. W., Dalton P. D. Melt electrospinning. Chem. An Asian J. **6**, 44–56 (2011).
50. Katz, R. United States: UNITED STATES. Eur. J. Polit. Res. Polit. Data Yearbook **54**, 309–315 (2016).
51. Owens, J. E. & Scheinberg, S. P. Apparatus for collecting fibrous material in sheet form, USA, Patent No. 3490115 (1970).

52. Lin, T. Needleless electrospinning: A practical way to mass production of nanofibers. J Text. Sci. Eng. **2**, 2–4 (2012).
53. Rangkupan, R. & Reneker, D. H. Electrospinning process of molten polypropylene in vacuum. J. Met. Mater. Miner. **12**, 81–87 (2003). https://doi.org/10.4172/2165-8064.1000e109
54. Martin, G. E., Cockshott, I. D., & Fildes, F. J. T. Fibrillar lining for prosthetic device USA, Patent No 4044404A (1977).
55. Thien V. How. Synthetic vascular grafts, and methods of manufacturing such grafts, USA, Patent No. 4552707A (1985).
56. Yang, J. et al. Electrospun Janus nanofibers loaded with a drug and inorganic nanoparticles as an effective antibacterial wound dressing. Mater. Sci. Eng. C **111**, 110805 (2020).
57. Zhu, C. et al. Novel antibacterial fibers of amphiphilic N-halamine polymer prepared by electrospinning. Polym. Adv. Technol. **30**, 1386–1393 (2019).
58. Augustine, R. et al. Titanium Nanorods Loaded PCL Meshes with enhanced blood vessel formation and cell migration for wound dressing applications. Macromol. Biosci. **7,** 1900058 (2019).
59. Grumezescu, A. M. et al. Electrospun polyethylene terephthalate nanofibers loaded with silver nanoparticles: Novel approach in anti-infective therapy. J. Clin. Med. **8**, 1039 (2019).
60. Karagoz, S. et al. Synthesis of Ag and TiO_2 modified polycaprolactone electrospun nanofibers (PCL/TiO_2-Ag NFs) as a multifunctional material for SERS, photocatalysis and antibacterial applications. Ecotoxicol. Environ. Saf. **188**, 109856 (2020).
61. Mukheem, A. et al. Fabrication and characterization of an electrospun PHA/graphene silver nanocomposite scaffold for antibacterial applications. Materials (Basel) **11**, 1–15 (2018).
62. Kumar, N. S. et al. Electrospun polyurethane and soy protein nanofibres for wound dressing applications. Let Biotechnology **12**, 94–98 (2017).
63. Howard, D., Buttery, L. D., Shakesheff, K. M. & Roberts, S. J. Tissue engineering: Strategies, stem cells and scaffolds. J. Anat. **213**, 66–72 (2008).
64. Tiwari, S. K., Thakur, A. K., Adhikari, A. D., Zhu, Y. & Wang, N. Current research of graphene-based nanocomposites and their application for supercapacitors. Nanomaterials **10**, 2046 (2020).
65. Alberti, T.B., Coelho, D.S., de Prá, M., Maraschin, M. and Veleirinho, B., 2020. Electrospun PVA nanoscaffolds associated with propolis nanoparticles with wound healing activity. Journal of Materials Science, 55(23), pp.9712-9727.
66. Zhang, C., et al. The surface grafting of graphene oxide with poly(ethylene glycol) as a reinforcement for poly(lactic acid) nanocomposite scaffolds for potential tissue engineering applications. J. Mech. Behav. Biomed. Mater. **53,** 403–413 (2016).
67. Safikhani, M. M., Zamanian, A. & Ghorbani, F. Synergistic effects of retinoic acid and graphene oxide on the physicochemical and in-vitro properties of electrospun polyurethane scaffolds for bone tissue engineering. e-Polymers **17,** 362–371 (2017). https://doi.org/10.1515/epoly-2016-0304.
68. He, F. et al. A novel layer-structured scaffold with large pore sizes suitable for 3D cell culture prepared by near-field electrospinning. Mater. Sci. Eng. C **86**, 18-27 (2018).
69. Kadakia, P. U. et al. Comparison of silk fibroin electrospun scaffolds with poloxamer and honey additives for burn wound applications. J. Bioact. Compat. Polym. **33**, 79-94 (2017).
70. Su, S. et al. Coaxial and emulsion electrospinning of extracted hyaluronic acid and keratin based nanofibers for wound healing applications. Eur. Polym. J. **142**, 110158 (2021).
71. Mahoney, C., Conklin, D., Waterman. J. & Bhattarai, N. Electrospun nanofibers of poly (ε-caprolactone)/depolymerized chitosan for respiratory tissue engineering Applications. J. Biomater. Sci. Polym. Ed. **27**, 611–625 (2016).
72. Ayyar, M., Mani, M. P., Jaganathan, S. K. & Rathanasamy, R. Preparation , characterization and blood compatibility assessment of a novel electrospun nanocomposite comprising polyurethane and ayurvedic-indhulekha oil for tissue engineering applications. J. Biomed. Eng. **63**, 245-253 (2018).
73. Cesur. S. et al. Preparation and characterization of electrospun polylactic acid / sodium alginate / orange oyster shell composite nanofiber for biomedical application. J. Aust. Ceram. Soc. **56,** 533-543 (2019).

74. So, H. S. et al. Novel lavender oil and silver nanoparticles simultaneously loaded onto polyurethane nano fi bers for wound-healing applications. Int. J. Pharm. **569**, 118590 (2019).
75. Khan A. S. et al. Fabrication and in vivo evaluation of hydroxyapatite/carbon nanotube electrospun fibers for biomedical/dental application. Mater. Sci. Eng. C. **80**, 387-396 (2017)
76. Korani, S. et al. Application of nanotechnology to improve the therapeutic benefits of statins. Drug Discov. Today **24**, 567-574 (2018).
77. Nekounam, H., Allahyari, Z., Gholizadeh, S., Mirzaei, E., Shokrgozar, M.A. and Faridi-Majidi, R., 2020. Simple and robust fabrication and characterization of conductive carbonized nanofibers loaded with gold nanoparticles for bone tissue engineering applications. Materials Science and Engineering: C, 117, p.111226.
78. Abbas, W. A., Sharafeldin, I. M., Omar, M. M. & Allam, N.K. Nanoscale advances applications : computational and experimental insights. Nanoscale Adv.**2** 1512–1522 (2020).
79. Adhikari, S.P., Pant, H.R., Mousa, H.M., Lee, J., Kim, H.J., Park, C.H. and Kim, C.S., 2016. Synthesis of high porous electrospun hollow TiO2 nanofibers for bone tissue engineering application. Journal of industrial and engineering chemistry, 35, pp.75-82.
80. de Castro, J.G., Rodrigues, B.V., Ricci, R., Costa, M.M., Ribeiro, A.F., Marciano, F.R. and Lobo, A.O., 2016. Designing a novel nanocomposite for bone tissue engineering using electrospun conductive PBAT/polypyrrole as a scaffold to direct nanohydroxyapatite electrodeposition. RSC advances, 6(39), pp.32615-32623.
81. Santocildes-Romero, M.E., Hadley, L., Clitherow, K.H., Hansen, J., Murdoch, C., Colley, H.E., Thornhill, M.H. and Hatton, P.V., 2017. Fabrication of electrospun mucoadhesive membranes for therapeutic applications in oral medicine. ACS applied materials & interfaces, 9(13), pp.11557-11567.
82. Suryavanshi. A. et al. Magnesium oxide nanoparticle-loaded polycaprolactone composite electrospun fiber scaffolds for bone – soft tissue engineering applications: in-vitro and in-vivo evaluation Magnesium oxide nanoparticle-loaded polycaprolactone composite electrospun fiber. Biomed. Mater. **5**, 055011 (2017).
83. Zarei, M. & Karbasi, S. Evaluation of the effects of multiwalled carbon nanotubes on electrospun poly (3-hydroxybutirate) scaffold for tissue engineering applications. J. Porous Mater. **25**, 259-272 (2017).
84. Li, P. et al. Applying electrospun gelatin / poly (lactic acid-co-glycolic acid) bilayered nanofibers to fabrication of meniscal. J. Nanosci. Nanotechnol. **16**, 4718–4726 (2016).
85. Rychter, M. et al. Cilostazol-loaded poly (ε -caprolactone) electrospun drug delivery system for cardiovascular applications. Pharm. Res. **35**, 1-20 (2018).
86. Dubsky. M., Lesny, P., Jirkovska. A. & Munzarova. M. Nanofibers prepared by needleless electrospinning technology as scaffolds for wound healing. J. Mater. Sci. Mater. Med. **23**, 931–941(2012).
87. Maslakci, N.N., Ulusoy, S., Uygun, E., Çevikbaş, H., Oksuz, L., Can, H.K. and Oksuz, A.U., 2017. Ibuprofen and acetylsalicylic acid loaded electrospun PVP-dextran nanofiber mats for biomedical applications. Polymer Bulletin, 74(8), pp.3283-3299.
88. Morsy. R., Hosny, M., Reisha, F. & Elnimr, T. Developing and physicochemical evaluation of cross-linked electrospun gelatin glycerol nanofibrous membranes for medical applications. J. Mol. Struct. **1135**. 222-237 (2017).
89. Komur, B. et al. Starch/PCL composite nanofibers by co-axial electrospinning technique for biomedical applications. Biomed. Eng.**16** Online 1–13 (2017).
90. Li, X. et al. Electrospinning-based strategies for battery. Materials. Adv. Energy Mater. **11**, 2000845 (2020).
91. 1703237:1https://doi.org/10.1002/smll.201703237 Zeng, Z., Zhang, W., Liu, Y., Lu, P. and Wei, J., 2017. Uniformly electrodeposited α-MnO2 film on super-aligned electrospun carbon nanofibers for a bifunctional catalyst design in oxygen reduction reaction. Electrochimica Acta, 256, pp.232–240.
92. Monaca, A. et al. Electrospun ceramic nanofibers as 1D solid electrolytes for lithium batteries. Electrochem. Commun. **104**, 106483 (2019).

93. Zhuang, R. et al. Synthesis and characterization of electrospun molybdenum dioxide – carbon nanofibers as sulfur matrix additives for rechargeable lithium – sulfur battery applications. J. Nanotechnol. **9**, 262–270 (2018).
94. Angel, N., Vijayaraghavan, S.N., Yan, F. and Kong, L., 2020. Electrospun cadmium selenide nanoparticles-loaded cellulose acetate fibers for solar thermal application. Nanomaterials, 10(7), p.1329.
95. Imran, M., Haider, S., Ahmad, K., Mahmood, A. and Al-Masry, W.A., 2017. Fabrication and characterization of zinc oxide nanofibers for renewable energy applications. Arabian Journal of Chemistry, 10, pp.S1067-S1072.
96. Kwon, Y. et al. Electrospun CuS/PVP Nanowires and superior near-infrared filtration efficiency for thermal shielding applications. ACS Appl Mater Interfaces **11**, 6575–6580 (2019).
97. Ponnamma, D., Aljarod, O., Parangusan, H. and Al-Maadeed, M.A.A., 2020. Electrospun nanofibers of PVDF-HFP composites containing magnetic nickel ferrite for energy harvesting application. Materials Chemistry and Physics, 239, p.122257.
98. Cheon S. et al. High-performance triboelectric nanogenerators based on electrospun polyvinylidene fluoride – silver nanowire composite nanofibers. Adv. Funct. Mater. **28**, 703778 (20181).
99. Li, Z., Zhao, Y., Wang, X., Sun, Y., Zhao, Z., Li, Y., Zhou, H. and Chen, Q., 2018. Cost analysis of perovskite tandem photovoltaics. Joule, 2(8), pp.1559-1572.
100. Kang, S. B. et al. Enhanced piezoresponse of highly aligned electrospun poly (vinylidene fluoride) nano fibers. Nanotechno. **28**, 395402 (2017).
101. Szewczyk, P. K. et al. Enhanced piezoelectricity of electrospun polyvinylidene fluoride fibers for energy harvesting. ACS Appl. Mater. Interfaces **12**, 13575-13583(2020).
102. Karumuthil, S. C. et al. Electrospun poly(vinylidene fl uoride-tri fl uoroethylene)-based polymer nanocomposite fibers for piezoelectric nanogenerators. ACS Appl. Mater. Interfaces **11**, 40180-40188 (2019).
103. Adhikary, P., Biswas, A. & Mandal, D. Improved sensitivity of wearable nanogenerators made of electrospun Eu^{3+} doped P (VDF-HFP)/graphene composite nano fibers for self-powered voice recognition. Nanotechnology **27**, 495501 (2016).
104. Fuh, Y. K., Wang, B. S. & Liu, B. J. Near field sequentially electrospun three-dimensional piezoelectric fibers arrays for self-powered sensors of human gesture recognition. Nano Energy **30**, 677-683 (2016).
105. Gualandi, C., Celli, A., Zucchelli, A. and Focarete, M.L., 2014. Nanohybrid materials by electrospinning. Organic-Inorganic Hybrid Nanomaterials, pp.87–142.
106. Kotatha, D. et al. Preparation and Characterization of electrospun gelatin nanofibers for use as nonaqueous electrolyte in electric double-layer capacitor. J. Nanotechnol., 2501039 (2019).
107. Pant, B., Park, M. & Park, S. TiO_2 NPs Assembled into a Carbon Nanofiber Composite Electrode by a One-Step Electrospinning Process for Supercapacitor Applications. Polymers **11**, 811 (2019).
108. Dai, Y. et al. Ceramic nanofibers fabricated by electrospinning and their applications in catalysis, environmental science, and energy technology. Polym. Adv. Technol. **22**, 326–338 (2011).
109. Ghorai, S., Sarkar, A. K., Panda, A. B. & Pal, S. Effective removal of Congo red dye from aqueous solution using modified xanthan gum/silica hybrid nanocomposite as adsorbent. Bioresour. Technol. **144**, 485–491 (2013).
110. Thomas, S. Chitin nanowhisker (ChNW) -functionalized electrospun PVDF membrane for enhanced removal of Indigo carmine. Carbohydr. Polym. **165**, 115-122 (2017).
111. Oya, N. et al. Encapsulation of living bacteria in electrospun cyclodextrin ultrathin fibers for bioremediation of heavy metals and reactive dye from wastewater. Colloids Surf. B Biointerfaces **161**, 169-176 (2017).
112. Qin, X. & Subianto, S. Electrospun nanofibers for filtration applications. Electrospun Nanofibers, Woodhead Publishing, 449–466 (2017).

113. Vinh, N. D & Kim, H. Electrospinning fabrication and performance evaluation of polyacrylonitrile nanofiber for air filter applications. Appl. Sci. **6**, 235 (2016).
114. Koena, M., Ojijo, V., Sadiku, R. & Sinha, S. Development of bacterial-resistant electrospun polylactide membrane for air fi ltration application : Effects of reduction methods and their loadings. Polym. Degrad. Stab. **178**, 109205 (2020).
115. Dung, N. V. et al. CuO nanofibers prepared by electrospinning for gas sensing application : effect of copper salt concentration. J. Nanosci. Nanotechnol. **19**, 7910-7918 (2016).
116. Kumar, M. et al. Mixture of PLA-PEG and Biotinylated Albumin enables Immobilization of Avidins on Electrospun Fibers. J. Biomed. Mater. Res. A. **105**, 356-362 (2017).
117. Leonardi, S. G. et a.l A comparison of the ethanol sensing properties of α -iron oxide nanostructures prepared via the sol-gel and electrospinning techniques. Nanotechnology **27**, 075502 (2016).
118. Almasian, A., Chizari Fard, G., Parvinzadeh Gashti, M., Mirjalili, M. and Mokhtari Shourijeh, Z., 2016. Surface modification of electrospun PAN nanofibers by amine compounds for adsorption of anionic dyes. *Desalination and Water Treatment*, *57*(22), pp.10333-10348.
119. DM Follmann, H., F Naves, A., A Araujo, R., Dubovoy, V., Huang, X., Asefa, T., Silva, R. and N Oliveira, O., 2017. Hybrid materials and nanocomposites as multifunctional biomaterials. Current pharmaceutical design, 23(26), pp. 3794–3813.
120. Ghosh, A., Nayak, A.K. and Pal, A., 2017. Nano-particle-mediated wastewater treatment: A review. Current Pollution Reports, 3(1), pp.17-30.
121. Sapountzi,. E. et al. Gold nanoparticles assembly on electrospun poly (vinyl alcohol)/ poly (ethyleneimine / glucose oxidase nanofibers for ultrasensitive electrochemical glucose biosensing. Sens. Actuators B Chem. **238**, 392–401(2017).
122. Yang, C. et al. Acta Biomaterialia Electrospun pH-sensitive core – shell polymer nanocomposites fabricated using a tri-axial process. ACTA Biomater. **35**, 77-86 (2016).
123. Deitzel, J. M. et al, Generation of polymer nanofibers through electrospinning. (No ARL-TR-1989) Army Res. Lab. Aberdeen Proving Gr. Md, (1999).
124. Lee, S., & Obendorf, S. K. Developing protective textile materials as barriers to liquid penetration using melt-electrospinning. J. Appl. Polym. Sci. **102**, 3430–3437 (2006).
125. Serbezeanu, D. et al. Preparation and characterization of thermally stable polyimide membranes by electrospinning for protective clothing applications. Text. Res. J. **85**, 1763–1775 (2015).
126. Schreuder-Gibson, H. et al. Protective textile materials based on electrospun nanofibers. J Adv. Mater. **34**, 44–55 (2002).
127. Gibson, P. W., Schreuder-Gibson, H. L. & Rivin, D. Electrospun fiber mats: Transport properties. AIChE J. **45**, 190–195 (1999).
128. Gashti, M. P., Pakdel, E. & Alimohammadi, F. Nanotechnology-based coating techniques for smart textiles. Active Coating for Smart Textiles, Woodhead Publishing, 243–268 (2016).
129. Gharehaghaji, A. A. Nanotechnology in Sport Clothing. Materials in Sports Equipment, Woohead Publishing, 521–568 (2019).
130. Liu, R. & Ji, D. Electrospun nano fi bers for personal protection in mines. Chem. Eng. J. **404**, 126558 (2021).
131. Angelica, D. et al. A novel nanocomposite for food packaging developed by electrospinning and electrospraying. Food Packag. Shelf Life **20**, 100314 (2019).
132. Rovira, F., Mas, L. C., Lorena, J. & Mayorga, C. Antimicrobial nanocomposites and electrospun coatings based on poly (3-hydroxybutyrate- co -3-hydroxyvalerate) and copper oxide nanoparticles for active packaging and coating applications. J. Appl. Polym. Sci**. 135,** 45673 (2018).
133. Ebrahimi. S., Fathi, M. & Kadivar, M. Production and characterization of chitosan-gelatin nano fi bers by nozzle- less electrospinning and their application to enhance edible film's properties. Food Packag. Shelf Life **22**, 100387 (2019).
134. Cherpinski, A., Cabedo, L. & Lagaron, J. M. Post-processing optimization of electrospun sub- micron poly (3-hydroxybutyrate) fibers to obtain continuous films of interest in food packaging applications. Food Addit. Contam. Part A **34**, 1817-1830 (2017).

135. Sharma, G., Rastogi, S. & Kandasubramanian, B. Smart Electrospun Materials. Electrospun Materials and Their Allied Applications, 351–378 (2020).
136. Scaffaro, R., Lopresti, F., Marino, A. & Nostro, A. Antimicrobial additives for poly (lactic acid) materials and their applications : current state and perspectives.A ppl. Microbiol. Biotechnol. **102**, 7739–7756 (2018).
137. Castro-Enríquez, D. D. et al. Preparation, characterization and release of urea from wheat gluten electrospun membranes. Materials **5**, 2903–2916 (2012).
138. Johnson, B.A. Agriculture and Nanotechnology. Ward and Dutta. University of Wisconsin-Madison (2006).
139. Prasad, R., Kumar, V. & Kumar, M. Nanotechnology: Food and environmental paradigm. Nanotechnology Food Environment Paradigm. 1–344 (2017).
140. Corradini, E., De Moura, M. R. & Mattoso, L. H. C. A preliminary study of the incorparation of NPK fertilizer into chitosan nanoparticles. Express Polym. Lett. **4**, 509–515 (2010).
141. Bulus. E., Sakarya, B. G. & Yakuphanoglu, F. Production and characterization of novel nature-friendly organic fertilizer covers based on nanotechnology for the agricultural sector. J. Mater. Electron, Devices **5**, 12–16 (2010).
142. Bisotto-de-oliveira, R. et al. Nanofibers as a vehicle for the synthetic attactant TRIMEDLURE to be used for ceratitis capitata wied : (Diptera , Tethritidae) capture. J. Res. Updates Polym..**3**, 40–47 (2014).
143. Castañeda, L. M. F., Genro, C., Roggia, I. & Bender, S. S. Innovative rice seed coating (Oryza Sativa) with polymer nanofibres and microparticles using the electrospinning method. J. Res. Updates Polym. **3**, 33–39 (2014).
144. Bansal, P. Water-Based Polymeric Nanostructures for Agricultural Applications PhD diss., PhD Thesis. Philipps-Universität Marburg (2010).
145. Topçu, G., Savac, U., Genç, A. & Turan, S. Polarized emission from $CsPbBr_3$ nanowire embedded-electrospun PU fibers. Nanotechnology **29**, 135202 (2018).
146. Hu, Z. et al. Composite film polarizer based on the oriented assembly of electrospun nano fibers. Nanotechnology **27**, 135301 (2016).
147. Sanfelice, R.C., Mercante, L.A., Pavinatto, A., Tomazio, N.B., Mendonça, C.R., Ribeiro, S.J., Mattoso, L.H. and Correa, D.S., 2017. Hybrid composite material based on polythiophene derivative nanofibers modified with gold nanoparticles for optoelectronics applications. Journal of Materials Science, 52(4), pp.1919-1929.
148. Nada, A. A. et al. Mesoporous $ZnFe_2O_4$ @ TiO_2 nanofibers prepared by electrospinning coupled with PECVD as highly performing photocatalytic materials mesoporous $ZnFe_2O_4$ @ TiO_2 nanofibers prepared by electrospinning coupled with PECVD as Highly performing photocatalytic . J. Phys. Chem. C **121**, 24669-24677 (2017).
149. Jafarzadeh, A., Sohrabnezhad, S., Zanjanchi, M. A. & Arvand, M. Microporous and mesoporous materials synthesis and characterization of thiol-functionalized MCM-41 nano fibers and its application as photocatalyst. Microporous Mesoporous Mater. **236**, 109–119 (2016).
150. Al, Z., et al. Enhancement of electronic and optical properties of enhancement of electronic and optical properties of ZnO / Al_2O_3 nanolaminate coated electrospun .Nanofibers. J. Phys. Chem. C **120**, 5124-5132 (2016).

Metal-Oxide Semiconductor Nanomaterials as Alternative to Carbon Allotropes for Third-Generation Thin-Film Dye-Sensitized Solar Cells

Muhammad Sufyan, Umer Mehmood, Sadia Yasmeen, Yasir Qayyum Gill, Muhammad Sadiq, and Mohsin Ali

Abstract In photovoltaic generations of a solar cell, thin-film dye-sensitized solar cells have a significant role in clean energy production due to low-cost, easy fabrication process, and maximum efficiencies even in low-intensity radiations from the sun in a cloudy environment. The charge transportation mechanism, electron diffusion and movement, charge collection efficiency, charge recombination reactions, and thc electron path length influence the DSSC performance. All these factors are linked with the photoanode material. Porosity, surface area, composition, and architecture are the key parameters which should be considered for the material selection of DSSC photoanode. This chapter comprises three parts. In the first section, a brief introduction to photovoltaic technologies, working mechanism of DSSC, and structure of DSSC will be discussed in detail. The role of metal oxide semiconductor materials in DSSCs and their types will be discussed in second part. Finally, the morphology, modification of semiconductor materials, and their effect on the photovoltaic properties of light-harvesting devices will be discussed in detail in the last part.

Keywords Metal oxides · Semiconductors · Nanoparticles · Photoanode · Thin film · Energy efficiency

1 Introduction

Photovoltaic cells (PV) are the best technology to convert solar radiations into electrical power because they generate no harmful by-product as well as provide benefit to be installed in remote areas where grid installation is difficult. Third-generation PV cells are flexible in nature, provide ease during fabrication, and low-temperature solution-based methods are used. Perovskite has achieved the highest efficiency but it creates problems in humid environment and unstable oxygen presence. Polymer solar cell (PSC) efficiencies are lower due to complicated fabrication

M. Sufyan · U. Mehmood (✉) · S. Yasmeen · Y. Q. Gill · M. Sadiq · M. Ali
Polymer and Process Engineering (PPE) Department, University of Engineering & Technology (UET) Lahore, Lahore, Pakistan
e-mail: umermehmood@uet.edu.pk

S. Sahoo et al. (eds.), *Defect Engineering of Carbon Nanostructures*, Advances in Material Research and Technology, https://doi.org/10.1007/978-3-030-94375-2_9

techniques, making them unfavorable for commercialization and practical applications [1]. DSSCs emerge on top in the third-generation solar cells due to ease of fabrication, low cost [2, 3], and stable working in diffused sunlight and cloudy environment [4]. However, its efficiency is less than a conventional silicon-based solar cell, but it reached up to 15% higher as compared to other generation solar cells.

The overall efficiency of DSSC is dependent on photoelectrode. Photoanode is a combination of three parts: conductive substrate, semiconductor metal oxide material, and dye. Mesoporous semiconductor oxide layer provides the surface area for sensitizer loading, which is supported by a conductive glass substrate. Main role of the photoanode is to provide a pathway for photoexcited electrons from chromophore to substrate surface through conduction band of the semiconductor. In order to enhance the overall efficiency, photoanode has to build some key characteristics: (1) higher surface area for maximum sensitizer loading; (2) a smooth path for the facilitation of electrons from dye to external circuit; (3) optimum porosity of material for maximum diffusion of the dye and electrolyte; (4) higher photo-corrosion resistance; (5) high light-scattering ability, surface roughness, and ability to accept electrons; and (6) good adhesion with glass substrate. All these parameters have shown a direct link with photoanode material.

2 DSSC Structure and Working Mechanism

The structure of DSSCs is made up of photoanode/photoelectrode (PE), the counter-electrode (CE), and electrolyte. Photoanode consists of glass or flexible (polymer) substrate on which the conductive layer is coated, a mesoporous semiconductor layer film, and dye. Counter-electrode is fabricated by depositing the catalyst on the substrate. The electrolyte is sandwiched between the two electrodes to complete the assembly of DSSC. When light falls on the dye (photosensitizer) through a transparent substrate, it becomes excited by absorbing photons of energy. The photoexcitation of dye molecules causes the electrons to move from lower energy level (HOMO, highest occupied molecular orbitals) to higher energy level (LUMO, lowest unoccupied molecular orbitals). Because of the difference in energy levels, electrons are diffused and dispersed into the conduction band of the mesoporous semiconductor and become collected on the conductive substrate. The electrons emit through photoanode to the external load and proceed toward the counter-electrode. Oxidized dye is regenerated through redox reaction with the electrolyte. At the same time, electrolyte is oxidized and the electron-deficient ions move toward the CE where redox reaction occurs to regenerate the electrolyte. With this step the working cycle of DSSC is completed, as shown in Fig. 1 [5].

2.1 *Metal Oxide Semiconductors as Photoanode Materials*

Proper material selection for photoanode is critical for an efficient DSSC. Many semiconductor materials having an energy bandgap higher than 3 eV, i.e., ZnO, TiO_2, SnO_2, $SrTiO_3$, Zn_2SnO_4, WO_3, and Nb_2O_5, have been tested as suitable material for the photoanode because they retain resistance against photo-corrosion and have excellent conductive nature. Metal oxides other than TiO_2 and ZnO proved less efficient material because they have faster charge recombination, poor dye intake absorption, lower isoelectric point (SnO_2), smaller pore diameter and reduced dye loading site (Nb_2O_5), positive conduction band edge (WO_3), shorter electron diffusion length (Zn_2SnO_4), and lower ionization potential and bandgap ($SrTiO_3$) [6, 7]. That is why only ZnO and TiO_2 as photoanode materials are being discussed in the material section.

2.1.1 ZnO

One of the earliest materials used for a dye-sensitized solar cell is ZnO. Its bandgap (3.37 eV) and conduction band edge are nearer to anatase TiO_2. The higher electron mobility of ZnO helps in fast electron transportation. But the chemical stability of ZnO is lower than TiO_2, which can enhance its dissolution both in an acidic and basic environments. Park et al. applied shell of SiO_2 on ZnO colloids to improve the chemical stability. The efficiency of 5.2% was achieved using SiO_2-modified structure of ZnO, compared with initial ZnO, the leaching of Zn^{2+} ions in acidic solution is less than 1% [8]. At the interface of ZnO/dye, agglomerates of complexes (Zn^{2+}/dye) are formed, leading to poor efficiency of the DSSC. Agglomerates of Zn^{2+}/dye complexes act as an insulating layer when the dye injects electrons into the semiconductor layer. Iso-electric point of ZnO is high [9, 10], which makes it more prone toward acidic dyes attack due to its essential nature [11]. To avoid agglomerate formation of complexes, shell structure is applied using a buffer layer of suitable materials like SiO_2 [12], Al_2O_3, and TiO_2 to protect the core of ZnO [13]. These materials are very useful shell materials, and their role is to minimize the formation of the Zn^{2+}/dye complex. Greene et al. used TiO_2 as a shell material during the growth of nanorod morphology. They achieved a value of 2.25% efficiency as compared to 0.45% for simple ZnO-NWs [14].

The best efficiency of 7.5% is achieved till now with ZnO hierarchically assembled nanocrystallites using the spray pyrolysis method [15]. ZnO NP synthesized by Akthar and co-worker gives PCE value of 4.10% when employed as photoanode of DSSC [16]. One of the main advantages of using ZnO semiconductor as photoanode is that one can easily develop desired nanostructured morphologies. Surface-to-volume ratio varies in different morphologies; higher surface-to-volume ratio leads toward more dye absorption, which harvests more radiations to convert into electrical energy. In Table1, different novel morphologies of ZnO have been reported with their efficiencies.

Table 1 Different novel morphologies used as photoanode of DSSC

Sr. no	Nanostructures	FF	V_{oc} (V)	J_{sc} ($mAcm^{-2}$)	η (%)	References
1	Nanoplates	0.41	0.554	8.4	1.90	[17]
2	Spindle-shaped	0.45	0.79	5.10	1.82	[18]
3	Nanoflowers	0.53	0.65	5.50	1.90	[19]
4	Nanoflakes	0.50	0.63	11.6	3.64	[20]
5	Hexagonal clubs	0.65	0.58	11.5	4.28	[21]
6	Cauliflower	0.55	0.66	6.08	2.18	[22]
7	Microspheres	0.61	0.51	14.73	5.16	[23]
8	Caterpillar-like	0.50	0.69	15.20	5.20	[24]
9	Nanospikes decorated sheets	0.60	0.68	6.07	2.51	[25]
10	Disk-like	52.50	0.69	6.92	2.49	[26]

2.1.2 TiO_2

Physical and chemical properties of TiO_2 nanomaterial are much more suitable and stable as the photoanode layer of DSSC compared to other semiconductor materials [27, 28]. An increment in surface area of 2000 times is reported using nanosized TiO_2 photoelectrode compared with flattered photoanodes [29, 30]. Parameters like particle size, surface area, porosity, thickness, and morphology of TiO_2 are influential in the overall performance of DSSC. The next section highlights the in-depth analysis of TiO_2 performance-measuring factors as an efficient photoanode material.

3 TiO_2 as Proficient Material for Photoanode

TiO_2 is a stable, non-toxic material with a high refractive index value of 2.5. It exists in three natural forms: rutile, anatase, and brooklite. Rutile and anatase are used extensively in different applications (paint, toothpaste, sunscreen, food). Rutile has a bandgap value of 3.05 $_{eV}$ and thermodynamically it is the most stable form of TiO_2. Anatase is the preferred form of TiO_2 which is used as photoanode of DSSCs owing to a higher bandgap value (3.2 eV), superior conduction band edge, and high electron diffusion coefficient [31, 32]. Whereas the bandgap value of brooklite is 3.28 eV, but the intricate synthesis process makes it unfavorable for application in DSSC. Accordingly, the anatase form of TiO_2 is massively applied as photoanode of DSSC.

3.1 Performance Evaluation Factors of TiO_2 Photoanode

TiO_2 layer is not just used to provide the electron movement path, but it can also be applied as a blocking layer, light absorption layer, and light-scattering layer on the photoanode of DSSC. The performance of PEs is based on the surface area, composition, and morphology of TiO_2. Light-harvesting efficiency (LHE) and light-scattering ability of TiO_2 thin film are poor owing to the nanometer size of particles. The objective of this section is to highlight the essential performance determining factors of PEs.

3.1.1 Particle Size and Surface Area

The surface area of the film decides the amount of dye intake, which in turn affects the number of electrons produced. The smaller surface area will lead to lower intake of the dye so the efficiency will be decreased because of smaller current density. The larger surface area will provide more grain boundaries due to smaller particle size, which acts as trapping site for the electrons. Figure 2 shows the relation of the particle size with the surface area and dye intake amount. It is shown that as the size of the particle increases, the quantity of dye adsorption also increases, and ultimately the current density will also become higher. On the other hand, the surface area decreases with the increase of particle size. So, a trade-off between particle size, surface area, and dye intake must be done to get maximum efficiency [33]. Particles produced with different synthesis techniques have different surface areas and show divergent results for every method. So the synthesis routes also affect the efficiency of the photoanode [34]. Anatase TiO_2 particles of 15–20 nm produced with hydrothermal method when applied as photoanode layer reported an efficiency of 5.2% [35]. 3D aerogels provide higher surface area and porosity, which is produced by S. Alvin et al. using the micro-wave-assisted sol–gel method. A 14.5 nm size aerogel of TiO_2, when employed as photoanode gives PCE of 5.2% with 15.18 mA/cm^2 current density and 0.62 V open-circuit voltage [36].

3.1.2 Thickness of TiO_2 Film

The optimum thickness of the semiconductor oxide layer is also vital to get enhanced efficiency of a DSSC. The efficiency of the cell depends on (1) amount of dye adsorbed, (2) electron transportation, and (3) minimum recombination reactions. All these parameters are linked with the thickness of the photoanode [37]. Chen et al. used a TiO_2 layer of varied thickness on separate anodes with different diameter-sized beads. It is found that the highest efficiency is reported around 13 μm film thickness [38]. Kao et al. used films of 0.5 to 2 μm thickness as photoanode and reported that the efficiency increases from 1.9 to 2.9% when the film thickness reaches up to 1.5 μm and then the efficiency starts decreasing from 1.5 to 2 μm (2.6%). First, the efficiency

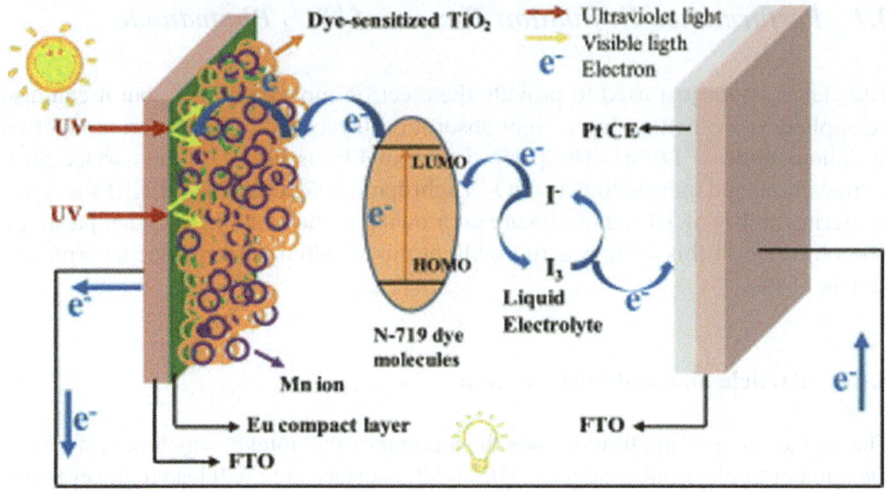

Fig. 1 Schematic diagram of DSSC structure and working mechanism (Reproduced with permission from Ref. [5])

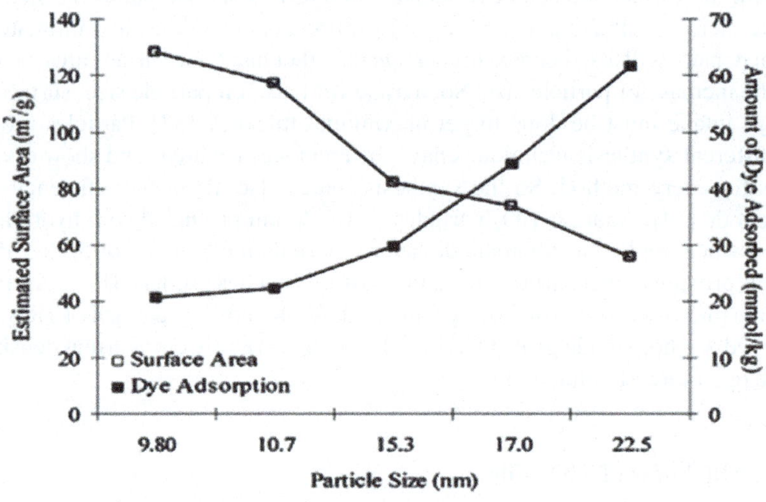

Fig. 2 Schematic relation of particle size with surface area and dye adsorbed by Chou et al. (Reproduced with permission from Ref. [33])

increases with thickness due to higher absorption of the dye molecules. After the optimum thickness, a further increase in the film thickness (2 μm) leads toward a decrease in the cell efficiency due to low transmittance and reduced intensity of light that reaches the dye [39]. Kumari et al. published the highest efficiency (6.07%) of a DSSC at an optimized thickness of 12.73 μm. They applied layers of varied thickness around 5.57–20.65 μm with the doctor-blade technique. At this optimum thickness

Table 2 Effect of scattering materials on DSSC photoanode

Sr. no	Improvement method	Core material	Improvement material	PCE (%) of core sample	PCE (%) after improvement	References
1	Scattering layer	TiO_2 NPs	Hierarchical TiO_2 sphere	6.80	9.37	[41]
2		TiO_2 NPs	Large rutile particles	7.2	9.00	[42]
3		TiO_2 NPs	Large TiO_2 particles	8.03	11.05	[43]
4		TiO_2 NPs	TiO_2 hollow sphere	7.79	9.43	[44]
5		TiO_2 NPs	Hierarchical TiO_2 beads	6.19	8.84	[45]
6		TiO_2 NPs	TiO_2 hierarchical flowers	6.59	9.08	[46]
7		TiO_2 NPs	SnO_2 hollow microspheres	7.29	9.53	[47]
8	Scattering layer and nanoarchitecture	TiO_2 NPs	TiO_2 nanooctahedra	5.76	8.76	[48]
9		TiO_2 nanosheets	TiO_2 spheres	5.39	7.72	[49]
10		TiO_2 NPs	TiO_2 hierarchical nanowires and TiO_2 hierarchical spheres	7.62	11.01	[50]
11		TiO_2 NFs with smaller diameter	TiO_2 NFs with large diameter	7.14	8.40	[51]
12		TiO_2 NPs	Hollow TiO_2 NPs and TiO_2 spindles	5.52	8.65	[52]

and surface area with better crystallization, the highest injection rate of electrons is possible from excited dye to the conduction band of TiO_2. Further increment in the thickness from 12.73 μm to 20.65 μm will enhance the surface area but the electron transport resistance increases at a higher order, and fast recombination of electrons occurs at TiO_2/electrolyte interface. Moreover, the electron lifetime decreases after the optimum thickness, and the recombination lifetime increases [40] (Table 2).

3.1.3 Doping Effect

The crystal structure of TiO_2 contains oxygen absences, which can create electron–hole pairs in the DSSCs [53]. These holes can react with the dye or extinguish it owing to its oxidizing nature, and the iodide ions of electrolyte will scavenge them, which will ultimately decrease the life span of the DSSC [54]. TiO_2 NPs can absorb UV light, which causes degradation, and that is why doping is done to modify the electronic properties of TiO_2 [55]. Free charges are enhanced with the doping (insertion of small impurity into base element) material to uplift the host element's conductivity. N-type dopants donate the electrons and p-type materials give holes to the host material. So, generally, weak or trapping points of the electrons and the electronic assembly of host (TiO_2) material are affected with the doping impurity. TiO_2 has positive (Ti^{4+}) and negative (O^{2-}) ions, and both can be replaced depending on the type of dopant. Ti^{4+} electronic cloud falls in 3d bands which constitutes conduction band (CB), 2p bands of O_2 constitute edge, and valance band (VB). Cationic impurity will make an impact on the CB, and anions will affect the VB energy levels [56]. The atomic radius of both the doping impurity and the base element should be analogous to get higher efficiency. Multi-dopants can also be used to get/decrease the combined effect of each impurity. Performance of the photoanode may deteriorate if the quantity of contaminants is not optimized [57–59]. Doping has two main types that are efficient for DSSCs: (1) nitrogen (non-metal) doping and (2) metal doping.

Nitrogen (Non-Metal) Doping

Nitrogen is a very effective non-metallic element for doping of the TiO_2 material. Improvement of the photoelectric properties and replacement of the oxygen deficiencies in TiO_2 is made through nitrogen doping by making the core material visible light active to boost the overall conversion efficiency of DSSCs. An increment in IPCE is seen, and Ma et al. and his co-workers achieved a conversion efficiency of 8% through N-doped nanocrystalline TiO_2 photoanode. N-doped TiO_2 photoelectrode has also shown greater stability as compared to pure TiO_2 and P25 [54, 60].

Performance is varied when the preparation technology of N-doped TiO_2 is changed. Technologies like sputtering and implantation [61], sintering under N-atmosphere [62], and wet methods have been very useful to deposit N-doped TiO_2. Wet methods to synthesize N-doped TiO_2 using urea, ammonia, and triethylamine as nitrogen dopants have been well studied. An efficiency of 8.32% is achieved, which is approximately 36% greater when ammonia is used as N-dopant owing to its higher dye intake and smooth path for the flow of electrons by N-doping [63, 64].

Metal Doping

The transition metals and rare earth metals have been used widely to ease the electron transportation in TiO_2 film. The addition of transition metal into the host material will give the advantages of prolonged band edges, upgraded fermi level, and amplified electron density of energy states. P-type doping effect befalls when rare metals (mostly tri-valent positive ions) are applied as dopants resulting in higher energy level and boosted the efficiency of DSSCs. Wu et al. and his co-workers discovered that light-harvesting efficiency amended/enhanced due to the luminous process along with the energy level perfection. Many ways have been used to introduce dopant into the lattice of TiO_2, but the simple mixing method is the most common. Dopant and host material can mix physically and disperse properly. TiO_2 can be in the form of paste or suspension, which is easy to deposit on the substrate. To deposit prepared

Table 3 Improvement of DSSC using doping of various materials

Sr. no	Dopant material	PCE (%) before doping	PCE (%) after doping	References
1	N-doped TiO_2 NPs	6	8	[60]
2	1 wt% Nb-doped TiO_2 NTs	2.40	3.21	[66]
3	2.5 mol% Nb-doped TiO_2 NPs	5.08	7.41	[80]
4	W-doped TiO_2 NWs	4.14	8.71	[81]
5	0.5 mol% Sn-doped TiO_2 NPs	7.45	8.31	[56]
6	$Y_{0.78}Yb_{0.20}Er_{0.02}F_3$-doped TiO_2 NPs	5.84	7.90	[82]
7	0.1% Ce-doped TiO_2	7.2	7.65	[83]
8	10% Cr-doped TiO_2	6.86	8.40	[84]
9	50 ppm Cr-doped TiO_2	5.54	6.35	[85]
10	F-doped TiO_2	7.25	8.07	[86]
11	S-doped TiO_2 NFs	1.54	4.27	[87]
12	$S_{0.75}$-doped TiO_2 particles	5.56	6.91	[88]
13	1 mol% Sb-doped TiO_2	7.36	8.13	[89]
14	3% Ni-doped TiO_2 NPs	2.93	4.00	[90]
15	NaYF4:Yb^{3+}, Er^{3+}-doped TiO_2 NPs	6.71	7.65	[91]
16	4 mol% Sm^{3+}/Y^{3+}-doped TiO_2 NPs	3.48	4.09	[92]

materials, techniques like sol–gel, hydrothermal, electrochemical, electrospinning, atomic layer deposition, anodization, and thermal oxidation have been widely used [65, 66]. Table 3 lists the effect of various materials on DSSC improvement.

3.1.4 Effect of Nanocomposites of TiO_2

The efficiency of pristine TiO_2 is good, but it is still low for high recital applications. The control of synthesis to produce different morphologies is also an issue. For example, nanotubes and nanorods control their dimensions which needs very precise reaction conditions [67]. The nanocomposite is an efficient way to make high-performance photoanode for DSSC applications. The materials used for nanocomposites are carbon allotropes (carbon black, graphene, and carbon nanotubes), noble metals (gold and silver), and transition metal oxides. These materials offer healthier/good electrical conductivity, faster movement of electrons, and higher stability toward electrolytes [68]. In this section, nanocomposites of TiO_2 with carbon nanoparticles, graphene, carbon nanotubes, silver, gold, metal oxides, and their effects on photoanode of the DSSC have been discussed in detail. The PV performance-measuring parameters of the nanocomposite-based DSSCs are listed in Table4.

Photoanode of TiO_2/Carbon Black Nanocomposite

In composite manufacturing, during the composite making, two layers (internal and external) are formed in its structure. When carbon black nanoparticles are used in nanocomposites, they are applied in an external layer of TiO_2/carbon black film. The higher conduction band of carbon black would ultimately enhance the conduction band energy level of TiO_2/carbon black nanocomposite. This composite, when applied as a PE film in the DSSC, gives higher open-circuit voltage. Ting et al. achieved a V_{OC} of 0.6 V, which ultimately improved the PCE of DSSC [69].

Carbon black performance is enhanced after thermal treatment of its surface. Kang et al. employed nanocomposite of TiO_2 and thermally treated carbon black as photoanode of DSSC and achieved 31% higher efficiency [70]. The production method of the carbon black nanoparticles also has a significant role in determining its performance in every application. Hydrothermally produced carbon black particles and TiO_2 nanocomposite, when used as PE of DSSC, achieved the PCE of 3.38% as compared to pristine TiO_2-based DSSC (2.49%) [71]. Three different sizes of carbon black (250 nm, 500 nm, 700 nm) particles are synthesized by Yang et al. through the hydrothermal method. The highest efficiency of 7.2% is achieved by using a nanocomposite of 15wt% (500 nm) carbon black/TiO_2 as a photoanode film in the DSSC, more elevated than that of a standard TiO_2-based cell (5.6%) [72]. When employed in TiO_2/carbon black unified PE film, carbon black extracted from glucose gives 11.6% higher efficiency than the conventional DSSCs [73].

Nanocomposite of Carbon Nanotubes/TiO_2

Interfaces of photoanode with the dye and electrolyte are the most sensitive points in DSSCs where recombination reactions can occur [74]. 1D nanostructure is the most promising solution to avoid the charge recombination reactions, which provides a higher surface area, faster electron flow path, and superior mechanical properties. Carbon nanotubes (SWCNTs and MWCNTs) are extensively used 1D nanostructure which offer higher motion, smooth transportation, and excellent electrical conductivity [75–77]. Figure 3 shows the improved PCE of DSSC using CNTs/TiO_2 composite. Loading, position, and dispersion of CNTs are the main parameters to make composite for high-performance application. Direct contact is developed between the CNTs and electrolytes at higher loading, which enhances the recombination reactions [78]. The potential of TiO_2 becomes more positive at CNTs loading, which boosts the electron injection from an excited state to a conduction band [79].

CNTs have less hydrophilic groups on its surface, which are necessary for proper distribution [94]. Therefore, the surface of CNTs is modified through functionalization to improve the dispersibility. Guai et al. observed that the semiconducting

Table 4 Improvement of DSSC through nanocomposites formation

Sr. no	Improvement method	Core material	Material of improvement	PCE (%) before improvement	PCE (%) after improvement	References
1	TiO_2 nanocomposites with carbon particles	TiO_2 paste	1 wt% carbon black	4.87	5.65	[70]
2		TiO_2 NPs	15wt% spherical carbon particles	5.55	7.20	[72]
3		TiO_2 NRs	Carbon fibers	0.76	1.28	[108]
4		TiO_2 NPs	Nanoporous carbon	2.49	3.38	[71]
5	TiO_2 nanocomposite with carbon nanotubes	TiO_2 NFs	SWCNTs	2.9	4.8	[109]
6		TiO_2 NFs	CNTs and graphene	4.66	6.11	[110]
7		TiO_2 NPs	MWCNTs	6.51	7.27	[111]
8		Titania particles	Acid-treated SWCNTs	3.55	3.85	[112]
9		TiO_2 NRs	0.1% MWCNTs	5.63	10.2	[113]
10		3wt% Cr-doped TiO_2	0.025wt% CNTs	5.16	7.47	[114]
11		Porous titania	0.32wt% MWCNTs	3.55	3.85	[115]
12		TiO_2 NPs	Multilayer MWCNTs	5.32	7.53	[116]
14		TiO_2 paste	Functionalized MWCNTs	6.10	7.95	[78]
15		TiO_2 NPs	0.3% MWCNTs	5.86	9.05	[117]
16	TiO_2 nanocomposite with grapheme	TiO_2 NTs	1% Graphene	6.25	8.67	[118]
17		TiO_2 NPs	Graphene	5.52	6.49	[119]
18		TiO_2 NPs	Functionalized graphene	5.80	8.13	[107]
19		TiO_2 NPs	Graphene nanosheets	6.49	8.07	[120]
20		TiO_2 NPs	0.75wt% Graphene oxide nanosheets	4.01	4.65	[121]

(continued)

Table 4 (continued)

Sr. no	Improvement method	Core material	Material of improvement	PCE (%) before improvement	PCE (%) after improvement	References
21		TiO_2 NPs	Graphene	3.11	4.28	[122]
22		P25 particles	Graphene	2.58	7.25	[123]
23		TiO_2 NPs	Graphene	5.01	6.97	[106]
24		TiO_2 Nanosheets	Graphene	4.61	5.77	[105]
25		TiO_2 NPs	Graphene	3.11	4.28	[122]
26		TiO_2 NPs	Graphene	5.80	8.13	[107]
27		TiO_2 NPs	Graphene	5.3	7.1	[124]
28	Plasmon resonance effect of Au and Ag	TiO_2 NPs	0.7% Au@SiO_2 NPs	9.3	10.2	[125]
29		TiO_2 NPs	Core–shell 0.1% Ag@TiO_2 NPs	7.8	9.00	[126]
30		TiO_2 NPs	Ag NPs	7.1	8.9	[127]

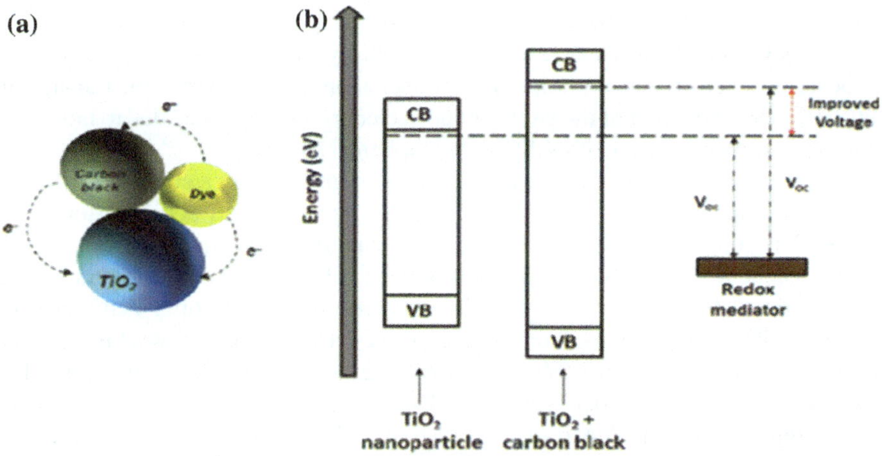

Fig. 3 **a** Triangular schematic of the TiO_2 particles, carbon black, and dye and **b** possible mechanism for the V_{OC} improvement of TiO_2 by adding carbon black (reproduced with permission from Ref. [68])

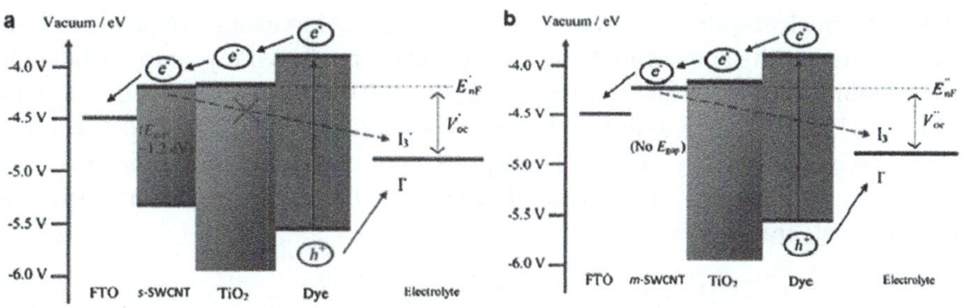

Fig. 4 Energy illustration of **a** semiconducting SWCNTs, **b** metallic SWCNTs composite with TiO_2 nanoparticles (Reproduced with permission from Ref. [93])

SWCNTs achieved higher value of PCE than metallic SWCNTs [93] due to their non-continuous band structure than zero bandgap metallic SWCNTs/and zero bandgap [95]. Charge recombination reactions at the interface can be reduced using a 3D network structure, which provides a faster track for injected electrons to move into PE film. Figure 4 shows that higher energy barriers boost the performance of DSSC by restricting the backward flow of electrons.

Graphene/TiO_2 Nanocomposite-Based Photoanode

Graphene is anticipated as an appropriate additive due to the 2D network structure and tunable bandgap. Moreover, it holds a large surface area (2630 m^2 g^{-1}) [96], higher electrical conductivity, excellent mechanical properties, quicker electron flow

[97–99], and mobility of 10^6 cm^2V^{-1} s^{-1} [100]. Graphene minimizes the recombination reactions and interfaces resistance of electrolyte and photoanode; meanwhile, it boosts the electron transportation from the semiconducting layer to the translucent substrate. The collection of the photo-induced electrons is increased due to improved light-harvesting efficiency of DSSC with graphene [101–104].

Efficiency increased by 25% using 0.75wt% of graphene in a composite of TiO_2 nanosheet/graphene[105]. The 2D structure of graphene gives better performance in DSSC as compared to 1D nanotubes. Graphene links TiO_2 nanoparticles, which is why graphene can easily capture photo-induced electrons and inject them toward the substrate [106]. Compared to CNTs, graphene sheets also offer better contact area with TiO_2 nanoparticles; therefore, graphene-based DSSCs show a higher value of PCE and fill factor [106, 107]. Additionally, graphene has a work function of −4.42 eV [103], which lies between TiO_2 (4.4 eV) and FTO (4.7 eV) work functions [2]. Therefore, graphene imparts a bridging role, and a stepwise injection of electrons occurs from the conduction band of TiO_2 to the transparent conductive substrate.

Except for the materials mentioned above for TiO_2 composite, attention has been diverted toward new emerging materials to apply in the DSSC photoanode. Yu et al. developed a modified composite of TiO_2 with graphite carbon nitride (g-C_3N_4) for the photoanode of DSSC. They achieved 28% more efficiency owing to barrier properties of g-C_3N_4 in electron recombination reactions [128]. Harvesting efficiency of light can be increased by converting harmful UV and near-infrared rays into visible light using luminescent materials coating layer [129, 130]. TiO_2 hybrid composite with SiO_2 [131], SiC [132], Li_2SiO_3 [133], CdS [134], CaF_2 [135], and hydrogels [62] is applied as photoanode of DSSC and gives improved conversion efficiencies. In Table 4, many nanocomposite materials and their efficiencies, before and after the improvement, are mentioned.

3.1.5 Effects of TiO_2 Morphologies

Each morphology has its own aspect ratio and properties, which determine the performance of the photoanode. In this section, the effect of nanosized architectures of TiO_2 on the photovoltaic performance of DSSCs has been discussed in detail.

One-Dimensional (1D) Nanostructures of TiO_2

1D nanostructures of TiO_2, which are extensively applied as photoanode material in DSSCs, are nanofibers, nanowires, nanorods, and nanotubes. These 1D architectures provided ease in electron transportation and boost the charge collection efficiency of the working electrodes.

Nanofibers (NFs)

1D nanofiber morphology of TiO_2 displayed severe light captivation/absorption in the visible region. Nanofibers improve the electron transportation capacity and reduce the recombination of electrons and holes [136, 137]. Electrospinning and sol–gel methods are mostly used to synthesize nanofibers. Cao et al. [138] used electro-spun way to fabricate TiO_2 nanofibers of hierarchical nanorod-branched type. They have used these fibers in the photoanode of DSSC and achieved an efficiency of 6.26%, owing to the enhancement of short-circuit current density (J_{sc}). Prakash Joshi and his co-workers made a composite of NFs/NPs of TiO2 through an electrospinning technique for the photoanode of DSSCs. Composite produced at 15% composition of TiO_2 NFs gave maximum PE efficiency of 8.8% [139]. Electrospinning and sol–gel techniques are compositely used by Surawut et al. [140] to synthesize nanofibers of TiO_2. Highly crystalline, one-dimensional anatase nanofibers with an average 250 nm diameter are produced after calcination. They found a conversion efficiency of 10.3% at a wavelength of 540 nm and quantum efficiency (QE) of 85%. Won Hu Jung and his co-workers use a modified electrospinning process to fabricate the highly porous NFs. The sol–gel process helps to make a spinning solution. A highly crystalline anatase structure NFs is produced, which possessed thickness of 80–150 nm, a surface area of 103.3 m^2/g, 80.5% maximum porosity, and a PCE of 4.6% which is higher than typical TiO_2 NFs [141]. Yang et al. synthesized the nanofibers using electrospinning method. These nanofibers were used as an intermediate layer between NP/NF/ZrO_2 and used as photoanode of DSSC, which gave the power conversion efficiency of 6.72% [51].

TiO_2 Nanowires/nanorods

1D TiO_2 nanostructures act as a light diffuser and help largely in scattering of the light. The output performance of solar cells made up of 1D TiO_2 nanostructure strictly depends on their crystalline phase, dimension, and surface morphologies. Nanowire and nanorod are also 1D nanomaterials, which are extensively used in DSSC applications. They provide outer walls the surface area for loading of the dye [143–147]. 1D nanorods have a large surface area, and improved light diffusion proportion that promotes the dye loading and light absorption. N. Sriharan et al. synthesized TiO_2 nanorods by sol–gel method, which shows the PCE of 6.5% at J_{sc} value of 15.23 $mAcm^{-2}$ and a FF of 0.63 [148]. Rutile TiO_2 nanorod arrays are grown on FTO substrate by Lin et al., and selective etching is applied through the hydrothermal treatment process to make nanorods porous as shown in Fig. 5. When these larger surface area nanorods are used in DSSC, a photoconversion efficiency of 7.91% is accomplished [149]. The energy barrier at the interface of semiconductor and electrolytes can be enhanced by using TiO_2 nanowire, which also reduces the surface traps that act as recombination centers [150, 151]. Surface area is the crucial factor in achieving a higher dye intake, which can be amplified by increasing the length of 1D nanomaterials [152]. Multiple hydrothermal steps are used to grow ultra-long anatase-type TiO_2 nanowires by Kuang et al. [153]. Nanowires of 50 μm

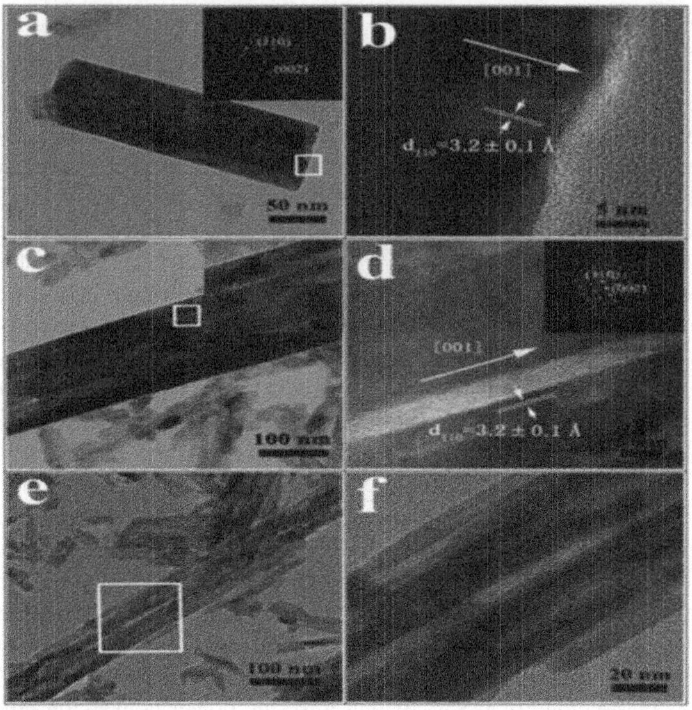

Fig. 5 TEM and HRTEM images of **a**, **b** unetched TiO_2 NRs, **c**–**f** etched TiO_2 NRs (Reproduced with permission from Ref. [149])

size length are attained, when employed as photoanode of DSSC, an efficiency of 9.40% is achieved, as shown in Fig. 6. This efficiency is higher than both short TiO_2 nanowires and P25 TiO_2 nanoparticles (7.23%)-based DSSCs. Surfactant-assisted process is used to make the network structure of TiO_2 nanowires at low temperature. The cell shows an IPCE of 9.3%, which is higher than that of P25 nanoparticle film [151]. To reduce the recombination reactions, porous nanowires are treated with $TiCl_4$ treatment, which helps the morphology boost the dye intake by three times. The effective conversion efficiency of 7.34% is claimed by increasing the length of nanowires [154]. The mutual effect of nanowires and nanoparticles is reported for the photoanode of DSSC. In the first part, nanowires are fabricated on the FTO substrate using the hydrothermal method. Then, in the second part, TiO_2 nanoparticles are introduced into the nanowires using the spin coating technique. PCE of 3.8% is achieved and compared with nanowires and nanoparticles separately. The attained efficiency is 2.2 times higher than the ordinary nanowires and 1.5 times higher than the nanoparticles-based photoanodes [155, 156].

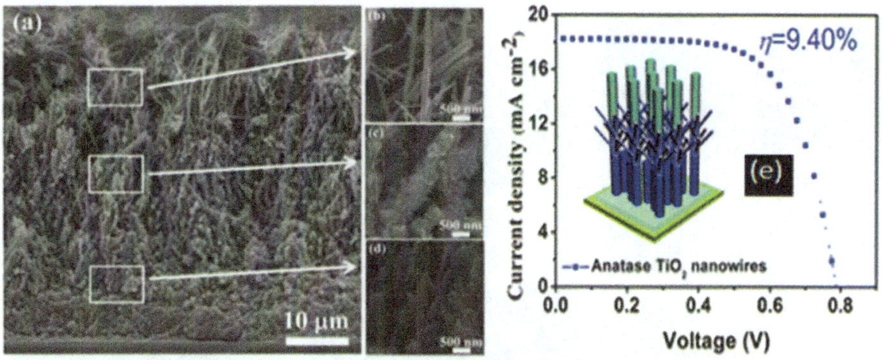

Fig. 6 **a** Cross-sectional SEM image of multilayer TiO_2 NWs, **b**, **c**, **d** SEM images of upper, intermediate, and bottom layer of TiO_2 NWs, respectively, **e** I-V curve of anatase multilayer TiO_2 nanowires (Reproduced with permission from Ref. [153])

TiO_2 Nanotubes (NTs)

The unique morphology of nanotubes offers a hollow cavity, which leads toward a higher active surface area compared to other 1D nanomaterials. NTS provides a smooth path and fewer inter-crystalline connections, contrary to NPS, which helps in more light absorption and faster transportation of electrons [157, 158]. Reduction in recombination reactions occurs owing to the free-flowing movement of electrons, and hence the electrical conductivity is boosted. In an early study, the sputtering technique is used to grow TiO_2 nanotubes on Ti foil, which was already deposited on FTO substrate. The efficiency of DSSC using TiO_2 nanotubes is reported to be 4.7% [159]. H. Park et al. have done a post-treatment of TiC_4 that were already grown as nanotubes on the FTO substrate. They used surface-treated nanotubes as photoanode of the DSSC and attained an efficiency of 5.36% [160].

The porosity of nanotubes should be lower to get the maximum active surface area. To modify the geometry by optimizing the pore size of tubes, we can place more tubes on the FTO substrate, which will ultimately enhance the overall surface area. The role of NTs dimensions, i.e., the thickness of walls, length of NTs, plays a critical role in the performance of DSSCs [161–164]. Charge transportation and recombination reactions are influenced by the thickness of nanotubes. It should be optimum enough to minimize the charge clogging. The higher length of nanotubes would cause more dye intake, and ultimately the current density would be also rising. Therefore, length of the nanotubes must be higher but less than the diffusion length of electrons to get maximum efficiency of charge collection. In a recent study, the effect of nanotube length has been investigated on the photovoltaic performance of DSSCs. It reported an IPCE of 8.1% at 20.8-μm-long nanotube array [165]. Potentiostatic anodization technique is used to fabricate the TiO_2 nanotubes (TNT) array on Ti foil substrate. It was found that the TNT dimensions are dependent on the electrolyte concentration and anodization time, as shown in Fig. 7. When these TNT arrays are used in DSSC, different efficiencies are achieved at every length. Efficiencies of 2.93,

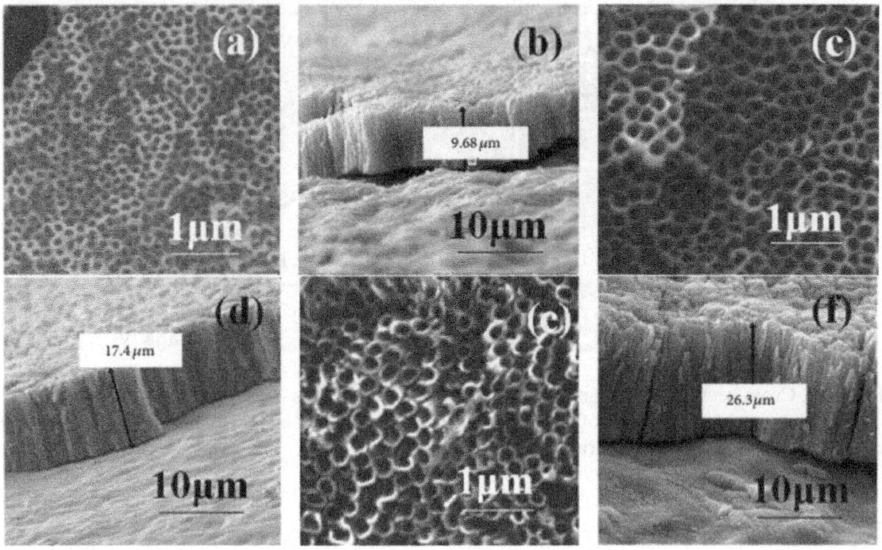

Fig. 7 SEM images of top views for **a** 1 h, **c** 3 **h**, and **e** 5 **h** and side views for **b** 1 h, **d** 3 **h**, and **f** 5 h, anodized time period, respectively, at 0.5wt% of $NH_4F + 2H_2O$ + EG (Reproduced with permission from Ref. [166])

3.46, and 4.09% are reported at 9.68 μm, 17.4 μm, and 26.3 μm lengths of nanotubes, respectively [166]. C. Wei-Chieh et al. used two-step anodization to develop single- and double-walled TiO_2 nanotube arrays. Single-walled TiO_2 nanotubes (SWTNTs) are produced in the first step of anodization, but for double-walled TiO_2 nanotubes (DWTNTs) the second step of anodization is compulsory. It is concluded that the average diameter (around 150 nm) of nanotubes remains the same, but the length of tubes increases (19.23–54 μm) with anodization time (2–6 h). The best value of IPCE is reported to be 6.9% at 4 h anodization time period for DWTNTs [167]. The hydrothermal method was used by Lee et al. to fabricate TiO_2 NTs. The size and structure of NTs is controlled by optimizing the growth temperature. At 120 °C hydrothermal temperature, highly ordered crystalline phases of NTs are produced. Maximum efficiency of 2.41% is attained when DSSC is fabricated using these TiO_2 NTs [168].

Two-Dimensional (2D) TiO_2 Nanostructures

Other than 1D and hollow sphere nanostructure, novel morphologies in 2D shape have also been developed to enhance the performance of DSSCs. TiO_2 2D morphologies include nanosheets [87, 88] [169, 170], nanodisks [171], nanoflakes [172], nanoribbons [173], and nanoleaves [174], which helped to improve the management of light, fast flow of electrons, and higher amount of dye pickup. Biao Chen et al.

synthesized ultra-thin-sized single-crystalline anatase TiO_2 nanosheets via low-cost and two-step graphene-assisted hydrothermal process. To make hybrid photoanode, 2D micro-sized TiO_2 (MS-TiO_2) material and nanosized TiO_2 (NS-TiO_2) nanosheet are incorporated in the nanoparticles of TiO_2 (P25) and the effect of lateral size (micro and nano) is explored. The MS-TiO_2/P25 shows superior efficiency of 4.26% compared to NS-TiO_2/P25 (3.46%), which indicates that the nanosheets of ultrathin sized with higher lateral size are good to use in photovoltaic applications [175]. Chang-Soo Lee and his co-workers developed 2D disk-shaped TiO_2 particles by using a simple one-step process. TiO_2 disks of thickness around 1.5–3.5 μm were synthesized, when these disk-shaped TiO_2 particles were used in the photoanode of DSSC, an efficiency of 6.6% was achieved [171]. To design TiO_2 nanostructure with higher surface area and effective light-scattering properties is a challenge in DSSC research. Deepak et al. developed a leaf-like morphology of TiO_2 which poses dual functionality. When used these leaf-like mesostructures in DSSC, power conversion efficiency of 7.92% is reported as compared to P25 (6.50%) [174].

TiO_2 Hierarchical Structures

In recent years, hierarchical materials have been increasingly applied to create primary nanostructures, which is the base to synthesize secondary structures. These structures can be either spherical, three dimensional, or low dimensional [176]. The large surface area-to-volume ratio results from higher dye loading, more efficiency of light-harvesting action, fewer recombination reactions, and smooth diffusion of the electrolyte within the secondary structure [177, 178] (Table 5).

4 Conclusions

The current book chapter elaborately discussed some important parameters and working mechanism of DSSC. The role of metal-oxide semiconductors has been extensively demonstrated. Majorly, the TiO_2-based DSSC displayed superior performance in terms of PCE percentage. The impact of morphology of the semiconductive metal oxides on their DSSC performance has also been discussed. Overall, the book chapter signifies the role of metal-oxide semiconductors in the field of DSSC.

Table 5 Improvement of DSSC performance using TiO_2 nanoarchitectures

Sr. No	Morphologies		Synthesis method	J_{sc} ($mAcm^{-2}$)	V_{oc} (mV)	FF (%)	PCE (%)	References
1	Nanoparticles	Coral-like TiO_2	Sol – gel method	14.95	834	65.87	8.22	[179]
				16.12	752	75.22	9.4	[180]
		TiO_2nanocorals	Hydrothermal method	8.63	706	58	3.53	[181]
		Rutile rod-like TiO_2		7.55	700	60	3.16	[182]
		Nanoparticles	Arc discharge	–	–	–	–	[183]
		Anatase TiO_2 nanoparticles	Hydrothermal method	13	850	71.53	7.90	[184]
2	Nanofibers	TiO_2 nanofiber	Electrospinning method	14.03	774	61.93	6.72	[137]
		Wormhole-like mesoporous TiO_2 nanofiber		13.7	731	56.3	5.64	[142]
		TiO_2 nanofiber		-	-	-	-	[185]
3	Nanotubes	Chrysanthemum flower-like TiO_2	Hydrothermal method	20.1	710	54	7.77	[186]
		TiO_2 nanotube TNT	Anodization	11.32	712	49	4.09	[166]
		Bamboo-type TiO_2 nanotube		11.60	810	60	5.64	[187]
		Skein-like TiO_2 nanotube	Two-step chemical route	10.59	710	61	4.61	[188]
		Pine tree-like TiO_2 nanotubes	Hydrothermal method	17.7	740	62	8	[189]
		Double-walled TiO_2 nanotubes	Anodization	16.1	670	64	6.90	[167]
4	Nanorods	Single-crystalline Rutile TiO_2 nanorods	Hydrothermal method	15.23	680	63	6.5	[148]

(continued)

Table 5 (continued)

Sr. No	Morphologies		Synthesis method	J_{sc} ($mAcm^{-2}$)	V_{oc} (mV)	FF (%)	PCE (%)	References
		TiO_2 fusiform nanorod		10.13	686	57	4.68	[190]
5	Nanowires	3D anatase hierarchical hyacinth TiO_2	Hydrothermal method	10.47	791	74.05	6.13	[191]
		TiO_2 nanowire	Sol–gel and electropatterning	16	726.53	61	4.75	[192]
		Cactus-like branched TiO_2	Hydrothermal method	11.15	835	69	6.43	[193]
		Rutile TiO_2 nanowire	Solvothermal method	3.16	767	63	1.81	[194]
		Rutile TiO_2 nanowires	Hydrothermal method	17.38	687	74.7	8.9	[195]
		Double-sided brush-like TiO_2		12.18	780	59	5.61	[150]
6	Nanosheet	TiO_2 nanosheet	Hydrothermal method	8.78	680	71.4	4.26	[175]
7	Hollow sphere	Urchin-like TiO_2 hierarchical microsphere	Hydrothermal method	15.07	716	66.56	7.21	[196]
		Hyperbranched TiO_2 hollow sphere		19.65	771	65	9.84	[197]
		Lotus root-like mesoporous TiO_2	Template-free method	–	–	–	–	[198]
		Porous TiO_2 sphere	Electrospray and hydrothermal	15.2	700	66	7.2	[199]
		Hollow TiO_2 porous microsphere	Anisotropic etching method	13.03	800	68	7.05	[200]
		Popcorn-like TiO_2	Facile solution method	13.95	770	70	7.56	[201]

(continued)

Table 5 (continued)

Sr. No	Morphologies		Synthesis method	J_{sc} ($mAcm^{-2}$)	V_{oc} (mV)	FF (%)	PCE (%)	References
		Hierarchical rutile TiO_2 microsphere with branched nanorods	Solvothermal method	11.9	767	73	6.66	[202]
		Hierarchical TiO_2 sub-micro-flowers		15.7	770	53.3	6.4	[203]
		Monodisperse TiO_2 microsphere	Hydrothermal	16.24	762	68.1	8.43	[204]
		Mesoporous TiO_2 yolk–shell microsphere	Solvothermal method	18.84	769	76.10	11.03	[205]
		Jujube-like hierarchical TiO_2 microsphere	Hydrothermal	19.03	800	59	8.93	[206]
		Mesoporous TiO_2 microspheres	Facile and controllable interface-directed co-assembly approach	16	720	74	8.5	[207]
		TiO_2 submicrospheres	Solvothermal method	21.53	691	74.7	11.11	[208]
		Multi-shell porous TiO_2 hollow nanoparticles	Sol–gel method	16.52	770	73	9.4	[209]
		Multi-hollow TiO_2 nanosphere	Microemulsion-based approach	17.85	761	60.7	8.25	[210]
8	Nanoforest	3D TiO_2 nanoforest	Hydrothermal	7.30	787	68	3.93	[211]
9	Nanoflowers	TiO_2 nanoflower	Hydrothermal	5.77	650	48.66	1.83	[212]
10	Nanohexagon	TiO_2 nanohexagon	Electrochemical anodization	12.85	710	70.50	6.43	[213]
11	Nanocubes	TiO_2 cubic	Combustion method	12.3	710	70	6.11	[214]

(continued)

Table 5 (continued)

Sr. No	Morphologies		Synthesis method	J_{sc} ($mAcm^{-2}$)	V_{oc} (mV)	FF (%)	PCE (%)	References
		TiO_2 nanocubes	Hydrothermal	19.07	760	66.21	9.60	[215]
		Cubic TiO_2		20.83	670	70	9.77	[216]
		Cubic TiO_2	Solvothermal process	15.90	740	60	7.06	[217]
12	Disk-shaped TiO_2	Disk-shaped TiO_2	Sol–gel method	13	790	65	6.6	[171]
13	Nanoparallelepipeds	Nanoparallelepipeds	Hydrothermal	19.07	760	66.21	9.60	[215]
14	TiO_2 beads	Mesoporous TiO_2 beads	Sol–gel and solvothermal	16.6	746	72	9	[38]
15	Nanospindles	Hierarchical TiO_2 needle-like microstructures	Solvothermal	16.22	808	74.7	9.79	[218]
		TiO_2 nanospindles	Hydrothermal	16.4	760	66.8	8.32	[219]
16	Square	Hollow box TiO_2	Solid-state precursor method	13.4	758	60.1	6.1	[220]
17	Composites	Titania nanoparticle (TNPs) and Titania nanotube (TNTs)	Facile sol–gel and hydrothermal methods	12.20	729	61.05	5.43	[221]
		Composite TiO_2 hollow sphere and nanorods	Hydrothermal methods	17.56	748	69	9.06	[222]
		TiO_2 nanobranches and nanoparticles hybrids		13.53	690	58	5.3	[223]
		TiO_2 nanosheet-based microsphere	Sol–gel method	15.2	650		6.64	[224]

Acknowledgements The authors acknowledge the support provided by PPE Department, UET Lahore.

References

1. Etxebarria, I., Ajuria, J. & Pacios, R. Polymer:fullerene solar cells: materials, processing issues, and cell layouts to reach power conversion efficiency over 10%, a review. J. Photonics Energy**5**, 057214 (2015).
2. Hagfeldt, A., Boschloo, G., Sun, L., Kloo, L. & Pettersson, H. Dye-Sensitized Solar Cells. Chem. Rev.**110**, 6595–6663 (2010).
3. O'Regan, B. & Grätzel, M. A low-cost, high-efficiency solar cell based on dye-sensitized colloidal TiO_2 films. Nature**353**, 737–740 (1991).
4. Iwata, S., Shibakawa, S. ichiro, Imawaka, N. & Yoshino, K. Stability of the current characteristics of dye-sensitized solar cells in the second quadrant of the current–voltage characteristics. Energy Reports**4**, 8–12 (2018).
5. Akman, E. Enhanced photovoltaic performance and stability of dye-sensitized solar cells by utilizing manganese-doped ZnO photoanode with europium compact layer. J. Mol. Liq.**317**, 114223 (2020).
6. Ou, J. Z. et al. Erratum: Elevated temperature anodized Nb 2O 5-A photoanode material with exceptionally large photoconversion efficiencies (ACS Nano (2012) 6 (4045–4053) DOI: https://doi.org/10.1021/nn300408p). ACS Nano**6**, 5737 (2012).
7. Nazeeruddin, M. K., Liska, P., Moser, J., Vlachopoulos, N. & Grätzel, M. Conversion of Light into Electricity with Trinuclear Ruthenium Complexes Adsorbed on Textured TiO2 Films. Helv. Chim. Acta**73**, 1788–1803 (1990).
8. Shin, Y.-J., Lee, J.-H., Park, J.-H. & Park, N.-G. Enhanced Photovoltaic Properties of SiO_2 -treated ZnO Nanocrystalline Electrode for Dye-sensitized Solar Cell. Chem. Lett.**36**, 1506–1507 (2007).
9. Horiuchi, H. et al. Electron injection efficiency from excited N3 into nanocrystalline ZnO films: Effect of (N3-Zn2+) aggregate formation. J. Phys. Chem. B**107**, 2570–2574 (2003).
10. Keis, K., Lindgren, J., Lindquist, S. E. & Hagfeldt, A. Studies of the adsorption process of Ru complexes in nanoporous ZnO electrodes. Langmuir**16**, 4688–4694 (2000).
11. Parks, G. A. The Isoelectric Points of Solid Oxides, Solid Hydroxides, and Aqueous Hydroxo Complex Systems. Chem. Rev.**65**, 177–198 (1965).
12. Shin, Y. J., Lee, J. H., Park, J. H. & Park, N. G. Enhanced photovoltaic properties of SiO2-treated ZnO nanocrystalline electrode for dye-sensitized solar cell. Chem. Lett.**36**, 1506–1507 (2007).
13. Law, M. et al. ZnO-Al2O3 and ZnO-TiO2 core-shell nanowire dye-sensitized solar cells. J. Phys. Chem. B**110**, 22652–22663 (2006).
14. Greene, L. E., Law, M., Yuhas, B. D. & Yang, P. ZnO - TiO2 Core - Shell nanorod/P3HT solar cells. J. Phys. Chem. C**111**, 18451–18456 (2007).
15. Memarian, N. et al. Hierarchically assembled ZnO nanocrystallites for high-efficiency dye-sensitized solar cells. Angew. Chemie - Int. Ed.**50**, 12321–12325 (2011).
16. Akhtar, M. S. A Facile Synthesis of ZnO Nanoparticles and Its Application as Photoanode for Dye Sensitized Solar Cells. Sci. Adv. Mater.**7**, 1137–1142 (2015).
17. Akhtar, M. S., Khan, M. A., Jeon, M. S. & Yang, O. B. Controlled synthesis of various ZnO nanostructured materials by capping agents-assisted hydrothermal method for dye-sensitized solar cells. Electrochim. Acta**53**, 7869–7874 (2008).
18. Ameen, S., Akhtar, M. S., Seo, H. K., Kim, Y. S. & Shin, H. S. Influence of Sn doping on ZnO nanostructures from nanoparticles to spindle shape and their photoelectrochemical properties for dye sensitized solar cells. Chem. Eng. J.**187**, 351–356 (2012).

19. Jiang, C. Y., Sun, X. W., Lo, G. Q., Kwong, D. L. & Wang, J. X. Improved dye-sensitized solar cells with a ZnO-nanoflower photoanode. Appl. Phys. Lett.**90**, 3–6 (2007).
20. Mou, J., Zhang, W., Fan, J., Deng, H. & Chen, W. Facile synthesis of ZnO nanobullets/nanoflakes and their applications to dye-sensitized solar cells. J. Alloys Compd.**509**, 961–965 (2011).
21. Lee, C. P. et al. Synthesis of hexagonal ZnO clubs with opposite faces of unequal dimensions for the photoanode of dye-sensitized solar cells. Phys. Chem. Chem. Phys.**13**, 20999–21008 (2011).
22. Wang, Y., Cui, X., Zhang, Y., Gao, X. & Sun, Y. Preparation of Cauliflower-like ZnO Films by Chemical Bath Deposition: Photovoltaic Performance and Equivalent Circuit of Dye-sensitized Solar Cells. J. Mater. Sci. Technol.**29**, 123–127 (2013).
23. Li, Z. et al. Fabrication of hierarchically assembled microspheres consisting of nanoporous ZnO nanosheets for high-efficiency dye-sensitized solar cells. J. Mater. Chem.**22**, 14341–14345 (2012).
24. McCune, M., Zhang, W. & Deng, Y. High efficiency dye-sensitized solar cells based on three-dimensional multilayered ZnO nanowire arrays with 'caterpillar-like' structure. Nano Lett.**12**, 3656–3662 (2012).
25. Ameen, S., Shaheer Akhtar, M. & Shin, H. S. Growth and characterization of nanospikes decorated ZnO sheets and their solar cell application. Chem. Eng. J.**195–196**, 307–313 (2012).
26. Wang, J. X. et al. Synthesis of hierarchical porous ZNO disklike nanostructures for improved photovoltaic properties of Dye-Sensitized solar cells. J. Phys. Chem. C**114**, 13157–13161 (2010).
27. Kalyanasundaram, K. & Grätzel, M. Applications of functionalized transition metal complexes in photonic and optoelectronic devices. Coord. Chem. Rev.**177**, 347–414 (1998).
28. O'Regan, B., Grätzel, M. A. Low-cost, high-efficiency solar cell based on dye-sensitized colloidal TiO2 films. Nature**354**, 56–58 (1991).
29. Barbé, C. J. et al. Nanocrystalline Titanium Oxide Electrodes for Photovoltaic Applications. J. Am. Ceram. Soc.**80**, 3157–3171 (2005).
30. Fillinger, A. The Adsorption Behavior of a Ruthenium-Based Sensitizing Dye to Nanocrystalline TiO[sub 2] Coverage Effects on the External and Internal Sensitization Quantum Yields. J. Electrochem. Soc.**146**, 4559 (1999).
31. Chai, S., Lau, T., Dayou, J., Sipaut, S. & Mansa, R. F. Development in Photoanode Materials for High Efficiency Dye Sensitized Solar Cells. vol. 4 (2014).
32. Valencia, S., Marín, J. M. & Restrepo, G. Study of the Bandgap of Synthesized Titanium Dioxide Nanoparticules Using the Sol-Gel Method and a Hydrothermal Treatment. Open Mater. Sci. J.**4**, 9–14 (2010).
33. Chou, T. P., Zhang, Q., Russo, B., Fryxell, G. E. & Cao, G. Titania particle size effect on the overall performance of dye-sensitized solar cells. J. Phys. Chem. C**111**, 6296–6302 (2007).
34. Bagheri, S., Mohd Hir, Z. A., Yousefi, A. T. & Abdul Hamid, S. B. Progress on mesoporous titanium dioxide: Synthesis, modification and applications. Microporous Mesoporous Mater.**218**, 206–222 (2015).
35. Muniz, E. C. et al. Synthesis and characterization of mesoporous TiO2 nanostructured films prepared by a modified sol-gel method for application in dye solar cells. Ceram. Int.**37**, 1017–1024 (2011).
36. Alwin, S., Shajan, X. S., Karuppasamy, K. & Warrier, K. G. K. Microwave assisted synthesis of high surface area TiO2 aerogels: A competent photoanode material for quasi-solid dye-sensitized solar cells. Mater. Chem. Phys.**196**, 37–44 (2017).
37. Hochbaum, A. I. & Yang, P. Semiconductor nanowires for energy conversion. Chem. Rev.**110**, 527–546 (2010).
38. Chen, Y. et al. Effect of mesoporous TiO2 bead diameter in working electrodes on the efficiency of dye-sensitized solar cells. ChemSusChem**4**, 1498–1503 (2011).
39. Kao, M. C., Chen, H. Z., Young, S. L., Kung, C. Y. & Lin, C. C. The effects of the thickness of TiO2 films on the performance of dye-sensitized solar cells. Thin Solid Films**517**, 5096–5099 (2009).

40. Dissanayake, L. & Thotawatthage, C. A. The effect of TiO2 photo anode film thickness on photovoltaic properties of dye-sensitized solar cells The effect of TiO 2 photoanode film thickness on photovoltaic properties of dye-sensitized solar cells. (2016) doi:https://doi.org/10.4038/cjs.v45i1.7362.
41. Park, Y. C. et al. Size-tunable mesoporous spherical TiO2 as a scattering overlayer in high-performance dye-sensitized solar cells. J. Mater. Chem.**21**, 9582–9586 (2011).
42. Sun, X. et al. Mixed P25 nanoparticles and large rutile particles as a top scattering layer to enhance performance of nanocrystalline TiO 2 based dye-sensitized solar cells. Appl. Surf. Sci.**337**, 188–194 (2015).
43. Miao, Q., Wu, L., Cui, J., Huang, M. & Ma, T. A New Type of Dye-Sensitized Solar Cell with a Multilayered Photoanode Prepared by a Film-Transfer Technique. Adv. Mater.**23**, 2764–2768 (2011).
44. Koo, H. J. et al. Nano-embossed hollow spherical TiO2 as bifunctional material for high-efficiency dye-sensitized solar cells. Adv. Mater.**20**, 195–199 (2008).
45. Huang, F., Chen, D., Zhang, X. L., Caruso, R. A. & Cheng, Y. B. Dual-function scattering layer of submicrometer-sized mesoporous TiO 2 beads for high-efficiency dyesensitized solar cells. Adv. Funct. Mater.**20**, 1301–1305 (2010).
46. Wu, W. Q., Xu, Y. F., Rao, H. S., Su, C. Y. & Kuang, D. Bin. A double layered TiO2 photoanode consisting of hierarchical flowers and nanoparticles for high-efficiency dye-sensitized solar cells. Nanoscale**5**, 4362–4369 (2013).
47. Dong, Z. et al. Quintuple-shelled Sno2 hollow microspheres with superior light scattering for high-performance dye-sensitized solar cells. Adv. Mater.**26**, 905–909 (2014).
48. Yan, K., Qiu, Y., Chen, W., Zhang, M. & Yang, S. A double layered photoanode made of highly crystalline TiO2 nanooctahedra and agglutinated mesoporous TiO2 microspheres for high efficiency dye sensitized solar cells. Energy Environ. Sci.**4**, 2168–2176 (2011).
49. Gao, Z. et al. Application of hierarchical TiO2 spheres as scattering layer for enhanced photovoltaic performance in dye sensitized solar cell. CrystEngComm**15**, 3351–3358 (2013).
50. Wu, W. Q., Xu, Y. F., Rao, H. S., Su, C. Y. & Kuang, D. Bin. Multistack integration of three-dimensional hyperbranched anatase titania architectures for high-efficiency dye-sensitized solar cells. J. Am. Chem. Soc.**136**, 6437–6445 (2014).
51. Yang, L. & Leung, W. W. F. Application of a bilayer TiO2 nanofiber photoanode for optimization of dye-sensitized solar cells. Adv. Mater.**23**, 4559–4562 (2011).
52. Wang, G., Zhu, X. & Yu, J. Bilayer hollow/spindle-like anatase TiO2 photoanode for high efficiency dye-sensitized solar cells. J. Power Sources**278**, 344–351 (2015).
53. Asahi, R., Morikawa, T., Ohwaki, T., Aoki, K. & Taga, Y. Visible-light photocatalysis in nitrogen-doped titanium oxides. Science (80-.).**293**, 269–271 (2001).
54. Mrowetz, M., Balcerski, W., Colussi, A. J. & Hoffmann, M. R. Oxidative power of nitrogen-doped TiO 2 photocatalysts under visible illumination. J. Phys. Chem. B**108**, 17269–17273 (2004).
55. Hoye, R. L. Z., Musselman, K. P. & Macmanus-Driscoll, J. L. Research update: Doping ZnO and TiO2 for solar cells. APL Mater.**1**, (2013).
56. Duan, Y. et al. Sn-doped TiO 2 photoanode for dye-sensitized solar cells. J. Phys. Chem. C**116**, 8888–8893 (2012).
57. Ko, K. H., Lee, Y. C. & Jung, Y. J. Enhanced efficiency of dye-sensitized TiO2 solar cells (DSSC) by doping of metal ions. J. Colloid Interface Sci.**283**, 482–487 (2005).
58. Zhang, J. C. et al. N, S-doped TiO_2 anode effect on performance of dye-sensitized solar cells. J. Phys. Chem. Solids**72**, 1239–1244 (2011).
59. Berglund, S. P., Hoang, S., Minter, R. L., Fullon, R. R. & Mullins, C. B. Investigation of 35 elements as single metal oxides, mixed metal oxides, or dopants for titanium dioxide for dye-sensitized solar cells. J. Phys. Chem. C**117**, 25248–25258 (2013).
60. Ma, T., Akiyama, M., Abe, E. & Imai, I. High-efficiency dye-sensitized solar cell based on a nitrogen-doped nanostructured titania electrode. Nano Lett.**5**, 2543–2547 (2005).
61. Tian, H. et al. Retarded charge recombination in dye-sensitized nitrogen-doped tio 2 solar cells. J. Phys. Chem. C**114**, 1627–1632 (2010).

62. Fu, H., Zhang, L., Zhang, S., Zhu, Y. & Zhao, J. Electron spin resonance spin-trapping detection of radical intermediates in N-doped TiO2-assisted photodegradation of 4-chlorophenol. J. Phys. Chem. B**110**, 3061–3065 (2006).
63. Etacheri, V., Seery, M. K., Hinder, S. J. & Pillai, S. C. Highly visible light active TiO2-xNx heterojunction photocatalysts. Chem. Mater.**22**, 3843–3853 (2010).
64. Guo, W., Shen, Y., Boschloo, G., Hagfeldt, A. & Ma, T. Influence of nitrogen dopants on N-doped TiO2 electrodes and their applications in dye-sensitized solar cells. Electrochim. Acta**56**, 4611–4617 (2011).
65. Nah, Y.-C., Paramasivam, I. & Schmuki, P. Doped TiO2 and TiO2 Nanotubes: Synthesis and Applications. ChemPhysChem**11**, 2698–2713 (2010).
66. Yang, M. et al. Nb doping of TiO2 nanotubes for an enhanced efficiency of dye-sensitized solar cells. Chem. Commun.**47**, 2032–2034 (2011).
67. Low, F. W. & Lai, C. W. Recent developments of graphene-TiO2 composite nanomaterials as efficient photoelectrodes in dye-sensitized solar cells: A review. Renew. Sustain. Energy Rev.**82**, 103–125 (2018).
68. Batmunkh, M., Biggs, M. J. & Shapter, J. G. Carbonaceous Dye-Sensitized Solar Cell Photoelectrodes. Adv. Sci.**2**, 1400025 (2015).
69. Ting, C. C. & Chao, W. S. Efficiency improvement of the DSSCs by building the carbon black as bridge in photoelectrode. Appl. Energy**87**, 2500–2505 (2010).
70. Kang, S. H., Kim, J. Y., Kim, Y. K. & Sung, Y. E. Effects of the incorporation of carbon powder into nanostructured TiO2 film for dye-sensitized solar cell. J. Photochem. Photobiol. A Chem.**186**, 234–241 (2007).
71. Kim, D. Y. et al.The photovoltaic efficiencies on dye sensitized solar cells assembled with nanoporous carbon/tio 2 composites. J. Ind. Eng. Chem.**18**, 1–5 (2012).
72. Yang, G. et al. Light scattering enhanced photoanodes for dye-sensitized solar cells prepared by carbon spheres/TiO2 nanoparticle composites. Curr. Appl. Phys.**11**, 376–381 (2011).
73. Jang, Y. J., Jang, Y. H. & Kim, D. H. Carbohydrate-Derived Carbon Sheaths on TiO_2 Nanoparticle Photoanodes for Efficiency Enhancement in Dye-Sensitized Solar Cells. Part. Part. Syst. Charact.**30**, 1030–1033 (2013).
74. Nissfolk, J., Fredin, K., Hagfeldt, A. & Boschloo, G. Recombination and transport processes in dye-sensitized solar cells investigated under working conditions. J. Phys. Chem. B**110**, 17715–17718 (2006).
75. Cai, J. et al. Enhanced conversion efficiency of dye-sensitized solar cells using a CNT-incorporated TiO2 slurry-based photoanode. AIP Adv.**5**, 027118 (2015).
76. Liu, Y., Zhang, J., Cheng, Y. & Jiang, S. P. Effect of Carbon Nanotubes on Direct Electron Transfer and Electrocatalytic Activity of Immobilized Glucose Oxidase. ACS Omega**3**, 667–676 (2018).
77. Peng, H., Sun, X., Weng, W. & Fang, X. Electronic Polymer Composite. in Polymer Materials for Energy and Electronic Applications 107–149 (Elsevier, 2017). doi:https://doi.org/10.1016/b978-0-12-811091-1.00004-5.
78. Benetti, D. et al. Functionalized multi-wall carbon nanotubes/TiO2 composites as efficient photoanodes for dye sensitized solar cells. J. Mater. Chem. C**4**, 3555–3562 (2016).
79. Anusorn Kongkanand, Rebeca Martínez Domínguez, and & Kamat*, P. V. Single Wall Carbon Nanotube Scaffolds for Photoelectrochemical Solar Cells. Capture and Transport of Photogenerated Electrons. (2007) doi:https://doi.org/10.1021/NL0627238.
80. Kim, S. G. et al. Nb-doped TiO2 nanoparticles for organic dye-sensitized solar cells. RSC Adv.**3**, 16380 (2013).
81. Archana, P. S., Gupta, A., Yusoff, M. M. & Jose, R. Tungsten doped titanium dioxide nanowires for high efficiency dye-sensitized solar cells. Phys. Chem. Chem. Phys.**16**, 7448–7454 (2014).
82. Wu, J. et al. Enhancement of the Photovoltaic Performance of Dye-Sensitized Solar Cells by Doping Y0.78Yb0.20Er0.02F3 in the Photoanode. Adv. Energy Mater.**2**, 78–81 (2012).
83. Zhang, J., Peng, W., Chen, Z., Chen, H. & Han, L. Effect of cerium doping in the TiO 2 photoanode on the electron transport of dye-sensitized solar cells. J. Phys. Chem. C**116**, 19182–19190 (2012).

84. Kim, C., Kim, K. S., Kim, H. Y. & Han, Y. S. Modification of a TiO2 photoanode by using Cr-doped TiO 2 with an influence on the photovoltaic efficiency of a dye-sensitized solar cell. J. Mater. Chem.**18**, 5809–5814 (2008).
85. Xie, Y. et al. Improved performance of dye-sensitized solar cells by trace amount Cr-doped TiO2 photoelectrodes. J. Power Sources**224**, 168–173 (2013).
86. Yang, S. et al. Improved efficiency of dye-sensitized solar cells applied with F-doped TiO2 electrodes. J. Fluor. Chem.**150**, 78–84 (2013).
87. Mahmoud, M. S. et al. Demonstrated photons to electron activity of S-doped TiO2 nanofibers as photoanode in the DSSC. Mater. Lett.**225**, 77–81 (2018).
88. Sun, Q. et al. Sulfur-doped TiO 2 nanocrystalline photoanodes for dye-sensitized solar cells. in Journal of Renewable and Sustainable Energy vol. 4 023104 (American Institute of PhysicsAIP, 2012).
89. Wang, M. et al. Improved photovoltaic performance of dye-sensitized solar cells by Sb-doped TiO 2 photoanode. Electrochim. Acta**77**, 54–59 (2012).
90. Sakthivel, T., Kumar, K. A., Senthilselvan, J. & Jagannathan, K. Effect of Ni dopant in TiO2 matrix on its interfacial charge transportation and efficiency of DSSCs. J. Mater. Sci. Mater. Electron.**29**, 2228–2235 (2018).
91. Zhu, G., Wang, H., Zhang, Q. & Zhang, L. Enhanced photovoltaic performance of dye-sensitized solar cells based on NaYF4:Yb3+, Er3+-incorporated nanocrystalline TiO2 electrodes. J. Colloid Interface Sci.**451**, 15–20 (2015).
92. Qin, Y. et al. Performance improvement of dye-sensitized solar cell by introducing Sm3 +/Y3 + co-doped TiO2 film as an efficient blocking layer. Thin Solid Films**631**, 141–146 (2017).
93. Guai, G. H., Li, Y., Ng, C. M., Li, C. M. & Chan-Park, M. B. TiO2 Composing with Pristine, Metallic or Semiconducting Single-Walled Carbon Nanotubes: Which Gives the Best Performance for a Dye-Sensitized Solar Cell. ChemPhysChem**13**, 2566–2572 (2012).
94. Munkhbayar, B. et al. Influence of dry and wet ball milling on dispersion characteristics of the multi-walled carbon nanotubes in aqueous solution with and without surfactant. Powder Technol.**234**, 132–140 (2013).
95. Dang, X. et al. Virus-templated self-assembled single-walled carbon nanotubes for highly efficient electron collection in photovoltaic devices. Nat. Nanotechnol.**6**, 377–384 (2011).
96. Balandin, A. A. et al. Superior Thermal Conductivity of Single-Layer Graphene. Nano Lett.**8**, 902–907 (2008).
97. Geim, A. K. & Novoselov, K. S. The rise of graphene. Nat. Mater.**6**, 183–191 (2007).
98. Sun, Y., Wu, Q. & Shi, G. Graphene based new energy materials. Energy Environ. Sci.**4**, 1113 (2011).
99. Geim, A. K. Graphene: status and prospects. Science**324**, 1530–4 (2009).
100. Bonaccorso, F., Sun, Z., Hasan, T. & Ferrari, A. C. Graphene photonics and optoelectronics. Nat. Photonics**4**, 611–622 (2010).
101. Sacco, A. et al. Investigation of transport and recombination properties in graphene/titanium dioxide nanocomposite for dye-sensitized solar cell photoanodes. Electrochim. Acta**131**, 154–159 (2014).
102. Song, J. et al. Enhancement of Photogenerated Electron Transport in Dye-Sensitized Solar Cells with Introduction of a Reduced Graphene Oxide-TiO2 Junction. Chem. - A Eur. J.**17**, 10832–10837 (2011).
103. Tang, Y. B. et al. Incorporation of graphenes in nanostructured TiO2 films via molecular grafting for dye-sensitized solar cell application. ACS Nano**4**, 3482–3488 (2010).
104. Batmunkh, M., Dadkhah, M., Shearer, C. J., Biggs, M. J. & Shapter, J. G. Incorporation of graphene into SnO 2 photoanodes for dye-sensitized solar cells. Appl. Surf. Sci.**387**, 690–697 (2016).
105. Fan, J., Liu, S. & Yu, J. Enhanced photovoltaic performance of dye-sensitized solar cells based on TiO 2 nanosheets/graphene composite films. J. Mater. Chem.**22**, 17027–17036 (2012).
106. Yang, N., Zhai, J., Wang, D., Chen, Y. & Jiang, L. Two-Dimensional Graphene Bridges Enhanced Photoinduced Charge Transport in Dye-Sensitized Solar Cells. ACS Nano**4**, 887–894 (2010).

107. Chen, T., Hu, W., Song, J., Guai, G. H. & Li, C. M. Interface Functionalization of Photoelectrodes with Graphene for High Performance Dye-Sensitized Solar Cells. Adv. Funct. Mater.**22**, 5245–5250 (2012).
108. Guo, W. et al. Rectangular bunched rutile TiO 2 nanorod arrays grown on carbon fiber for dye-sensitized solar cells. J. Am. Chem. Soc.**134**, 4437–4441 (2012).
109. Macdonald, T. J. et al. A TiO_2 Nanofiber-Carbon Nanotube-Composite Photoanode for Improved Efficiency in Dye-Sensitized Solar Cells. ChemSusChem**8**, 3396–3400 (2015).
110. Yen, M. Y. et al. Preparation of graphene/multi-walled carbon nanotube hybrid and its use as photoanodes of dye-sensitized solar cells. Carbon N. Y.**49**, 3597–3606 (2011).
111. Kilic, B. et al. Preparation of Carbon Nanotube/TiO2 Mesoporous Hybrid Photoanode with Iron Pyrite (FeS2) Thin Films Counter Electrodes for Dye-Sensitized Solar Cell. Sci. Rep.**6**, 27052 (2016).
112. Golobostanfard, M. R. & Abdizadeh, H. Influence of carbon nanotube wall thickness on performance of dye sensitized solar cell with hierarchical porous photoanode. Microporous Mesoporous Mater.**191**, 74–81 (2014).
113. Yang, L. & Leung, W. W. F. Electrospun TiO2 nanorods with carbon nanotubes for efficient electron collection in dye-sensitized solar cells. Adv. Mater.**25**, 1792–1795 (2013).
114. Massihi, N., Mohammadi, M. R., Bakhshayesh, A. M. & Abdi-Jalebi, M. Controlling electron injection and electron transport of dye-sensitized solar cells aided by incorporating CNTs into a Cr-doped TiO2 photoanode. Electrochim. Acta**111**, 921–929 (2013).
115. Golobostanfard, M. R. & Abdizadeh, H. Hierarchical porous titania/carbon nanotube nanocomposite photoanode synthesized by controlled phase separation for dye sensitized solar cell. Sol. Energy Mater. Sol. Cells**120**, 295–302 (2014).
116. Anjidani, M., Milani Moghaddam, H. & Ojani, R. Binder-free MWCNT/TiO2 multilayer nanocomposite as an efficient thin interfacial layer for photoanode of dye sensitized solar cell. Mater. Sci. Semicond. Process.**71**, 20–28 (2017).
117. Chan, Y. F., Wang, C. C., Chen, B. H. & Chen, C. Y. Dye-sensitized TiO2 solar cells based on nanocomposite photoanode containing plasma-modified multi-walled carbon nanotubes. Prog. Photovoltaics Res. Appl.**21**, 47–57 (2013).
118. Tang, B. & Hu, G. Two kinds of graphene-based composites for photoanode applying in dye-sensitized solar cell. J. Power Sources**220**, 95–102 (2012).
119. Zhang, H. et al. Effects of TiO2 film thickness on photovoltaic properties of dye-sensitized solar cell and its enhanced performance by graphene combination. Mater. Res. Bull.**49**, 126–131 (2014).
120. Tsai, C. H., Fei, P. H. & Wu, W. C. Enhancing the efficiency and charge transport characteristics of dye-sensitized solar cells by adding graphene nanosheets to TiO2 working electrodes. Electrochim. Acta**165**, 356–364 (2015).
121. Wang, P., He, F., Wang, J., Yu, H. & Zhao, L. Graphene oxide nanosheets as an effective template for the synthesis of porous TiO 2 film in dye-sensitized solar cells. in Applied Surface Science vol. 358 175–180 (Elsevier B.V., 2015).
122. Zhu, M., Li, X., Liu, W. & Cui, Y. An investigation on the photoelectrochemical properties of dye-sensitized solar cells based on graphene-TiO2 composite photoanodes. J. Power Sources**262**, 349–355 (2014).
123. He, Z. et al. Nanostructure control of graphene-composited TiO 2 by a one-step solvothermal approach for high performance dye-sensitized solar cells. Nanoscale**3**, 4613–4616 (2011).
124. Chen, L. et al. Enhanced photovoltaic performance of a dye-sensitized solar cell using graphene-TiO2 photoanode prepared by a novel in situ simultaneous reduction-hydrolysis technique. Nanoscale**5**, 3481–3485 (2013).
125. Choi, H., Chen, W. T. & Kamat, P. V. Know thy nano neighbor. Plasmonic versus electron charging effects of metal nanoparticles in dye-sensitized solar cells. ACS Nano**6**, 4418–4427 (2012).
126. Qi, J., Dang, X., Hammond, P. T. & Belcher, A. M. Highly efficient plasmon-enhanced dye-sensitized solar cells through metal@oxide core-shell nanostructure. ACS Nano**5**, 7108–7116 (2011).

127. Jeong, N. C., Prasittichai, C. & Hupp, J. T. Photocurrent enhancement by surface plasmon resonance of silver nanoparticles in highly porous dye-sensitized solar cells. Langmuir**27**, 14609–14614 (2011).
128. Xu, J. et al. G-C3N4 modified TiO2 nanosheets with enhanced photoelectric conversion efficiency in dye-sensitized solar cells. J. Power Sources**274**, 77–84 (2015).
129. Bella, F., Griffini, G., Gerosa, M., Turri, S. & Bongiovanni, R. Performance and stability improvements for dye-sensitized solar cells in the presence of luminescent coatings. J. Power Sources**283**, 195–203 (2015).
130. Griffini, G. et al. Multifunctional Luminescent Down-Shifting Fluoropolymer Coatings: A Straightforward Strategy to Improve the UV-Light Harvesting Ability and Long-Term Outdoor Stability of Organic Dye-Sensitized Solar Cells. Adv. Energy Mater.**5**, 1401312 (2015).
131. Yuan, S., Tang, Q., He, B., Men, L. & Chen, H. Transmission enhanced photoanodes for efficient dye-sensitized solar cells. Electrochim. Acta**125**, 646–651 (2014).
132. Gondal, M. A., Ilyas, A. M. & Baig, U. Facile synthesis of silicon carbide-titanium dioxide semiconducting nanocomposite using pulsed laser ablation technique and its performance in photovoltaic dye sensitized solar cell and photocatalytic water purification. Appl. Surf. Sci.**378**, 8–14 (2016).
133. Kim, J. T., Lee, S. H. & Han, Y. S. Enhanced power conversion efficiency of dye-sensitized solar cells with Li 2 SiO 3 -modified photoelectrode. Appl. Surf. Sci.**333**, 134–140 (2015).
134. Sabet, M., Salavati-Niasari, M. & Amiri, O. Using different chemical methods for deposition of CdS on TiO2 surface and investigation of their influences on the dye-sensitized solar cell performance. Electrochim. Acta**117**, 504–520 (2014).
135. Wang, Z. et al. Titanium dioxide/calcium fluoride nanocrystallite for efficient dye-sensitized solar cell. A strategy of enhancing light harvest. J. Power Sources**275**, 175–180 (2015).
136. Kane, S. N., Mishra, A. & Dutta, A. K. Electrospinning Titanium Dioxide (TiO2) nanofiber for dye sensitized solar cells based on Bryophyta as a sensitizer. J. Phys. Conf. Ser.**755**, 3–10 (2016).
137. Yang, X., Zhao, L., Lv, K., Dong, B. & Wang, S. Enhanced efficiency for dye-sensitized solar cells with ZrO 2 as a barrier layer on TiO 2 nanofibers. Appl. Surf. Sci.**469**, 821–828 (2019).
138. Cao, Y., Dong, Y. J., Feng, H. L., Chen, H. Y. & Kuang, D. Bin. Electrospun TiO2 nanofiber based hierarchical photoanode for efficient dye-sensitized solar cells. Electrochim. Acta**189**, 259–264 (2016).
139. Joshi, P. et al. Composite of TiO2 nanofibers and nanoparticles for dye-sensitized solar cells with significantly improved efficiency. Energy Environ. Sci.**3**, 1507–1510 (2010).
140. Chuangchote, S., Sagawa, T. & Yoshikawa, S. Efficient dye-sensitized solar cells using electrospun Ti O2 nanofibers as a light harvesting layer. Appl. Phys. Lett.**93**, 2012–2015 (2008).
141. Jung, W. H., Kwak, N. S., Hwang, T. S. & Yi, K. B. Preparation of highly porous TiO 2 nanofibers for dye-sensitized solar cells (DSSCs) by electro-spinning. Appl. Surf. Sci.**261**, 343–352 (2012).
142. Lin, Y. P., Chen, Y. Y., Lee, Y. C. & Chen-Yang, Y. W. Effect of wormhole-like mesoporous anatase TiO 2 nanofiber prepared by electrospinning with ionic liquid on dye-sensitized solar cells. J. Phys. Chem. C**116**, 13003–13012 (2012).
143. Mesoporous titania–vertical nanorod films with interfacial engineering for high performance dye-sensitized solar cells - IOPscience.
144. Zukalová, M. et al. Nanofibrous TiO2 improving performance of mesoporous TiO 2 electrode in dye-sensitized solar cell. J. Nanoparticle Res.**15**, 1–8 (2013).
145. Wang, J. et al. Improved morphology and photovoltaic performance in TiO2 nanorod arrays based dye sensitized solar cells by using a seed layer. J. Alloys Compd.**551**, 82–87 (2013).
146. Huang, Q., Zhou, G., Fang, L., Hu, L. & Wang, Z. S. TiO2 nanorod arrays grown from a mixed acid medium for efficient dye-sensitized solar cells. Energy Environ. Sci.**4**, 2145–2151 (2011).
147. Zhang, Z., Hu, Y., Qin, F. & Ding, Y. DC sputtering assisted nano-branched core-shell TiO 2 /ZnO electrodes for application in dye-sensitized solar cells. Appl. Surf. Sci.**376**, 10–15 (2016).

148. Sriharan, N., Ganesan, N. M., Kang, M., Kungumadevi, L. & Senthil, T. S. Improved photoelectrical performance of single crystalline rutile TiO2 nanorod arrays incorporating α-alumina for high efficiency dye-sensitized solar cells. Mater. Lett.**237**, 204–208 (2019).
149. Lv, M. et al. Optimized porous rutile TiO2 nanorod arrays for enhancing the efficiency of dye-sensitized solar cells. Energy Environ. Sci.**6**, 1615–1622 (2013).
150. Zha, C. et al. Double-sided brush-shaped TiO2 nanostructure assemblies with highly ordered nanowires for dye-sensitized solar cells. ACS Appl. Mater. Interfaces**6**, 122–129 (2014).
151. Motonari Adachi, * et al. Highly Efficient Dye-Sensitized Solar Cells with a Titania Thin-Film Electrode Composed of a Network Structure of Single-Crystal-like TiO2 Nanowires Made by the "Oriented Attachment" Mechanism. (2004) doi:https://doi.org/10.1021/JA048068S.
152. Wang, W. et al. Effects of low pressure plasma treatments on DSSCs based on rutile TiO 2 array photoanodes. Appl. Surf. Sci.**324**, 143–151 (2015).
153. Wu, W. Q., Xu, Y. F., Su, C. Y. & Kuang, D. Bin. Ultra-long anatase TiO2 nanowire arrays with multi-layered configuration on FTO glass for high-efficiency dye-sensitized solar cells. Energy Environ. Sci.**7**, 644–649 (2014).
154. Wu, W. Q. et al. Hierarchical oriented anatase TiO2 nanostructure arrays on flexible substrate for efficient dye-sensitized solar cells. Sci. Rep.**3**, 1–7 (2013).
155. Sun, P. et al. Rutile TiO2 nanowire array infiltrated with anatase nanoparticles as photoanode for dye-sensitized solar cells: Enhanced cell performance via the rutile-anatase heterojunction. J. Mater. Chem. A**1**, 3309–3314 (2013).
156. Sun, P. et al. Bilayer TiO2 photoanode consisting of a nanowire-nanoparticle bottom layer and a spherical voids scattering layer for dye-sensitized solar cells. New J. Chem.**39**, 4845–4851 (2015).
157. Mor, G. K., Varghese, O. K., Paulose, M., Shankar, K. & Grimes, C. A. A review on highly ordered, vertically oriented TiO2 nanotube arrays: Fabrication, material properties, and solar energy applications. Solar Energy Materials and Solar Cells vol. 90 2011–2075 (2006).
158. Zhou, Q., Fang, Z., Li, J. & Wang, M. Applications of TiO2 nanotube arrays in environmental and energy fields: A review. Microporous and Mesoporous Materials vol. 202 22–35 (2015).
159. Flores, I. C. et al. Dye-sensitized solar cells based on TiO2 nanotubes and a solid-state electrolyte. J. Photochem. Photobiol. A Chem.**189**, 153–160 (2007).
160. Park, H. et al. Fabrication of dye-sensitized solar cells by transplanting highly ordered TiO2 nanotube arrays. in Solar Energy Materials and Solar Cells vol. 95 184–189 (North-Holland, 2011).
161. Roy, P., Kim, D., Lee, K., Spiecker, E. & Schmuki, P. TiO2 nanotubes and their application in dye-sensitized solar cells. Nanoscale**2**, 45–59 (2010).
162. Kim, D., Ghicov, A. & Schmuki, P. TiO2 Nanotube arrays: Elimination of disordered top layers ("nanograss") for improved photoconversion efficiency in dye-sensitized solar cells. Electrochem. commun.**10**, 1835–1838 (2008).
163. Zhu, K., Neale, N. R., Miedaner, A. & Frank, A. J. Enhanced charge-collection efficiencies and light scattering in dye-sensitized solar cells using oriented TiO2 nanotubes arrays. Nano Lett.**7**, 69–74 (2007).
164. Jennings, J. R., Ghicov, A., Peter, L. M., Schmuki, P. & Walker, A. B. Dye-sensitized solar cells based on oriented TiO2 nanotube arrays: Transport, trapping, and transfer of electrons. J. Am. Chem. Soc.**130**, 13364–13372 (2008).
165. Liu, Z. & Misra, M. Dye-sensitized photovoltaic wires using highly ordered TiO2 nanotube arrays. ACS Nano**4**, 2196–2200 (2010).
166. Kuo, C. G., Yang, C. F., Hwang, L. R. & Huang, J. S. Effects of titanium oxide nanotube arrays with different lengths on the characteristics of dye-sensitized solar cells. Int. J. Photoenergy**2013**, (2013).
167. Chen, W. C., Yeh, M. H., Lin, L. Y., Vittal, R. & Ho, K. C. Double-Wall TiO2 Nanotubes for Dye-Sensitized Solar Cells: A Study of Growth Mechanism. ACS Sustain. Chem. Eng.**6**, 3907–3915 (2018).
168. Lee, C. H., Kim, K. H., Jang, K. U., Park, S. J. & Choi, H. W. Synthesis of TiO_2 Nanotube by Hydrothermal Method and Application for Dye-Sensitized Solar Cell. Mol. Cryst. Liq. Cryst.**539**, 125/[465]-132/[472] (2011).

169. Yu, J., Fan, J. & Lv, K. Anatase TiO2 nanosheets with exposed (001) facets: Improved photoelectric conversion efficiency in dye-sensitized solar cells. Nanoscale**2**, 2144–2149 (2010).
170. Zhu, H. et al. Growth of TiO 2 nanosheet-array thin films by quick chemical bath deposition for dye-sensitized solar cells. Appl. Phys. A Mater. Sci. Process.**105**, 769–774 (2011).
171. Lee, C. S., Kim, J. K., Lim, J. Y. & Kim, J. H. One-step process for the synthesis and deposition of anatase, two-dimensional, disk-shaped TiO2 for dye-sensitized solar cells. ACS Appl. Mater. Interfaces**6**, 20842–20850 (2014).
172. Shanmugam, M., Jacobs-Gedrim, R., Durcan, C. & Yu, B. 2D layered insulator hexagonal boron nitride enabled surface passivation in dye sensitized solar cells. Nanoscale**5**, 11275–11282 (2013).
173. Lin, J. et al. A Bi-layer TiO2 photoanode for highly durable, flexible dye-sensitized solar cells. J. Mater. Chem. A**3**, 4679–4686 (2015).
174. Deepak, T. G. et al. Cabbage leaf-shaped two-dimensional TiO2 mesostructures for efficient dye-sensitized solar cells. RSC Adv.**4**, 27084–27090 (2014).
175. Chen, B. et al. Graphene Oxide-Assisted Synthesis of Microsized Ultrathin Single-Crystalline Anatase TiO2 Nanosheets and Their Application in Dye-Sensitized Solar Cells. ACS Appl. Mater. Interfaces**8**, 2495–2504 (2016).
176. Li, X., Yu, J. & Jaroniec, M. Hierarchical photocatalysts. Chemical Society Reviews vol. 45 2603–2636 (2016).
177. Chen, H. Y., Kuang, D. Bin & Su, C. Y. Hierarchically micro/nanostructured photoanode materials for dye-sensitized solar cells. Journal of Materials Chemistry vol. 22 15475–15489 (2012).
178. Zhang, Q. & Cao, G. Hierarchically structured photoelectrodes for dye-sensitized solar cells. J. Mater. Chem.**21**, 6769–6774 (2011).
179. Sanjay, P., Deepa, K., Madhavan, J. & Senthil, S. Performance of TiO2 based dye-sensitized solar cells fabricated with dye extracted from leaves of Peltophorum pterocarpum and Acalypha amentacea as sensitizer. Mater. Lett.**219**, 158–162 (2018).
180. Bahramian, A. High conversion efficiency of dye-sensitized solar cells based on coral-like TiO2 nanostructured films: Synthesis and physical characterization. Ind. Eng. Chem. Res.**52**, 14837–14846 (2013).
181. Mali, S. S., Betty, C. A., Bhosale, P. N., Patil, P. S. & Hong, C. K. From nanocorals to nanorods to nanoflowers nanoarchitecture for efficient dye-sensitized solar cells at relatively low film thickness: All Hydrothermal Process. Sci. Rep.**4**, 2–9 (2014).
182. Su, C. et al. Preparation and characterization of pure rutile TiO2 nanoparticles for photocatalytic study and thin films for dye-sensitized solar cells. J. Nanomater.**2011**, (2011).
183. Fang, F., Kennedy, J., Manikandan, E., Futter, J. & Markwitz, A. Morphology and characterization of TiO 2 nanoparticles synthesized by arc discharge. Chem. Phys. Lett.**521**, 86–90 (2012).
184. Liu, L., Yu, X. M., Zhang, B., Meng, S. X. & Feng, Y. Q. Synthesis of nano-TiO2 assisted by diethylene glycol for use in high efficiency dye-sensitized solar cells. Chinese Chem. Lett.**28**, 765–770 (2017).
185. Hussian, H. A. R. A., Hassan, M. A. M. & Agool, I. R. Synthesis of titanium dioxide (TiO2) nanofiber and nanotube using different chemical method. Optik (Stuttg).**127**, 2996–2999 (2016).
186. Guo, H. et al. Facile synthesis of chrysanthemum flowers-like TiO2 hierarchical microstructures assembled by nanotube for high performance dye-sensitized solar cells. Org. Electron.**55**, 97–105 (2018).
187. Luan, X., Guan, D. & Wang, Y. Facile synthesis and morphology control of bamboo-type TiO 2 nanotube arrays for high-efficiency dye-sensitized solar cells. J. Phys. Chem. C**116**, 14257–14263 (2012).
188. Mojaddami, M., Mohammadi, M. R. & Madaah Hosseini, H. R. Improved Efficiency of Dye-Sensitized Solar Cells Based on a Single Layer Deposition of Skein-Like TiO 2 Nanotubes . J. Am. Ceram. Soc.**97**, 2873–2879 (2014).

189. Roh, D. K., Chi, W. S., Jeon, H., Kim, S. J. & Kim, J. H. High efficiency Solid-state Dye-sensitized solar cells assembled with hierarchical anatase pine Tree-like TiO2 nanotubes. Adv. Funct. Mater.**24**, 379–386 (2014).
190. Fan, K., Zhang, W., Peng, T., Chen, J. & Yang, F. Application of TiO2 fusiform nanorods for dye-sensitized solar cells with significantly improved efficiency. J. Phys. Chem. C**115**, 17213–17219 (2011).
191. Liu, Y. Y. et al. One-step hydrothermal fabrication of three dimensional anatase hierarchical hyacinth-like TiO2 arrays for dye-sensitized solar cells. Thin Solid Films**683**, 42–48 (2019).
192. Tański, T. et al. Study of dye sensitized solar cells photoelectrodes consisting of nanostructures. Appl. Surf. Sci.**491**, 807–813 (2019).
193. Wu, W. Q. et al. Morphology-controlled cactus-like branched anatase TiO2 arrays with high light-harvesting efficiency for dye-sensitized solar cells. J. Power Sources**260**, 6–11 (2014).
194. Wei, Z., Yao, Y., Huang, T. & Yu, A. Solvothermal growth of well-aligned TiO2 nanowire arrays for dye-sensitized solar cell: Dependence of morphology and vertical orientation upon substrate pretreatment. Int. J. Electrochem. Sci.**6**, 1871–1879 (2011).
195. Li, H. et al. Ultralong Rutile TiO2 Nanowire Arrays for Highly Efficient Dye-Sensitized Solar Cells. ACS Appl. Mater. Interfaces**8**, 13384–13391 (2016).
196. Li, Y. et al. Au nanoparticle-decorated urchin-like TiO2 hierarchical microspheres for high performance dye-sensitized solar cells. Electrochim. Acta**293**, 230–239 (2019).
197. Marandi, M. & Bayat, S. Facile fabrication of hyper-branched TiO2 hollow spheres for high efficiency dye-sensitized solar cells. Sol. Energy**174**, 888–896 (2018).
198. Liu, Y. et al. Synthesis of "lotus root"-like mesoporous titanium dioxide and its effects on UV response to aconitine release. J. Alloys Compd.**777**, 285–293 (2019).
199. Hu, B. & Liu, B. Dye-sensitized solar cells fabricated by the TiO 2 nanostructural materials synthesized by electrospray and hydrothermal post-treatment. Appl. Surf. Sci.**358**, 412–417 (2015).
200. Pan, H., Qian, J., Cui, Y., Xie, H. & Zhou, X. Hollow anatase TiO2 porous microspheres with V-shaped channels and exposed (101) facets: Anisotropic etching and photovoltaic properties. J. Mater. Chem.**22**, 6002–6009 (2012).
201. Chen, Y. Z., Wu, R. J., Lin, L. Y. & Chang, W. C. Novel synthesis of popcorn-like TiO2 light scatterers using a facile solution method for efficient dye-sensitized solar cells. J. Power Sources**413**, 384–390 (2019).
202. Zhang, J., He, X., Zhu, M., Guo, Y. & Li, X. The preparation of hierarchical rutile TiO2 microspheres constructed with branched nanorods for efficient dye-sensitized solar cells. J. Alloys Compd.**747**, 729–737 (2018).
203. Xu, L. et al. Hierarchical submicroflowers assembled from ultrathin anatase TiO2 nanosheets as light scattering centers in TiO2 photoanodes for dye-sensitized solar cells. J. Alloys Compd.**776**, 1002–1008 (2019).
204. Ma, C. et al. Monodisperse TiO 2 microspheres assembled by porous spindles for high performance dye-sensitized solar cells. Colloids Surfaces A Physicochem. Eng. Asp.**538**, 94–99 (2018).
205. Li, Z. Q. et al. Mesoporous TiO2 Yolk-Shell Microspheres for Dye-sensitized Solar Cells with a High Efficiency Exceeding 11%. Sci. Rep.**5**, 1–8 (2015).
206. Beall, C., Piipari, K., Al-qassab, H. & Smith, M. A. Alkali-Corrosion Synthesis and Excellent DSSC Performance of a Novel Jujube-like Hierarchical TiO2 Microspheres. Biochem. J. 0–14 (2010).
207. Zhao, T. et al. Monodisperse mesoporous TiO2 microspheres for dye sensitized solar cells. Nano Energy**26**, 16–25 (2016).
208. Li, Z. Q. et al. Fine Tuning of Nanocrystal and Pore Sizes of TiO 2 Submicrospheres toward High Performance Dye-Sensitized Solar Cells. ACS Appl. Mater. Interfaces**7**, 22277–22283 (2015).
209. Hwang, S. H., Yun, J. & Jang, J. Multi-shell porous TiO2 hollow nanoparticles for enhanced light harvesting in dye-sensitized solar cells. Adv. Funct. Mater.**24**, 7619–7626 (2014).

210. Jiang, J., Gu, F., Shao, W. & Li, C. Fabrication of spherical multi-hollow TiO 2 nanostructures for photoanode film with enhanced light-scattering performance. Ind. Eng. Chem. Res.**51**, 2838–2845 (2012).
211. Lin, C. M. et al. Multi-step hydrothermally synthesized TiO 2 nanoforests and its application to dye-sensitized solar cells. Mater. Chem. Phys.**135**, 723–727 (2012).
212. Ma, J., Ren, W., Zhao, J. & Yang, H. Growth of TiO2nanoflowers photoanode for dye-sensitized solar cells. J. Alloys Compd.**692**, 1004–1009 (2017).
213. Javed, H. M. A. et al. Investigation on the surface modification of TiO2 nanohexagon arrays based photoanode with SnO2 nanoparticles for highly-efficient dye-sensitized solar cells. Mater. Res. Bull.**109**, 21–28 (2019).
214. Umale, S., Sudhakar, V., Sontakke, S. M., Krishnamoorthy, K. & Pandit, A. B. Improved efficiency of DSSC using combustion synthesized TiO2. Mater. Res. Bull.**109**, 222–226 (2019).
215. Amoli, V. et al. Tailored Synthesis of Porous TiO2 Nanocubes and Nanoparallelepipeds with Exposed {111} Facets and Mesoscopic Void Space: A Superior Candidate for Efficient Dye-Sensitized Solar Cells. ACS Appl. Mater. Interfaces**7**, 26022–26035 (2015).
216. Chae, J. & Kang, M. Cubic titanium dioxide photoanode for dye-sensitized solar cells. J. Power Sources**196**, 4143–4151 (2011).
217. Shuang, Y., Hou, Y., Zhang, B. & Yang, H. G. Impurity-free synthesis of cube-like single-crystal anatase TiO2 for high performance dye-sensitized solar cell. Ind. Eng. Chem. Res.**52**, 4098–4102 (2013).
218. Li, Z. Q. et al. Solvothermal Synthesis of Hierarchical TiO2 Microstructures with High Crystallinity and Superior Light Scattering for High-Performance Dye-Sensitized Solar Cells. ACS Appl. Mater. Interfaces**9**, 32026–32033 (2017).
219. Qiu, Y., Chen, W. & Yang, S. Double-Layered Photoanodes from Variable-Size Anatase TiO2 Nanospindles: A Candidate for High-Efficiency Dye-Sensitized Solar Cells. Angew. Chemie**122**, 3757–3761 (2010).
220. He, X., Li, X. & Zhu, M. The application of hollow box TiO2 as scattering centers in dye-sensitized solar cells. J. Power Sources**333**, 10–16 (2016).
221. Jeyaraman, A. R. et al. Enhanced solar to electrical energy conversion of titania nanoparticles and nanotubes-based combined photoanodes for dye-sensitized solar cells. Mater. Lett.**243**, 180–182 (2019).
222. Marandi, M., Bayat, S. & Naeimi Sani Sabet, M. Hydrothermal growth of a composite TiO2 hollow spheres/TiO2 nanorods powder and its application in high performance dye-sensitized solar cells. J. Electroanal. Chem.**833**, 143–150 (2019).
223. Kim, J. S. et al. Facile Preparation of TiO2 Nanobranch/Nanoparticle Hybrid Architecture with Enhanced Light Harvesting Properties for Dye-Sensitized Solar Cells. J. Nanomater.**2015**, (2015).
224. Wang, Y., Yang, W. & Shi, W. Preparation and characterization of anatase TiO2 nanosheets-based microspheres for dye-sensitized solar cells. Ind. Eng. Chem. Res.**50**, 11982–11987 (2011).

The manufacturer's authorised representative in the EU is Springer Nature Customer Service Centre GmbH, Europaplatz 3, 69115 Heidelberg, Germany. If you have any concerns regarding our products, please contact ProductSafety@springernature.com

Printed and bound by CPI Group (UK) Ltd, Croydon, CR0 4YY
28/09/2026
02224093-0004